Geological Prior Information: Informing Science and Engineering

Special Publication reviewing procedures

The Society makes every effort to ensure that the scientific and production quality of its books matches that of its journals. Since 1997, all book proposals have been refereed by specialist reviewers as well as by the Society's Books Editorial Committee. If the referees identify weaknesses in the proposal, these must be addressed before the proposal is accepted.

Once the book is accepted, the Society has a team of Book Editors (listed above) who ensure that the volume editors follow strict guidelines on refereeing and quality control. We insist that individual papers can only be accepted after satisfactory review by two independent referees. The questions on the review forms are similar to those for *Journal of the Geological Society*. The referees' forms and comments must be available to the Society's Book Editors on request.

Although many of the books result from meetings, the editors are expected to commission papers that were not presented at the meeting to ensure that the book provides a balanced coverage of the subject. Being accepted for presentation at the meeting does not guarantee inclusion in the book.

Geological Society Special Publications are included in the ISI Index of Scientific Book Contents, but they do not have an impact factor, the latter being applicable only to journals.

More information about submitting a proposal and producing a Special Publication can be found on the Society's web site: www.geolsoc.org.uk.

It is recommended that reference to all or part of this book should be made in one of the following ways:
CURTIS, A. & WOOD, R. (eds) 2004. *Geological Prior Information: Informing Science and Engineering*. Geological Society, London, Special Publications, **239**.
BURGESS, P. M. & EMERY, D. J. 2004. Sensitive dependence, divergence and unpredictable behaviour in a stratigraphic forward model of a carbonate system. *In*: CURTIS, A. & WOOD, R. (eds) 2004. *Geological Prior Information: Informing Science and Engineering*. Geological Society, London, Special Publications, **239**, 77–93.

GEOLOGICAL SOCIETY SPECIAL PUBLICATION NO. 239

Geological Prior Information: Informing Science and Engineering

EDITED BY

ANDREW CURTIS

Schlumberger Cambridge Research, and the University of Edinburgh, UK

and

RACHEL WOOD

Schlumberger Cambridge Research, and the University of Cambridge, UK

2004

Published by

The Geological Society

London

THE GEOLOGICAL SOCIETY

The Geological Society of London (GSL) was founded in 1807. It is the oldest national geological society in the world and the largest in Europe. It was incorporated under Royal Charter in 1825 and is Registered Charity 210161.

The Society is the UK national learned and professional society for geology with a worldwide Fellowship (FGS) of 9000. The Society has the power to confer Chartered status on suitably qualified Fellows, and about 2000 of the Fellowship carry the title (CGeol). Chartered Geologists may also obtain the equivalent European title, European Geologist (EurGeol). One fifth of the Society's fellowship resides outside the UK. To find out more about the Society, log on to www.geolsoc.org.uk.

The Geological Society Publishing House (Bath, UK) produces the Society's international journals and books, and acts as European distributor for selected publications of the American Association of Petroleum Geologists (AAPG), the American Geological Institute (AGI), the Indonesian Petroleum Association (IPA), the Geological Society of America (GSA), the Society for Sedimentary Geology (SEPM) and the Geologists' Association (GA). Joint marketing agreements ensure that GSL Fellows may purchase these societies' publications at a discount. The Society's online bookshop (accessible from www.geolsoc.org.uk) offers secure book purchasing with your credit or debit card.

To find out about joining the Society and benefiting from substantial discounts on publications of GSL and other societies worldwide, consult www.geolsoc.org.uk, or contact the Fellowship Department at: The Geological Society, Burlington House, Piccadilly, London W1J 0BG: Tel. +44 (0)20 7434 9944; Fax +44 (0)20 7439 8975; E-mail: enquiries@geolsoc.org.uk.

For information about the Society's meetings, consult *Events* on www.geolsoc.org.uk. To find out more about the Society's Corporate Affiliates Scheme, write to enquiries@geolsoc.org.uk.

Published by The Geological Society from:
The Geological Society Publishing House
Unit 7, Brassmill Enterprise Centre
Brassmill Lane
Bath BA1 3JN, UK
(*Orders*: Tel. +44 (0)1225 445046
Fax +44 (0)1225 442836)

Online bookshop: http://bookshop.geolsoc.org.uk

British Library Cataloguing in Publication Data
A catalogue record for this book is available from the British Library.

ISBN 1-86239-171-8

Typeset by Tradespools, Frome, UK

Printed by MPG Books Ltd, Bodmin, UK

Distributors

USA
AAPG Bookstore
PO Box 979
Tulsa
OK 74101-0979
USA
Orders: Tel. + 1 918 584-2555
Fax +1 918 560-2652
E-mail: bookstore@aapg.org

India
Affiliated East-West Press PVT Ltd
G-1/16 Ansari Road, Darya Ganj,
New Delhi 110 002
India
Orders: Tel. +91 11 2327-9113/2326-4180
Fax +91 11 2326-0538
E-mail: affiliat@vsnl.com

Japan
Kanda Book Trading Company
Cityhouse Tama 204
Tsurumaki 1-3-10
Tama-shi, Tokyo 206-0034
Japan
Orders: Tel. +81 (0)423 57-7650
Fax +81 (0)423 57-7651
Email: geokanda@ma.kcom.ne.jp

Contents

Applications

Preface

To some, the title of this collected volume, *Geological Prior Information*, might seem novel and somewhat baffling. This is because this discipline represents an emerging field within the geosciences, where methods and applications are still in their infancy. While the use of Bayesian analysis and prior information have long been established within both the sciences and humanities, in the Geosciences these methods have to date only rarely extended beyond the realm of geophysics.

Prior information is that which is provided as an *a priori* component of a solution to any problem of interest. That is, it comprises all information that pre-existed to the collection of any new or current data sets that were designed specifically to help solve the problem. Geological prior information takes many forms, ranging from basic assumptions of physics, chemistry, biology, and geology, to the design of the problem to be solved, and to the use of prior experience from previous studies in order to interpret new data and provide a solution.

Such information is always central to the creation of each solution and indeed to the formulation of the problem in the first place. This is especially true in the geological sciences in which typically available data sets rarely unambiguously constrain interpretations, so that much must be inferred from subjective intuition and past experience. It is often the case that geoscientists give short shrift to subjective information that they themselves have injected into their solutions, preferring instead to focus on the 'objective' components of their work. This is unnecessary: all results from all scientific disciplines contain judgements and assumptions, and indeed these are usually key to enabling the course of study in the first place, and hence are highly valuable and important components to the results and solutions found.

This special volume from the Geological Society of London is devoted to understanding how prior information of a geological nature can be harnessed in order to solve problems. As such, it spans a wide range of conceptual and analytical research in several aspects of geological prior information: how this can be captured, quantified, ascribed an associated uncertainty or reliability, and also how geological prior information might be used to provide solutions to both practical and theoretical problems, both within or beyond the discipline of geology.

The importance of using such knowledge in the most effective way possible cannot be overestimated. For example, historically recent perturbations to the Earth's environmental and climatic systems appears to be so widespread and significant that it seems inevitable that we will alter geological processes, and hence future geologies, on human timescales. It also appears highly probable that the consequences of global climate change will, in some senses, be catastrophic, with all the attendant economic consequences (such as drowning of land areas, mass movement of populations, and disturbed weather patterns including an increased frequency of storms and flooding). Belief in the causal mechanisms governing climate change, in particular the anthropogenically-influenced component, will have a dramatic affect on how Society reacts to this impending crisis. In particular, the degree and nature of their belief will govern whether effective, preventative action is taken beforehand, in which case the rate of global change may slow, or whether remedial action will be taken so late that climatic and hence geological processes may continue to change rapidly and significantly. Even given the most accurate climate models, beliefs about the mitigation required contain many subjectivities (e.g. exactly what preventative action should be taken), and yet which outcome is chosen will result, amongst many other effects, in the creation of a quite different set of near surface geologies within our own lifetimes.

Hence, our prior, subjective beliefs contribute to stimulate actions (solutions to the problem at hand) that can change future geological outcomes. The same effect is seen in the more rapidly evolving social sciences such as Economics. For example, the purchase of shares is always stimulated by the belief that share prices will rise, and that same purchase will indeed drive share prices upwards. In such globally important cases as our climatic and geological future, as well as our collective and personal wealth, it is surely best to be explicit about the level of subjectivity that influences our decisions and solutions to problems. Such information can then be judged fairly and the course of action most likely to provide greatest benefit chosen.

Most of the papers in this volume resulted from a Discussion Meeting hosted by the Royal Astronomical Society at Burlington House on May 9th 2003; others are included to provide a more cohesive and comprehensive overview of this new area of research. Given the breadth of the contributions, we hope that this Special Publication will serve as both a review and reference text for the emerging field of geological prior information, and that academics, industrial geoscientists, and many readers outside the field of geology, will find some interest and stimulation.

Andrew Curtis, Edinburgh
Rachel Wood, Cambridge

Acknowledgements

First, we thank the Royal Astronomical Society at Burlington House, London, who generously funded the Discussion Meeting of which this volume is the outcome. We would also like to thank all participants and contributing authors for their timely submission of articles, and their assistance in the assembly of this volume. In addition, we are grateful to all reviewers of submissions who have given freely of their expertise and, through their constructive comments, have contributed significantly to this volume: Michelle Baddeley, Peter Burgess, Peter Clift, Nick Drinkwater, János Fodor, Cedric Griffiths, David Holmes, David Jeanette, Charles Jenkins, Peter Kaufman, Anthony Lomax, Ian Main, Alberto Malinverno, David McCormick, Louis Moresi, Bob Parker, Mike Prang, Cyril Pshenichny, Silja Renooij, Wim Spakman, Ru Smith, Albert Tarantola, Jeannot Trampert, Dan Teztlaff, Jonathan Turner, Ron Wehrens, and Paul Wright. We are grateful to Schlumberger for supporting this project. Finally, we thank Angharad Hills and Rob Holdsworth for their help in the final stages of publication.

Geological prior information and its applications to geoscientific problems

RACHEL WOOD[1,2] & ANDREW CURTIS[1,3]

[1]*Schlumberger Cambridge Research, High Cross, Madingley Road, Cambridge CB3 0EL, UK (e-mail: rwood@cambridge.oilfield.slb.com)*

[2]*Department of Earth Sciences, University of Cambridge, Downing Street, Cambridge CB2 3EQ, UK*

[3]*Grant Institute of Earth Science, School of GeoSciences, University of Edinburgh, West Mains Road, Edinburgh EH9 3JW, UK*

Abstract: Geological information can be used to solve many practical and theoretical problems, both within and outside of the discipline of geology. These include analysis of ground stability, predicting subsurface water or hydrocarbon reserves, assessment of risk due to natural hazards, and many others. In many cases, geological information is provided as an *a priori* component of the solution (i.e. information that existed before the solution was formed and which is incorporated into the solution). Such information is termed 'geological prior information'.

The key to the successful solution of such problems is to use only *reliable* geological information. In turn this requires that: (1) multiple geological experts are consulted and any conflicting views reconciled, (2) all prior information includes measures of confidence or uncertainty (without which its reliability and worth is unknown), and (3) as much information as possible is quantitative, and qualitative information or assumptions are clearly defined so that uncertainty or risk in the final result can be evaluated. This paper discusses each of these components and proposes a probabilistic framework for the use and understanding of prior information.

We demonstrate the methodology implicit within this framework with an example: this shows how prior information about typical stacking patterns of sedimentary sequences allows aspects of 2-D platform architecture to be derived from 1-D geological data alone, such as that obtained from an outcrop section or vertical well. This example establishes how the extraction of quantitative, multi-dimensional, geological interpretations is possible using lower dimensional data. The final probabilistic description of the multi-dimensional architecture could then be used as prior information sequentially for a subsequent problem using exactly the same method.

For decades, specific geological information has been transferred to other domains, helping solve applied and theoretical problems. These include, for example:

- Information about typical geological architectures in a particular environment allows subsurface properties to be estimated away from drilled observation wells in order to assess hydrocarbon or water reservoir potential (e.g. Mukerj *et al.* 2001; Schon 2004).
- Assessments of predicted geohazard risks contributes to the pricing of life and property insurance (e.g. Rosenbaum & Culshaw 2003).
- The distribution of ancient reef corals, combined with the relationship between coral growth and sea level, allows prediction of both past sea level oscillations and how future sea level change might affect the distribution of modern reefs (e.g. Potts 1984; Budd *et al.* 1998).
- Historical knowledge of slope stability in different geological environments allows proposed building projects to include appropriate remedial action against such risks (e.g. Selby 1982).
- Estimates of regional tectonic stability contribute to allowing toxic waste to be stored underground with minimal risk of future leakage (e.g. Pojasek 1980).

For each of these problems, geological information is not only provided as an *a priori* component of the solutions, but is central to their creation. This information pre-existed the formation of the solution, and so in this context is termed 'geological prior information' (GPI).

From: CURTIS, A. & WOOD, R. (eds) 2004. *Geological Prior Information: Informing Science and Engineering*. Geological Society, London, Special Publications, **239**, 1–14. 186239-171-8/04/$15.00

The special volume from the Geological Society of London, of which this paper is a part, is devoted to understanding the use of geological prior information in order to solve problems in both geological and other domains. As such, it spans research in several aspects of GPI: how GPI can be captured, quantified and ascribed an associated uncertainty or reliability, and then how this might be used to provide solutions to other problems. The purpose of research into the creation and use of GPI is therefore to make geology more accurate, useful and transferable to scientists and problem solvers.

In this paper, we begin by constructing a Bayesian, probabilistic framework that precisely defines GPI. We then define the various components of information that must be either assembled or assumed in order to solve problems using GPI, and demonstrate how uncertainty in such assembled information propagates to create uncertainty in the solutions found. This framework also allows a useful categorization of methods that use GPI to solve geoscientific problems: we demonstrate this by placing the papers of this volume into this framework and stating the principal ways in which each contributes to the field. We then present an example of the use of GPI: in this case, how it might be used to extract and extend geological information away from a single point in space at which data was collected. This example serves to illustrate definitions given in the probabilistic framework and highlights some of the issues and difficulties involved in using GPI.

A framework for using and analysing geological prior information

We begin with a short revision of some definitions of probabilities, since these are used frequently in this paper. In all that follows we use $P(Q)$ as a general notation for the probability of an event or hypothesis Q being true, or the value of the probability density function (p.d.f.) at Q if the values that Q can take are continuous. The p.d.f.s are always positive or zero, and they have the following relationships to probabilities Pr(...):

$$\Pr(a \leqslant X \leqslant b) = \int_a^b P(x)dx$$
$$\Pr(-\infty \leqslant X \leqslant \infty) = \int_{-\infty}^{\infty} P(x)dx = 1$$

for $b > a$. Respectively, these equations state first that the p.d.f. $P(X)$ contains all information necessary to calculate the probability of occurrence of any particular event concerning X (e.g. the event that X lies between a and b) – indeed this is the definition of a p.d.f. Second, p.d.f.s are normalized so that the total probability that X takes any value at all is 1.

We also make use of conditional probability notation $P(Q \mid R)$, meaning the probability that Q is true given that we know event or hypothesis R is true. Finally, we will make use of Bayes' rule, which is obeyed by all probability distributions. For any events or hypotheses Q and R it may be stated:

$$P(Q \mid R) = \frac{P(R \mid Q)P(Q)}{P(R)}. \quad (1)$$

Now, say we would like to answer a geoscientific question upon which geological information might have some bearing, and assume further that the question can be posed in such a way that its answer is given by constraining some model vector **m**. Model **m** might represent a predicted sea-level curve, a set of charges for insurance against geohazards, or the distribution of fluids in a subsurface reservoir, for instance. Examples of the types of pertinent background knowledge might include: knowledge of regional geological, diagenetic, tectonic or climatic history; inferences from previously acquired data; typical problems experienced when acquiring data; expected data uncertainties; generally accepted geological theory which might be assumed in a geological model, and existing methods that may be used to tackle different types of problems. We represent all such background knowledge by symbol B.

We will assume that B can be decomposed into geological knowledge G and other knowledge G', and that the geological knowledge can be decomposed further into quantitative and non-quantitative components, G_q and $G_{q'}$ respectively. This results in a description of all background knowledge by the partition $B = [G_q, G_{q'}, G']$.

Say a new dataset **d** is acquired which was designed to place more focused constraints on model **m**. Bayes' rule allows constraints on **m** both from the new data **d** and from pre-existing background information B to be quantified. This is useful because the relationships between **m**, **d** and B are often complicated, in which case it is not easy to calculate or weight the relative constraints offered by **d** and B in other ways. Since all data and background information will be uncertain to some extent, so our knowledge of **m** will always be uncertain. One way to express available information about **m** is to use a probability distribution. Substituting **m**, **d** and B

appropriately into Equation 1 we obtain:

$$P(\mathbf{m} \mid \mathbf{d}, B) = \frac{P(\mathbf{d} \mid \mathbf{m}, B)}{P(\mathbf{d} \mid B)} P(\mathbf{m} \mid B). \quad (2)$$

The totality of information about the model given sources **d** and B is called the *posterior distribution* and is described by the probability distribution on the left of Equation 2. This can be calculated for any particular model **m** by the product of two terms on the right of Equation 2. The first is called the *relative likelihood* term, and is the ratio of the probability of measuring data **d** if model **m** was true (numerator) to the probability when no knowledge about **m** is available (denominator). Notice that both the numerator and denominator must therefore implicitly incorporate information about the uncertainty in the measured data **d**. The second term on the right is called the *prior distribution* of model **m** and describes all information about **m** that existed prior to data **d** having been acquired (since data **d** does not feature in this term). Hence, this term includes all information about **m** from background information B alone.

At this point it helps to have a particular example in mind, and the following example will be illustrated in detail later in this paper. Consider the situation in which we would like to constrain the stacking patterns of sedimentary sequences or parasequences, observed on a laterally constrained but vertically extensive outcrop exposure surface in some particular location (assuming that the exposure is too restricted to reveal the stacking patterns directly). Let model **m** represent the various possible stacking patterns of an observed parasequence set. Hence, **m** may represent one of the following patterns: regressive (R), progradational (P), aggradational (A), or retrogradational (G). Pertinent background information B might include: regional geological history and age of the sequence, approximate location of the parasequence set within the overall succession, and the sedimentary composition. The prior information term $P(\mathbf{m} \mid B)$ then represents all possible constraints on the stacking pattern given only this pre-existing background information.

Now say the relative vertical thicknesses of the parasequence set observed on the vertical outcrop section were measured to provide a dataset **d**. In principle, this dataset contains some information about the stacking pattern that would be observed laterally (a particular, precise relationship is given later). The numerator of the relative likelihood term $P(\mathbf{d} \mid \mathbf{m}, B)$ is the probability that we could have recorded dataset **d** if we knew that a particular value of model **m** was true (e.g. if we knew that the stacking pattern was regressive) and given background information B. The denominator $P(\mathbf{d} \mid B)$ is the probability that data **d** could have been recorded with no knowledge of the true stacking pattern but still given background information B, and is a normalizing term which ensures that the integral of the posterior distribution equals unity.

To give a numerical demonstration, say we know from background information B that the age of deposition occurred during a period of general sea-level rise. Then *a priori* we may wish to assign a low probability that regression R occurred (i.e. $\mathbf{m} = \text{'R'}$ in reality) since this is often associated with a relative sea-level fall (less likely if sea level is generally rising), but equal probability that progradation P, aggradation A or retrogradation G occurred. Hence, for example, we might define the following prior distribution:

$$P(\mathbf{m} = \text{'R'} \mid B) = \frac{1}{10} \quad \text{and}$$
$$P(\mathbf{m} = \text{'P'} \mid B) = P(\mathbf{m} = \text{'A'} \mid B) = P(\mathbf{m} = \text{'G'} \mid B) = \frac{3}{10}.$$

These probabilities represent our subjective belief about how likely was each stacking pattern, given our previous field experience in similar areas, papers we had read about other people's observations, etc.

Say that the probabilities of recording data **d** given each of the possible values for **m** are:

$$P(\mathbf{d} \mid \mathbf{m} = \text{'R'}, B) = P(\mathbf{d} \mid \mathbf{m} = \text{'P'}, B) = P(\mathbf{d} \mid \mathbf{m} = \text{'G'}, B) = \frac{1}{5} \quad \text{and}$$
$$P(\mathbf{d} \mid \mathbf{m} = \text{'A'}, B) = \frac{2}{5}$$

(our example later will illustrate how such values can be calculated); i.e. the probability of the data having been recorded in an aggradational setting is twice that in each of the other settings.

Thus far we have completely defined the likelihood in the numerator, and the prior information term on the right of Equation 2. It is now possible to work out the other two terms in this equation: since the left of Equation 2 is a probability distribution it must integrate to unity (i.e. the sum of the probabilities for each of the possible values of model **m** – R, P, A and G – must be one). Hence, the denominator on the right must assume a value that normalizes the right side of the equation similarly. This gives (with terms in order $\mathbf{m}$ = R, P, A, then G):

$$\begin{aligned} P(\mathbf{d} \mid B) &= P(\mathbf{d} \mid \mathbf{m} = \text{`R'}, B) P(\mathbf{m} = \text{`R'} \mid B) \\ &\quad + P(\mathbf{d} \mid \mathbf{m} = \text{`P'}, B) P(\mathbf{m} = \text{`P'} \mid B) \\ &\quad + P(\mathbf{d} \mid \mathbf{m} = \text{`A'}, B) P(\mathbf{m} = \text{`A'} \mid B) \\ &\quad + P(\mathbf{d} \mid \mathbf{m} = \text{`G'}, B) P(\mathbf{m} = \text{`G'} \mid B) \\ &= \frac{1}{5} \cdot \frac{1}{10} + \frac{1}{5} \cdot \frac{3}{10} + \frac{2}{5} \cdot \frac{3}{10} + \frac{1}{5} \cdot \frac{3}{10} \\ &= \frac{13}{50}. \end{aligned}$$

Finally, we can work out the posterior distribution on the left of Equation 2:

$$\begin{aligned} P(\mathbf{m} &= [\mathrm{R}, \mathrm{P}, \mathrm{A}, \mathrm{G}] \mid \mathbf{d}, B) \\ &= \left[\frac{1}{13}, \frac{3}{13}, \frac{6}{13}, \frac{3}{13} \right], \end{aligned}$$

where the numbers in the square brackets on the right are the probabilities of each of the cases for $\mathbf{m}$ listed in the square brackets on the left. Thus, we show that the probability of aggradation A is still twice that of progradation P or retrogradation G, as was the case above when we considered only the likelihood distribution (data information). However, in the posterior distribution when prior information is also taken into account, the probability of aggradation A is shown to be *six* times that of regression R. This is due to the relatively low probability of regression given our prior knowledge that the sequence was deposited in a period of general sea-level rise.

Let us now expand the prior distribution in Equation 2 using the partition of background information B made above, to obtain:

$$P(\mathbf{m} \mid B) = P(\mathbf{m} \mid G_q, G_{q'}, G'). \tag{3}$$

Since probabilities are inherently quantitative, it is often difficult to use uncertain, qualitative information or hypotheses to constrain their distributions. For instance, in the prior information term (Equation 3), the value of the perceived probability of model $\mathbf{m}$ being true will be influenced by both quantitative and qualitative geological information G_q and $G_{q'}$, respectively. However, generally it will be more difficult to assess the exact, quantitative influence of $G_{q'}$ on this probability than the influence of G_q. Returning to the stacking pattern example for illustration, it is reasonably easy to assess the prior probability of each stacking pattern $\mathbf{m}$ given quantitative information G_q about the relative proportion of the various stacking patterns that have been observed in other outcrops – if we assume the truth of the qualitative hypothesis in $G_{q'}$ 'that these are good analogues for the outcrop in question'. It would be far more difficult to assess how that information on stacking patterns might be enshrined within a prior probability distribution if we also allowed within $G_{q'}$ that this hypothesis might be false. Although we have only discussed the prior distribution here, such arguments also apply to each probability term in Equation 2.

As a result, it is always the case that, in practice, it will be necessary to make an approximation of the posterior distribution by limiting the uncertainty in some qualitative information and hypotheses in $G_{q'}$ (i.e. by adding extra, assumed information) to give approximation $\hat{G}_{q'}$. This results in an approximation of the background information

$$\hat{B} = [G_q, \hat{G}_{q'}, G'] \tag{4}$$

and the following application of Bayes' rule:

$$\begin{aligned} P(\mathbf{m} \mid \mathbf{d}, B) &\simeq P(\mathbf{m} \mid \mathbf{d}, \hat{B}) \\ &= \frac{P(\mathbf{d} \mid \mathbf{m}, \hat{B})}{P(\mathbf{d} \mid \hat{B})} P(\mathbf{m} \mid \hat{B}). \end{aligned} \tag{5}$$

The accuracy with which $P(\mathbf{m} \mid \mathbf{d}, \hat{B})$ approximates $P(\mathbf{m} \mid \mathbf{d}, B)$ depends on the nature of the approximations made in the background information $\hat{B}$, and in particular on the truth of the assumptions added.

To summarize the discussion so far, all practical, calculable posterior distributions are approximations (Equation 5), all are conditional on background information, and in practice it is also always the case that some of this background information must be assumed to be true for expediency rather than because it is known to be so (Equation 4). This implies that all estimates of prior and posterior uncertainty are likely to be optimistic as they cannot account for the true range of uncertainty. The Bayesian method formalizes the fact that most science and inference is based partially on subjective prior belief. This forces users to be explicit about their subjective assumptions (rather than sweeping them under the carpet!)

We differentiate here between *static* and *dynamic* quantitative prior information. The former denotes quantitative information about the present-day (static) observed or inferred geology; the latter denotes quantitative information about either time-dependent (dynamic) geological processes, or geologies that existed at some point in the past (which implicitly requires the use of a dynamic model).

Static prior information about a succession might comprise statistics of the geological geometries observed in one or more outcrops from formations of analogous settings and origin. In the example given later we use a uniform prior distribution of typical parasequence architectures of platform carbonate formations, but any other, more accurate prior distribution could also be used.

Dynamic prior information might be derived, for example, from climatological data since climate affects parasequence formation by influencing carbonate production rate, and by affecting sea-level oscillations and hence accommodation space. Consequently, some have postulated that fourth-order parasequence geometries are observed to be less ordered in icehouse compared to greenhouse episodes (Lehrmann & Goldhammer 1999). An example of prior information derived from this argument is therefore as follows: given that a specific formation is carbonate (with no other information, e.g. of the geological age), the prior probability that deposition occurred during icehouse, transitional or greenhouse episodes is approximately 0.31, 0.19 or 0.5, respectively (calculated by integrating global carbonate accumulation rates provided by Bosscher & Schlager 1993, over time spanned by each episode as defined by Sandberg 1985). In the example given later we demonstrate the method using only static information. Note, however, that dynamic information can be treated in exactly the same way.

If prior information comes from multiple independent sources (e.g. static and dynamic) it can be combined as follows: let B_1 represent background information from, for example, static sources as described above, and B_2 represent background information from dynamic sources derived, for example, from a geological forward process model (e.g. Tetzlaff & Priddy 2001; **Burgess & Emery**; **Curtis & Wood**; **Tetzlaff**). Let us assume that the sources of background information B_1 and B_2 are independent. Then, $P(\mathbf{m} \mid B_1)$ denotes static prior information and $P(\mathbf{m} \mid B_2)$ denotes dynamic prior information on model $\mathbf{m}$. The correct way to combine these to derive the total resultant prior information $P(\mathbf{m} \mid B)$, where B is the total state of background information $[B_1, B_2]$, is to use Bayes' rule again making use of the independence of B_1 and B_2, resulting in:

$$P(\mathbf{m} \mid B) = \frac{P(\mathbf{m} \mid B_1)P(\mathbf{m} \mid B_2)}{P(\mathbf{m})}. \qquad (6)$$

Here $P(\mathbf{m})$ is called the null probability distribution and is the state of information about $\mathbf{m}$ when no background information at all is available. The null distribution describes the minimum possible information about the model $\mathbf{m}$ (Tarantola & Valette 1982). We assume that $P(\mathbf{m})$ is a uniform distribution and hence is constant with respect to $\mathbf{m}$.

Curtis & Lomax (2001) have shown that even weak prior information is often sufficient to make computationally tractable those problems that would otherwise be impossible to solve. Notice the implication of Equation 6: that no matter how weakly constrained the dynamic process model used, adding the extra information in B_2 cannot create more uncertainty than existed using static information alone. At worst, the dynamic model provides the minimum possible information so $P(\mathbf{m} \mid B_2) = P(\mathbf{m})$, in which case Equation 6 gives $P(\mathbf{m} \mid B) = P(\mathbf{m} \mid B_1)$. Thus, by using our method the addition of even weak dynamic (or additional static) information always improves knowledge of geological architecture, which in turn can render significantly improved computational efficiencies in constraining models from measured data.

A pitfall that might occur in practice when combining information using Equation 6 is that incompatible approximations to background knowledge B_1 and B_2 are made. For example, if the dynamic information in B_2 is produced using a geological process model, then that information may be obtained on the assumption that the process model is sufficiently detailed to represent reality. Thus, an approximation $\hat{B}_2$ to background information B_2 would be used (for further discussion, see **Curtis & Wood**). Such an assumption would not be required in order to incorporate geostatistical, static information in $P(\mathbf{m} \mid B_1)$, which could be measured directly from outcrops in the field; such information would therefore be conditional on different and incompatible approximations, $\hat{B}_1$, to the background information, B_1. Combining information $P(\mathbf{m} \mid \hat{B}_1)$ and $P(\mathbf{m} \mid \hat{B}_2)$ using Equation 6 directly would be incorrect since the two distributions are based on incompatible information. Typically this would again result in an optimistic assessment of the uncertainty in $P(\mathbf{m} \mid B)$. An example of one correct way of combining such sources of information would be to use the static, statistical information to constrain the range of process model outputs that should be considered (**Curtis & Wood**; **Tetzlaff**; **Wijns *et al.***). Thus, both static and dynamic sources of information are combined using the more restrictive assumptions implicit in the conjunction of $\hat{B}_1$ and $\hat{B}_2$ – that the outcrops used to measure static information were valid analogues, *and* that the geological process model assumptions were valid.

Towards a probabilistic framework

Geological prior information can now be defined more precisely as the field devoted to making geological background information G_q and $G_{q'}$ explicitly, or at least practically, available, and to using such information to solve geoscientific problems. As such, this field makes geology a useful tool to solve problems in other fields of geoscience.

We may use Equations 4 and 5 to create a two-step framework within which most work in the field can be described. Step 1 is defined as the quantification of existing geological knowledge, and is implicit within the partition on the right of Equation 4. As explained above, this is often necessary in order for uncertain, yet correctly calculated inferences to be made. Research in this area quantifies previously qualitative information, and thus, information is moved from $G_{q'}$ to G_q. Step 2 is defined to be the use of such information to solve other geoscientific problems, and is associated with the use of Equation 5 for Bayesian inference (or with the use of some other system of inference).

Table 1 summarises how each of the papers within this volume contributes to these two steps within this probabilistic framework and whether a paper presents a new method, application, or general discussion. We also differentiate between papers that contribute to either *static* or *dynamic* quantitative prior information within Step 1. The papers in this volume are therefore ordered roughly so as to begin with two introductory and discussion papers, of which this is one, then to progress logically through the various stages of Step 1 and then Step 2.

Notice that only two technical papers (**Pshenichny**; **Stephenson *et al.***) address the issue of calculating distribution $P(\mathbf{d} \mid B)$ in the denominator of Equation 2, and of these only **Stephenson *et al.***) addresses this directly. In most applications, this term, called the *evidence*, or *marginal likelihood*, serves merely to normalize the posterior distribution and is ignored in the calculation. This is usually because relative rather than absolute values of the posterior distribution are perceived to be sufficient. However, Malinverno (2000) and Malinverno & Briggs (2004) show how the value of this term can be used to determine the complexity of model **m** that is justified by the data available. In other words, this term quantifies the 'evidence' for any particular model parameterization, and allows the evidence for different possible model parameterizations to be compared.

Example: extraction of 2-D stratigraphic information from 1-D data

Parasequences forming over ~ 100 ka periods, nested within sequences that form over ~ 1–3 Ma, are usually considered to be the fundamental depositional units of marine platform systems, especially in carbonate successions. Yet

Table 1. *The contribution of each paper in this volume placed within a geological prior information framework (see main text for details).*

Contribution	STEP 1		STEP 2				Contribution Type		
	Quantify $G_{q'} \rightarrow G_q$ Static	Quantify $G_{q'} \rightarrow G_q$ Dynamic	Calculate Prior $P(\mathbf{m} \mid \hat{B})$	Calculate Evidence $P(\mathbf{d} \mid \hat{B})$	Calculate Likelihood $P(\mathbf{d} \mid \mathbf{m}, \hat{B})$	Calculate Posterior $P(\mathbf{m} \mid \mathbf{d}, B)$	Method	Application	Discussion
Wood & Curtis	*	*	*	*	*	*	*		*
Baddeley *et al.*	*		*						*
Verwer *et al.*	*							*	
Jones *et al.*	*							*	
Hodgetts *et al.*	*		*					*	
Burgess & Emery		*					*		
Tetzlaff		*					*		
Curtis & Wood	*	*	*				*		
Wijns *et al.*	*	*						*	
Pshenichny	*	*	*	*	*	*	*		
Bowden	*	*	*		*	*	*		
Shapiro *et al.*	*		*		*	*	*	*	
Stephenson *et al.*	*	*	*	*	*	*	*		
White					*		*	*	

the shape, size and stacking pattern of parasequences and sequences within any given basin, as well as details of their internal architectures and facies distribution, are usually poorly known in 2-D or 3-D. This is due either to the sparsity of outcrop or to lack of sufficient resolution afforded by remote-sensing techniques, such as seismic data (a good example of the latter is given in Tinker *et al.* 1999). It has been postulated, however, that under certain conditions the stratal order observed within any given parasequence includes vertical and lateral lithofacies changes that are at least semi-predictable at certain scales (e.g. Lehrmann & Goldhammer 1999), although this is far from accepted (see Wright 1992; Drummond & Wilkinson 1993; Wilkinson *et al.* 1997; Burgess 2001). There are, however, many general principles that govern degrees of predictability in sedimentary successions, and these are enshrined within the discipline of stratigraphy (e.g. Walther's law).

In the example given here, a method is described that uses GPI gained from sedimentary successions (with some additional, explicit assumptions) to obtain probabilistic information about the stratigraphic architecture of platform successions at the parasequence scale. We demonstrate the method on a simple model of transgressive system tracts (TSTs) and show that: (1) data describing only the depths of parasequence or sequence boundaries intersected by a 1-D data profile may be sufficient to provide significant spatial constraints on the far-field (2-D or 3-D) geology when combined with prior information, and (2) the probabilistic results obtained are consistent with those derived from independent geological reasoning.

Methodology

The method combines: (1) data from a single 1-D profile, that could be derived from either an outcrop stratigraphic section or well (henceforth we refer to both potential data sources as a *well*), (2) prior information about the 2-D (or 3-D) stratigraphic architecture that might be expected, and (3) knowledge of the relationships between the 1-D data profile and the 2-D architecture from which it is measured. Components (1) to (3) include all possible information pertinent to the problem considered, i.e. of estimating 2-D parasequence architecture from 1-D data profiles alone.

Using some additional assumptions, information in (1), (2) and (3) individually may be described by independent probability distributions. These distributions are combined using Bayes' rule. The result is a distribution that represents the state of knowledge that includes all possible information about the 2-D parasequence architecture conditional on the assumptions made.

In the example given here we will demonstrate the method on a 2-D representation of shallow platform parasequence architecture formed during a TST, as defined by a simple model (Fig. 1). Although highly simplified, this model may approximate the architecture of at least some real examples, and the probabilistic methodology established can be applied similarly to more complex, 2-D or 3-D geological models, and may also be extended to include a further variety of data types and distributions. That this model *may be* sufficient to explain architectures in question is a qualitative hypothesis, and hence is in $G_{q'}$. The explicit assumption that this *is* the case (i.e. that we may neglect the possibility that the model is insufficient) is made for convenience so that we need not consider the range of all other possible types of models, and results in the approximation $\hat{G}_{q'}$, and hence $\hat{B}$ in Equation 4.

Probability distributions

We now describe the three required probability distributions and the various components of the method:

1. Data distribution. Data from the well must include an estimate of the data uncertainty. This is used to define the *data probability distribution* (or simply the *data distribution*) $\rho(\mathbf{d})$ where $\mathbf{d}$ is the vector of data. In our example, $\mathbf{d}$ contains only the observed depths of intersection of a well with parasequence boundaries (Figs 1 and 2);

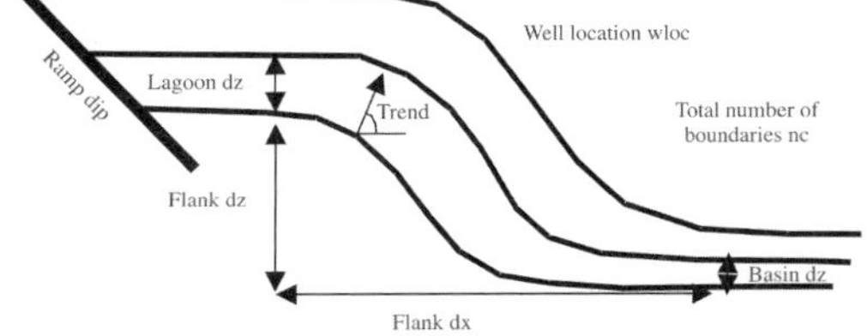

Fig. 1. Model and parameters describing the boundaries of parasequences within a transgressive systems tract (TST), using half a sine curve with horizontal terminations. All parameters are defined graphically except for nc (total number of boundaries), and wloc (horizontal location of well relative to geological architecture). The depositional trend is given as an angle to the horizontal. All parameters other than ramp dip are included in model vector **m**.

$\rho(\mathbf{d})$ describes uncertainties in those depth observations. Note that $\rho(\mathbf{d})$ is not explicitly included in Equation 5, but will be used below to construct the likelihood term.

2. Models and the prior distribution. We call the range of all possible geologies the *model space* M. Any particular geology is called a *model*, which we assume can be described by a vector $\mathbf{m}$. In our example, $\mathbf{m}$ comprises the seven parameters in Figure 1 (ramp dip is not included in $\mathbf{m}$), and although this model is particularly simple, it suffices to illustrate our method. Geological prior information includes only information about models that is independent of the current well data $\rho(\mathbf{d})$. Prior information is described by the *prior (probability) distribution* $P(\mathbf{m} \mid \hat{B})$ that describes the uncertainty in model parameter values given only the available, approximate, background information $\hat{B}$.

3. Model-data relationship and the likelihood function. In theory, the relationships between data $\mathbf{d}$ and models $\mathbf{m}$ can be described by an independent probability distribution (e.g. Tarantola & Valette 1982; Tarantola 1987). In practice, however, the relationship is usually used in the form of a *likelihood function*, a non-normalized probability distribution, since normalization can be difficult. This function describes the relative probability of occurrence of any geological model in the model space given the information contained in the current dataset alone, and therefore incorporates the data distribution $\rho(\mathbf{d})$. This function will be used below to construct the relative likelihood in Equation 5.

The likelihood is often calculated as follows: for any geological model $\mathbf{m}$ we assume that we can calculate synthetically data $\mathbf{d}_0 = \mathbf{f}(\mathbf{m})$ that would have been recorded if $\mathbf{m}$ represented the true geology ($\mathbf{d}_0$ is the *expected data*). These are calculated using modelling techniques or assumptions represented here by function $\mathbf{f}$. In our example, given particular model parameter values $\mathbf{m}$ we can define a set of parasequence boundaries, as shown in Figure 1. Function $\mathbf{f}(\mathbf{m})$ represents the calculation of the intersection depths $\mathbf{d}_0$ between the well and these 'synthetic' parasequences. We calculate the consistency of $\mathbf{d}_0$ with the measured current data distribution $\rho(\mathbf{d})$ by evaluating $\rho(\mathbf{d}_0)$. In turn, variations in $\rho(\mathbf{d}_0) = \rho(\mathbf{f}(\mathbf{m}))$ reflect the relative probability of occurrence of different models $\mathbf{m}$ given the current data distribution alone, and hence $\rho(\mathbf{f}(\mathbf{m}))$ defines the likelihood function.

Bayesian inference

We use Bayes' rule in Equation 5 as follows: term $P(\mathbf{d} \mid \mathbf{m}, \hat{B})$ is the (normalized) likelihood function described above, and $P(\mathbf{m} \mid \hat{B})$ is the prior distribution in the model space. In this example we estimate neither the value of $P(\mathbf{d} \mid \hat{B})$ (the probability of the measured data occurring at all under the assumptions inherent in our modelling) nor the normalization factor for the likelihood. That is because these are both constant with respect to model $\mathbf{m}$. Hence, in this example (as is commonly the case), Bayes' rule is used to estimate the posterior distribution $P(\mathbf{m} \mid \mathbf{d}, \hat{B})$ up to an unknown multiplicative constant. This is sufficient to find the best-fitting model, and the range of possible 2-D models that fit the 1-D data to any desired accuracy.

Application

In this example, we consider the placement of hypothetical vertical sections or wells within a succession of parasequences. Figure 2 shows the intersection of each successive parasequence boundary given at three horizontal locations, for both progradational and regressive phases of a cycle. Notice that, due to the similarity of shape and progressive horizontal offset of successive parasequences, a single, vertical section through multiple parasequences provides similar information to multiple sections through a *single* parasequence at different horizontal locations (Fig. 3). Intuitively then, intersection depths alone from a single, 1-D vertical well contain information about 2-D parasequence shape.

Also, different stacking patterns result in different patterns of depths of intersection between a vertical well and parasequence boundaries. For example, in the progradational situation in Figure 2a, parasequences become relatively thinner (i.e. intersection points become increasingly dense) towards the base of a section taken on the flank (right hand well). In the regressive situation in Figure 2b, the reverse is true. Hence, vertical intersection depth data also contain information about the parasequence stacking patterns. The proposed methodology allows the degree and nature of these constraints to be explored and quantified.

In this example the specific data distribution used was constructed by placing uniform uncertainties of $\pm$ 0.5 m on the synthetic intersection depth data from the right well of the regressive situation (Fig. 2b), representing, for example, uncertainty in the interpretation of parasequence boundaries given available well data.

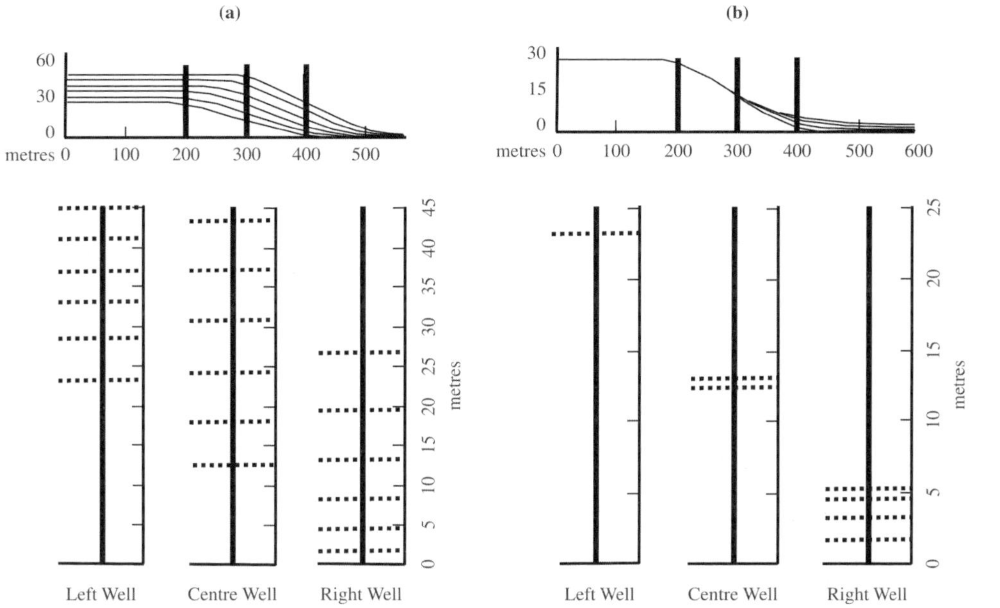

Fig. 2. Modelled parasequence successions. Each of the uppermost models shows the parasequence boundaries and three vertical sections or well locations. Dashed lines on the lower six plots show the depths of intersection of each well with parasequence boundaries. All dimensions are given in metres. (**a**) Progradational sequence where the depositional trend (trajectory) is 10 degrees basinwards. (**b**) Regressive sequence where the depositional trend (trajectory) is – 10 degrees basinwards.

Our static prior information consists of a joint uniform distribution over the parameters given in Figure 1, and this distribution is defined in Table 2. This includes examples of all possible stacking patterns (regressive, aggradational, progradational and retrogradational), and the parameters in Table 2 are derived from background information $\hat{B}$ that includes case studies in published literature (Bosscher & Schlager 1993; Lehrmann & Goldhammer 1999).

An example of the type of far-field information embedded intrinsically within the prior distribution is constructed as follows: the prior distribution is simply a uniform distribution with parameters given in Table 2. A Monte

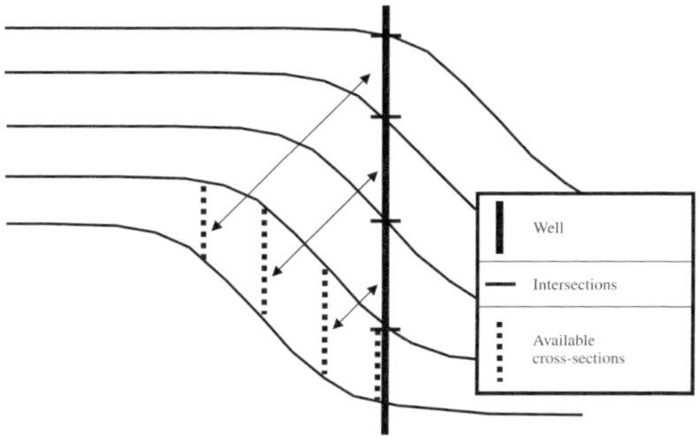

Fig. 3. Sections of different parasequences sampled at a well are approximately equivalent to sections of the same (e.g. lowermost) parasequence taken at successively horizontally offset locations. *Intersections*, parasequence boundaries; *available cross-sections*, horizontally offset points where parasequence thicknesses occur that are equivalent to those found in the well; double arrows, well section to which each available cross-section is similar.

Table 2. *Typical ranges of model parameter distributions appropriate for carbonate formations. Minimum and maximum values for each parameter are sufficient to define the joint uniform prior probability distribution. Distributions for all parameters other than ramp dip were used to illustrate the prior and posterior distributions in Figures 5 and 6. Published sources: Bosscher & Schlager (1993); Lehrmann & Goldhammer (1999).*

Parameter	Min. value	Max. value
Flank dx	25 m	1 km
Flank dz	0 m	50 m
Basin dz	0 m	2 m
Lagoon dz	1 m	10 m
Ramp dip	1 degree	5 degrees
Trend	– 50 degrees	180 degrees
Well location wloc	0 km	1 km
Number of boundaries	1	10

Carlo procedure was used to sample the prior distribution $P(\mathbf{m} \mid \hat{B})$ of Equation 5 across model parameters **m** (i.e. random samples were taken such that the distribution or density of the samples exactly matched $P(\mathbf{m} \mid \hat{B})$). Each Monte Carlo sample is a model consisting of seven parameters. For each, the parasequence flank gradient g, defined to be flank dz/flank dx (Fig. 1), is calculated, and a histogram of the gradient values is created. Figure 4a shows this histogram, which is an approximation to the (non-normalized) prior marginal probability distribution of parasequence flank gradient g. Gradient g is defined using parameters related to the geological architecture up to hundreds or thousands of metres laterally away from the well, since it depends on the lateral extent (length) of the parasequences. This length is not observable at the well. Hence, the non-uniform nature of the distribution in Figure 4 shows that the prior distribution contains significant far-field geological information that cannot be derived from well data alone.

A Monte Carlo procedure was then used to sample the posterior distribution of Equation 5 across model parameters **m** using Bayes' rule. Marginal histograms of these samples over each model parameter are shown in Figure 5. These are (non-normalized) approximations to the posterior probability distribution of each parameter given all prior and data-derived information. These histograms are highly non-uniform despite the fact that both prior and data distributions were uniform. This indicates that there is significant non-linearity in the model–data relationship since linear relationships would result in uniform posterior distributions.

Also, for around half of the parameters, both mean and the maximum likelihood parameter values proved to be poor indicators of the 'true' model parameter values used in Figure 2b. Hence, there is no easy way to infer a single, best estimate for each parameter: the complete 7-D posterior distribution should be considered in any subsequent interpretation.

The posterior marginal histogram over flank gradient g can be estimated by calculating flank dz/flank dx for each Monte Carlo sample of the posterior distribution (Fig. 4b). Differences between the prior and posterior histograms represent exactly the information gained by adding the well intersection depth data from the right-hand well in Figure 2b to the prior information (compare Fig. 4a & b). Clearly, adding only the depth data from a single well has reduced uncertainty in the distribution considerably since the distribution in Figure 4b is narrower than that in Figure 4a (the

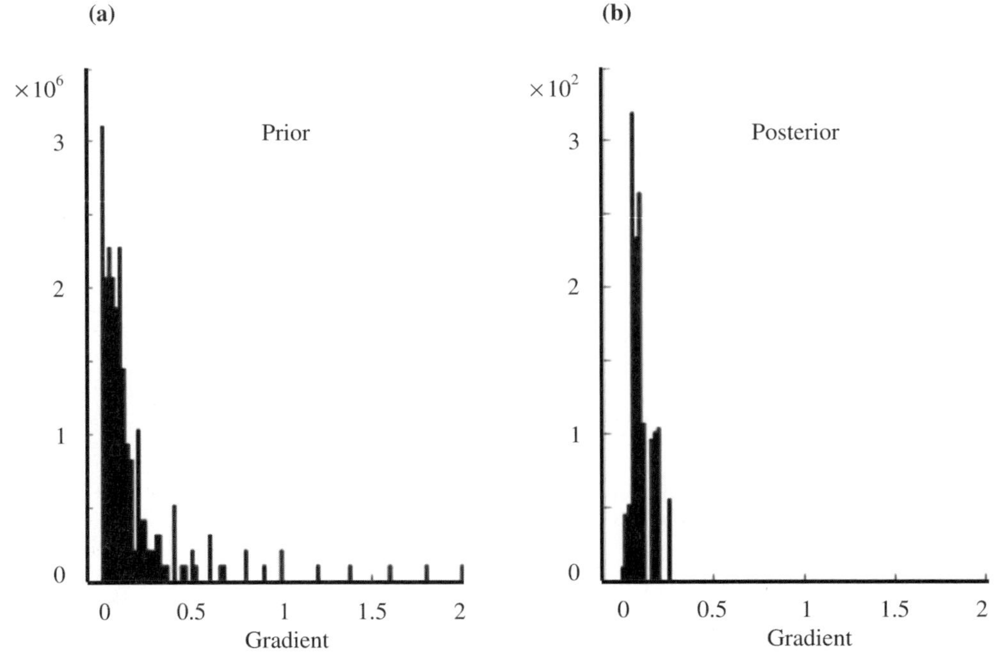

Fig. 4. Histograms over flank gradient g (defined in text): for (**a**) prior distribution; (**b**) posterior distribution.

range of possible gradients has been reduced). Also, the addition of the well data has shifted the maximum likelihood estimate of the gradient close to the value of 0.07 – the correct value for the true architecture shown in Figure 2b.

Similarly, marginal prior and posterior distributions over the various stacking patterns (retrogradational, progradational, aggradational, or regressive) can be calculated (Fig. 6). The stacking pattern is intrinsically a global parameter, which is not observable at the well since it is defined by geometries of the entire succession. Vertical depth data almost excludes the possibility that the true pattern was either progradational or aggradational. The result, that either regressive or retrogradational stacking patterns are almost equally likely, is consistent with the experience of field geologists: by examining a vertical section alone through either one of these patterns in the field, it is well known to be difficult to differentiate between the two.

Discussion of the example

In this example we have combined simple geological prior information (geometrical information about the possible distribution of parasequence or sequence shapes) with 1-D depth data to obtain significantly improved constraints on the parasequence architecture and stacking pattern away from a single outcrop or well (Figs 4, 5 & 6). These constraints were impossible to obtain from either the well data or the prior information alone, and the results in Figure 6 were shown to be consistent with the experience of field geologists.

The method in the example could be used similarly with a combination of more complex data types, and is easily generalized to include multiple outcrop sites or wells. For example, the use of vector data (e.g. dip and azimuth) along the vertical profile would constrain 3-D models in a similar way to the 2-D examples above.

In order to use Bayes' rule to combine information about the data with prior information about the model it was necessary to construct a model relating data and model parameters. The model used here was certainly an over-simplification for many geological scenarios, but serves as an example for the Bayesian method. In principle, it is straightforward to use a more realistic and complex model. For example, the introduction of more detailed stratal shape models or parameter prior information could extend the examples to low stand and high stand successions and result in significantly improved far-field interpretations. However, several studies have suggested that stratal patterns may be far less ordered than those predicted by our assumed model (see Wright 1992; Drummond & Wilkinson 1993; Wilkinson *et al.* 1997; Burgess 2001; Burgess *et al.* 2001; **Burgess & Emery**). This is clearly a contentious topic, and hence in principle our prior information $\hat{B}$ (that the model given in Fig. 1 holds true) might better be regarded as prior assumption, and should properly be assigned a significant uncertainty. This uncertainty would be manifest in the likelihood function, creating more broad posterior probability distributions. Thus, the example above also illustrates the common phenomenon in practical problems of under-estimating uncertainty in the solutions found.

Bayesian inference is explicitly a means to incorporate the prior beliefs of whoever is solving the problem, in a probabilistic, quantitative and rigorous way. The method forces the scientist to be as thorough and as objective as possible in specifying his/her uncertainties, but also implicitly recognizes the fact that all solutions to inference or inverse problems contain subjectivities (i.e. assumptions). It is false to assume that one can ever fully overcome this problem in any study. Provided that assumptions are made clear, however, the solution must simply be interpreted with their full acceptance.

While the assumptions made in our example may be overly stringent, the principle of obtaining information at distances away from the well remains proven as long as one has a model relating data and model parameters that can be believed with less than infinite uncertainty. If no such model exists then Bayes' rule simply results in a posterior distribution that equals the prior distribution (i.e. the model and data added nothing to our prior knowledge) as shown above using Equation 6. Replacing our simple model with a more realistic one in a particular situation would not, therefore, invalidate the method; it would almost certainly enhance it.

Despite the uncertainties and approximations discussed above, the conclusions of this example are that: (1) geological prior information and Bayes' rule allows data from a 1-D profile to be propagated laterally both quantitatively and probabilistically, and (2) quantitative, multidimensional, but uncertain far-field log interpretation is possible from a single 1-D profile of scalar data.

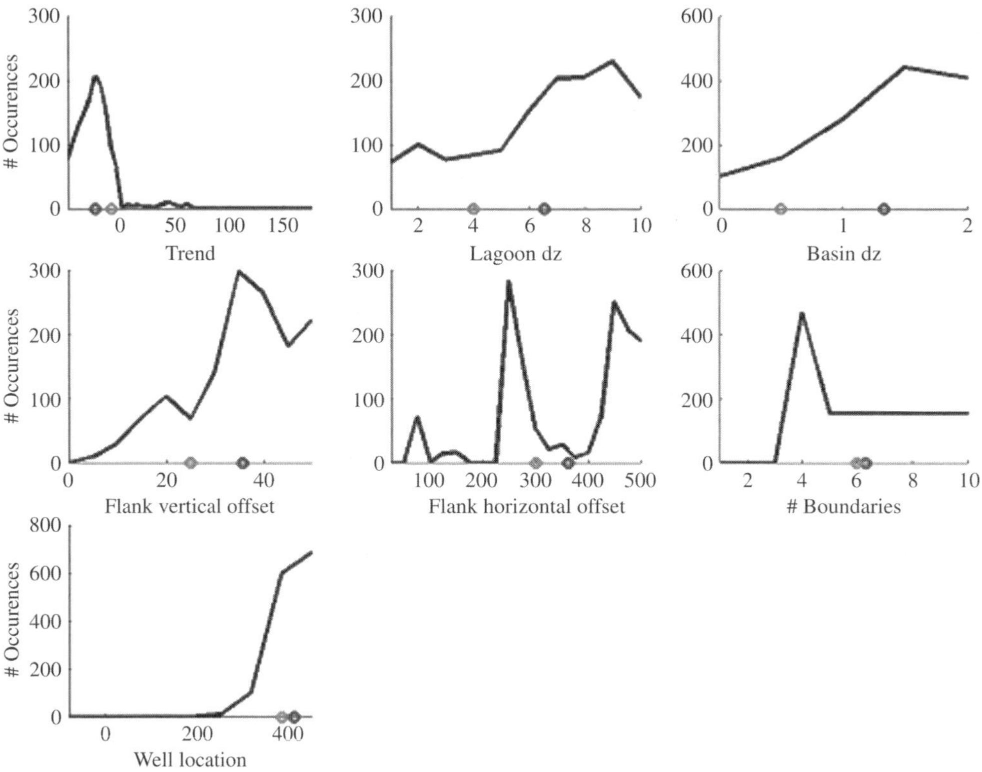

Fig. 5. 1-D marginal histograms for each of the parameters controlling both well location and stratigraphic scenario. Dark circles, mean value in each graph; light circles, the 'true' value in Figure 2.

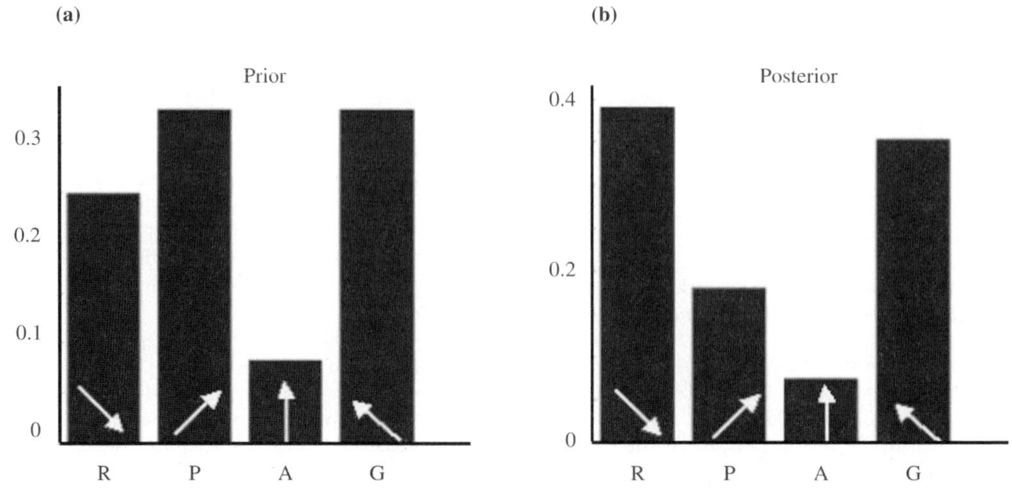

Fig. 6. Histograms over different depositional regimes for: (**a**) prior distribution and (**b**) posterior distribution. G, retrogradation; P, progradation; A, aggradational; R, regressive; arrows, example corresponding trend directions.

Conclusions

In this contribution, we have constructed a Bayesian probabilistic framework that clarifies the definition and applications of geological prior information. We introduce Bayesian inference as an explicit means to incorporate prior beliefs in a probabilistic, quantitative and rigorous way. This method forces a scientist to be as objective as possible in specifying uncertainties, but also implicitly recognizes that, while all solutions to inverse problems contain assumptions that can never be fully overcome, it is possible to interpret these in the light of their full acceptance.

Various components of information must be assembled or assumed in order to solve problems using geological prior information, and uncertainty in such assembled information will propagate to create uncertainty in the solutions found. We have created a two-step framework with which most work in the field can be described. Step 1 is defined as the quantification of geological knowledge; Step 2 is defined as the use of such information to solve other geoscientific problems, using Bayesian inference or another system of inference.

We also present an example of how Bayes' rule can be used to combine simple geological prior information with limited data to obtain significantly improved constraints on a solution by the construction of a model relating data and model parameters that can be believed with less than infinite uncertainty. These constraints were impossible to obtain from either data or prior information alone.

We thank Peter Burgess and Alberto Malinverno for their helpful reviews.

References

BADDELEY, M. C., CURTIS, A. & WOOD, R. A. 2004. An introduction to prior information derived from probabilistic judgements: elicitation of knowledge, cognitive bias and herding. *In:* CURTIS, A. & WOOD, R. (eds). 2004. *Geological Prior Information: Informing Science and Engineering*. Geological Society, London, Special Publications, **239**, 15–27.

BOSSCHER, H. & SCHLAGER, W. 1993. Accumulation rates of carbonate platforms. *Journal of Geology*, **10**, 345–55.

BOWDEN, R. A. 2004. Building confidence in geological models. *In:* CURTIS, A. & WOOD, R. (eds). 2004. *Geological Prior Information: Informing Science and Engineering*. Geological Society, London, Special Publications, **239**, 157–173.

BUDD, A. F., PETERSON, R. A. & MCNEILL, D. F. 1998. Stepwise faunal change during evolutionary turnover: a case study from the Neogene of Curacao, Netherlands Antilles. *Palaios*, **13**, 170–188.

BURGESS, P. M. 2001. Modeling carbonate sequence development without relative sea-level oscillations. *Geology*, **29**, 1127–1130.

BURGESS, P. M. & EMERY, D. J. 2004. Sensitive dependence, divergence and unpredictable behaviour in a stratigraphic forward model of a carbonate system. *In:* CURTIS, A. & WOOD, R. (eds). 2004. *Geological Prior Information: Informing Science and Engineering*. Geological Society, London, Special Publications, **239**, 77–93.

BURGESS, P. M., WRIGHT, V. P. & EMERY, D. 2001. Numerical forward modelling of peritidal carbonate parasequence development: implications for outcrop interpretation. *Basin Research*, **13**, 1–16.

CURTIS, A. & LOMAX, A. 2001. Prior information, sampling distributions and the curse of dimensionality. *Geophysics*, **66**, 372–378.

CURTIS, A. & WOOD, R. 2004. Optimal elicitation of probabilistic information from experts. *In:* CURTIS, A. & WOOD, R. (eds). 2004. *Geological Prior Information: Informing Science and Engineering*. Geological Society, London, Special Publications, **239**, 127–145.

DRUMMOND, C. N. & WILKINSON, B. H. 1993. On the use of cycle thickness diagrams as records of long-term sea level change during accumulation of carbonate sequences. *Journal of Geology*, **101**, 687–702.

HODGETTS, D., DRINKWATER, N. J., HODGSON, J., KAVANAGH, J., FLINT, S. S. & KEOGH, K. J. 2004. Three-dimensional geological models from outcrop data using digital data collection techniques: an example from the Tanqua Karoo depocentre, South Africa. *In:* CURTIS, A. & WOOD, R. (eds). 2004. *Geological Prior Information: Informing Science and Engineering*. Geological Society, London, Special Publications, **239**, 57–75.

JONES, R. R., MCCAFFREY, K. J. W., WILSON, R. W. & HOLDSWORTH, R. E. 2004. Digital field data acquisition: towards increased quantification of uncertainty during geological mapping. *In:* CURTIS, A. & WOOD, R. (eds). 2004. *Geological Prior Information: Informing Science and Engineering*. Geological Society, London, Special Publications, **239**, 43–56.

LEHRMANN, D. J. & GOLDHAMMER, R. K. 1999. Secular variation in parasequence and facies tacking patterns of platform carbonates: a guide to application of stacking pattern analysis in strata of diverse ages and settings. *In:* HARRIS, P. M., SALLER, A. H. & SIMO, J. A. (eds) *Advances in Carbonate Sequence Stratigraphy: Applications to Reservoirs, Outcrops and Models*. SEPM (Society for Sedimentary Geology) Special Publications, **63**, 187–225.

MALINVERNO, A. 2000. A Bayesian criterion for simplicity in inverse problem parameterization. *Geophysical Journal International*, **140**, 267–285.

MALINVERNO, A. & BRIGGS, V. A. (2004). Expanded uncertainty quantification in inverse problems: hierarchical Bayes and empirical Bayes. *Geophysics*, **69**, 1005–1016.

MUKERJ, T., JORSTAD, A., AVSETH, P., MAVCO, G. & GRANLI, J. R. 2001. Mapping lithofacies and pore-fluid probabilities in a North Sea reservoir: seismic inversions and statistical rock physics. *Geophysics*, **66**, 988–999.

PSHENICHNY, C. A. 2004. Classical logic and the problem of uncertainty. *In:* CURTIS, A. & WOOD, R. (eds). 2004. *Geological Prior Information: Informing Science and Engineering*. Geological Society, London, Special Publications, **239**, 111–126.

POJASEK, R. B. 1980. *Toxic and Hazardous Waste Disposal*. Ann Arbor Science, Ann Arbor.

POTTS, D. C. 1984. Generation times and Quaternary evolution of reef-building corals. *Paleobiology*, **10**, 48–58.

ROSENBAUM, M. S. & CULSHAW, M. G. 2003. Communicating the risks arising from geohazards. *Journal of the Royal Statistical Society' Series A (Statistics in Society)*, **166**, 261 –288.

SANDBERG, P. A. 1985. An oscillating trend in Phanerozoic non-skeletal carbonate mineralogy. *Nature*, **305**, 19–22.

SCHON, J. H. 2004. *Physical Properties of Rocks: Fundamentals and Principles of Petrophysics*. Pergamon Press. Handbook of Geophysical Exploration: Seismic Exploration Series, **18**.

SELBY, M. J. 1982. *Hillslope Materials and Processes*. Oxford University Press, Oxford.

SHAPIRO, N. M., RITZWOLLER, M. H., MARESCHAL, J. C. & JAUPART, C. 2004. Lithospheric structure of the Canadian Shield inferred from inversion of surface-wave dispersion with thermodynamic a *priori* constraints. *In:* CURTIS, A. & WOOD, R. (eds). 2004. *Geological Prior Information: Informing Science and Engineering*. Geological Society, London, Special Publications, **239**, 175–194.

STEPHENSON, J., GALLAGHER, K. & HOLMES, C. C. 2004. Beyond kriging: dealing with discontinuous spatial data fields using adaptive prior information and Bayesian partition modelling. *In:* CURTIS, A. & WOOD, R. (eds). 2004. *Geological Prior Information: Informing Science and Engineering*. Geological Society, London, Special Publications, **239**, 195–209.

TARANTOLA, A. 1987. *Inverse Problem Theory*. Xth edition. Elsevier Sciences Amsterdam.

TARANTOLA, A. & VALETTE, B. 1982. Inverse problems = Quest for information. *Journal of Geophysics*, **50**, 159–170.

TETZLAFF, D. M. 2004. Input uncertainty and conditioning in siliciclastic process modelling. *In:* CURTIS, A. & WOOD, R. (eds). 2004. *Geological Prior Information: Informing Science and Engineering*. Geological Society, London, Special Publications, **239**, 95–109.

TETZLAFF, D. M. & PRIDDY, G. 2001. Sedimentary process modeling, from academia to industry. *In:* MERRIAM, D. F. & DAVIS, J. C. (eds) *Geologic Modeling and Simulation: Sedimentary Systems*. Kluwer Academic/Plenum Publishers, New York, 45–69.

TINKER, S., CALDWELL, D., BRINTON, L., BRONDOS, M., CARLSON, J., COX, D., DEMIS, W., HAMMAN, J., LASKOWSKI, L., MILLER, K. & ZAHM, L. 1999. Sequence stratigraphy and 3-D modelling of a Pennsylvanian, distally steepened ramp reservoir: Canyon and Cisco Formations, South Dagger Draw field, New Mexico, USA. *In:* HENTZ, T. F. (ed.) *Advanced Reservoir Characterization for the 21st Century*. 19th Gulf Coast Section, Society of Economic Palaeontologists and Mineralogists Foundation Research Conference, Papers, 214–232.

VERWER, K., KENTER, J. E. M., MAATHUIS, B., & DELLA PORTA, G. 2004. Stratal patterns and lithofacies of an intact seismic-scale Carboniferous carbonate platform (Asturias, Northwestern Spain): a virtual outcrop model. *In:* CURTIS, A. & WOOD, R. (eds). 2004. *Geological Prior Information: Informing Science and Engineering*. Geological Society, London, Special Publications, **239**, 29–41.

WHITE, N. 2004. Using prior subsidence data to infer basin evolution. *In:* CURTIS, A. & WOOD, R. (eds). 2004. *Geological Prior Information: Informing Science and Engineering*. Geological Society, London, Special Publications, **239**, 211–224.

WIJNS, C., POULET, T., BOSCHETTI, F., DYT, C. & GRIFFITHS, C. M. 2004. Interactive inverse methodology applied to stratigraphic forward modelling. *In:* CURTIS, A. & WOOD, R. (eds). 2004. *Geological Prior Information: Informing Science and Engineering*. Geological Society, London, Special Publications, **239**, 147–156.

WILKINSON, B. H., DRUMMOND, C. N., ROTHMAN, E. D. & DIEDRICH, N. W. 1997. Stratal order in peritidal carbonate sequences. *Journal of Sedimentary Research*, **67**, 1068–1082.

WRIGHT, V. P. 1992. Speculations on the controls on cyclic peritidal carbonates: icehouse versus greenhouse eustatic controls. *Sedimentary Geology*, **76**, 1–5.

An introduction to prior information derived from probabilistic judgements: elicitation of knowledge, cognitive bias and herding

MICHELLE C. BADDELEY[1], ANDREW CURTIS[2,3] & RACHEL WOOD[2,4]
[1]*Faculty of Economics and Politics, Gonville & Caius College, Cambridge CB2 1TA, UK (e-mail: mb150@econ.cam.ac.uk)*

[2]*Schlumberger Cambridge Research, High Cross, Madingley Road, Cambridge CB3 0EL, UK*

[3]*Grant Institute of Earth Science, School of GeoSciences, University of Edinburgh, West Mains Road, Edinburgh EH9 3JW, UK*

[4]*Department of Earth Sciences, Cambridge University, Downing Street, Cambridge CB2 3EQ, UK*

Abstract: Opinion of geological experts is often formed despite a paucity of data and is usually based on prior experience. In such situations humans employ heuristics (rules of thumb) to aid analysis and interpretation of data. As a result, future judgements are bootstrapped from, and hence biased by, both the heuristics employed and prior opinion.

This paper reviews the causes of bias and error inherent in prior information derived from the probabilistic judgements of people. Parallels are developed between the evolution of scientific opinion on one hand and the limits on rational behaviour on the other. We show that the combination of data paucity and commonly employed heuristics can lead to herding behaviour within groups of experts. Elicitation theory mitigates the effects of such behaviour, but a method to estimate reliable uncertainties on expert judgements remains elusive.

We have also identified several key directions in which future research is likely to lead to methods that reduce such emergent group behaviour, thereby increasing the probability that the stock of common knowledge will converge in a stable manner towards facts about the Earth as it really is. These include: (1) measuring the frequency with which different heuristics tend to be employed by experts within the geosciences; (2) developing geoscience-specific methods to reduce biases originating from the use of such heuristics; (3) creating methods to detect scientific herding behaviour; and (4) researching how best to reconcile opinions from multiple experts in order to obtain the best probabilistic description of an unknown, objective reality (in cases where one exists).

Geological information is often partly based on personal opinion or judgement. The aim of any good theorist must be to form judgements as rationally as possible, the goal being to coincide with some unobservable, but nonetheless objective reality. For this to be possible, well-reasoned, probabilistic judgements must have the potential to guide the evolution of scientific thought. However, subjectivity inevitably biases opinion. This is particularly true in economics, where belief has a causal role, i.e. changes in belief can change the reality of the phenomena of interest (e.g. stock market share prices, Kahneman & Tversky 1982*a*). In the geosciences, human belief does not usually affect the underlying data-generating system, i.e. Earth processes (although there are exceptions, see Wood & Curtis (2004)). However, accepted or prior opinions of existing experts certainly affect the judgement of others, including future experts in the making. It is therefore desirable to understand typical human biases and errors that may be implicit within the opinion-forming cognitive processes of experts so that their effects can be reduced rather than propagated.

This paper has two purposes: firstly, it explores the cognitive issues surrounding prior information based on probabilistic judgements and the methods of acquiring such information reliably. Secondly, it describes areas requiring future research in order to reduce bias in judgements within the geosciences. These pur-

From: CURTIS, A. & WOOD, R. (eds) 2004. *Geological Prior Information: Informing Science and Engineering*. Geological Society, London, Special Publications, **239**, 15–27. 186239-171-8/04/$15.00

poses are considered in parallel throughout the paper.

The next section explains differences and parallels in the terminology and concepts used in different fields of research, in order that results from other disciplines (statistics, economics, psychology) can be understood within the geosciences. There follows an explanation of how subjective beliefs can be analysed using Bayesian analysis. This is followed by a discussion of probabilistic judgements and cognitive bias, from both an individual and a group perspective, and explains phenomena such as herding. A discussion of methods of eliciting expert opinion in such a way as to reduce the effects of such phenomena is then presented, followed by a particular geoscientific interpretation example in which expert opinion was sought. A summary of potentially fruitful and useful further research directions concludes the paper.

Limits on quantification

A basic distinction common to several frameworks of probability and uncertainty found in different academic disciplines (at least in statistics, geoscience and economics) is that between subjective and objective probability. This distinction is important for the rest of this paper because we generally discuss subjective probabilities, which describe opinions or beliefs. Consequently we begin by reconciling variations in terminology between these different fields and describing key results from each field.

Statistical literature makes the distinction between statistical probability and inductive probability (Carnap 1950; Bulmer 1979). A *statistical probability* is the limiting value of the relative frequency of an event over many trials. Statistical probability is therefore an empirical concept about some objective reality and can be verified via observation and experiment (Bulmer 1979, p. 4). Statistical probabilities or frequencies are usually associated with some *ex post* calculation and/or a complete knowledge of a data-generating process; they may therefore have little to do with fundamental forms of uncertainty emerging from incomplete knowledge. Classical or frequentist statistical approaches have tended to assume implicitly that probabilities are statistical.

In contrast, an *inductive probability* describes rational expectations of a future event. Inductive probabilities act as a guide to life and are formed even when an anticipated event is unprecedented; they therefore have no necessary association with frequency ratios. In contrast to statistical probabilities, inductive probabilities are to do with *ex ante* prediction; they are formed in the face of uncertainty and incomplete knowledge. In most areas of academic investigation, inductive probabilities are of greater practical importance than statistical probabilities because knowledge of an underlying objective reality is either limited or absent. With incomplete knowledge, statistical probabilities based upon past outcomes and an assumption of stationarity are often inappropriate to the analysis of expert judgement in complex situations, either in natural scientific (e.g. geoscience) or social scientific (e.g. economics) contexts.

A similar distinction made in the geosciences is that between statistical probabilities and knowledge-based or conceptual uncertainty (see Pshenichny 2004). Pshenichny defines conceptual probability as a measure of conceptual uncertainty – uncertainty that arises from incompleteness of knowledge – that is clearly associated with the inductive probabilities described above.

In analysing some of the limitations on quantification of economic probabilities, Keynes (1921) distinguishes between *Knightian risk* (the quantifiable risks associated with frequentist concepts) and *Knightian uncertainty* (which is unquantifiable). Under Knightian uncertainty people can say no more than that an event is probable or improbable; they cannot assign a number or ranking in their comparison of probabilities of different events. In the simplest terms the probabilities of Knightian risk and statistical/objective probabilities can be understood as being the same thing: Knightian risk events can easily be calculated using the frequency concepts associated with Classical statistical theory. These events tend to be repeatable and the outcome of a deterministic and immutable data-generating mechanism, such as an unloaded die or a lottery machine. In a world of Knightian risk and quantifiable uncertainty it may be easy to assess and monitor expert judgement just by understanding the mathematics of the data-generating process.

Keynes (1921) argues, however, that in only a limited number of cases can probabilities be quantified in a cardinal sense; in some cases, ordinal comparisons of probability may be possible, but often, particularly in the context of unique events, probabilities may not be quantifiable at all. In reality there may be little consensus in expert (or amateur) opinions, particularly in economic decision making. Keynes (1921) therefore argues that events characterized by Knightian uncertainty are more common than those characterized by Knightian risk, at least in the economic and social sphere.

Such issues are of particular importance in economics because much economic behaviour is forward looking, experiments may not be repeatable and conditions cannot be controlled. People often make subjective probability judgements about events that have not occurred before, for which the data-generating mechanism cannot be known. This makes the quantification and assessment of probabilities particularly problematical because it becomes impossible to match subjective probability judgements with an objective probability distribution. Also, endogeneity (i.e. the path a system takes, determined by events within the system) will limit the accuracy of probabilistic judgements of future events when beliefs about the future are affected by beliefs about the present. Shiller (2003) analyses such phenomena in the context of feedback theory, describing the endogeneity in belief formation: beliefs about the system determine the path of that system (e.g. stock prices go up because people believe they will go up). In this sort of world, no matter how much experts know, there are no objective probability distributions waiting to be discovered; probabilistic judgements will always concern subjective beliefs rather than an immutable reality.

These problems are more worrying for economists than for most natural scientists and certainly for most geoscientists, as there is a limit to what can be done to change existing geology and geological processes (an exception to this may exist in the area of climate change; see Wood & Curtis 2004). However, even though natural scientists often attempt to find out about an immutable, objective reality, the analysis of subjective probabilities is still of fundamental importance in the evolution of knowledge about natural phenomena under conceptual uncertainty. Lack of knowledge of the immutable reality limits the ability to match objective probability distributions and subjective probability assessments. Without knowledge of the mechanisms generating future outcomes, experts must rely on their subjective assessment of prior information.

Statistical probabilities rely on large datasets and assume an absence of subjectivity (Gould 1970). It is unlikely that frequentist approaches will have much resonance in analysing the elicitation of expert opinion on more complex geoscientific or economic issues; on such issues the experts whose opinion might usefully be sought are often few in number. Other approaches to quantification have focused on stochastic modelling strategies, using genetic algorithms and simulated annealing, and on chaos or catastrophe theory (Gleick 1987; Smith *et al.* 1992; Brock 1998; Sornette 2003). These approaches adopt the assumption of some underlying order that might not be immediately obvious but is nonetheless theoretically quantifiable. If this is true, then the statistical probabilities will coincide with judgements of probabilities as long as the procedures of logical inference adopted are correct (Pshenichny 2004). If experts are assumed to be consistent, rational and not prone to making systematic mistakes, then the distinction between conceptual probabilities and statistical probabilities disappears as uncertainty is reduced and as experts increase their knowledge of underlying data-generating processes. However, experts can never be assumed to possess such qualities, as we show below.

Subjective probabilities and Bayesian analysis

Subjective beliefs are important in a world of conceptual uncertainty, and subjective probabilities can be analysed more effectively within a Bayesian approach than within a Classical statistical approach. Bayesian analysis focuses on the subjective confidence that people have in a hypothesis about a single event and can be used to analyse the process by which subjective probabilities or judgements of confidence are updated as new information arrives.

Subjectivity can be thought of as a negative quality, particularly in science. However, the formation of subjective judgements is not necessarily problematical if these subjective judgements are derived in a consistent way (Cox 1946). If any given set of information always generates the same set of probability judgements, then judgements can be said to have formed in a systematic way. The recognition of this insight has made the old subjectivist v. frequentist debates somewhat redundant as focus has shifted towards Bayesian methods, and thus the sting of the term 'subjective' has been drawn.

The starting point in Bayesian analysis is the *prior* probability, which represents the odds that a researcher would place on a particular hypothesis before considering new data. This prior probability is combined with new data using Bayes' theorem, resulting in a *posterior* probability. Bayes shows that the posterior probability of a hypothesis is given by the product of the prior probability derived from all relevant background information (B), and the relative likelihood of the data having been recorded if the hypothesis were true (Lee 1997).

This theorem is best illustrated by means of an example. The posterior probability of finding oil (O), given particular new geological data (G) in addition to the background knowledge (B), is calculated as:

$$\mathrm{P}(O|G,B) = \frac{\mathrm{P}(G|O,B)\mathrm{P}(O|B)}{\mathrm{P}(G|B)},$$

where $\mathrm{P}(O|G,B)$ is the posterior probability of finding oil, given data G, $\mathrm{P}(O|B)$ is the prior probability of finding oil, conditioned on the model B, and $\mathrm{P}(G|O,B)/\mathrm{P}(G|B)$ is the relative likelihood of seeing data G if there is oil. The use of Bayes' theorem to augment geological prior information with new geophysical data is illustrated and developed in Wood & Curtis (2004).

The Bayesian approach differs in at least two significant ways from a Classical frequentist approach. First, the output, i.e. the posterior, is a probability density function, not a point estimate. In addition, it is directly related to the beliefs about a population parameter rather than being a sampling distribution of *estimates* of a population parameter (Kennedy 1998).

The virtue of a Bayesian approach in analysing expert judgement is that it captures the process by which subjective beliefs or degrees of confidences can be updated as new information arrives. Reckhow (2002), for example, argues that the Bayesian approach provides a systematic procedure for pooling and combining knowledge in order to make decisions. The posterior is combined with a loss function (or a utility function representing gains) and a decision is made such that the expected loss is minimized (maximizing expected gain).

There are, however, a number of problems with the Bayesian approach. Firstly, there are practical problems in its application; for example, in economics, there is often a paucity of data that can be used to quantify subjectively formed probability judgements (Kennedy 1998, p. 205). Also, human intuitive cognitive processes do not deal well with Bayesian concepts. Anderson (1998) argues that this is a consequence of the nature of memes (the cultural analogy of genes; see below). Anderson suggests that Bayesian approaches can be refined using the advantages of a frequentist approach, such as mental/visual imagery. For example, consistent methods should be developed: probabilistic information should be represented in graphical or pictorial form; more generally frequentist approaches should be adopted in the presentation of information; attention should be paid to devices for cognitive interpretation; and Bayesian analysts should develop conventions for graphic displays. In other words, some frequentist methods can be used effectively within a Bayesian framework so that human cognition will process subjective probabilities more effectively.

Probabilistic judgements and cognitive bias

A Bayesian approach assumes some sort of order in the process of forming subjective beliefs. As outlined below, recent research within cognitive psychology suggests that expert opinion may not be the outcome of rational, systematic calculation. Some of the most common errors that characterize human assessment and processing of probabilities will now be outlined.

In making probabilistic judgements, research has shown that most ordinary people make common mistakes in their judgements of probabilities (e.g. Anderson 1998). Experts are susceptible to similar biases, both on an individual basis and in terms of group biases. This is because of cognitive limitations in the processing ability of the human mind (Gould 1970; Tversky & Kahneman 1974; Anderson 1998).

The problem originates in the input format of data and in the algorithms used; if prompted by clear signals, the human brain is able to deal with probabilities effectively (Anderson 1998). For example, if students are asked to judge the probability of two coincident events within the context of a statistics class, then they will know what to do. However, outside their class, if they are confronted with a problem requiring probability judgements in a situation in which it is not obvious that this is what is required, they may make a judgement using instincts and intuition rather than statistical reasoning (see Kyburg 1997). The key sources of inconsistency emerge from either *individual bias* or *group bias*.

Individual bias

At least two main types of individual bias can be distinguished: motivational bias and cognitive bias (Skinner 1999). *Motivational bias* reflects the interests and circumstances of the experts (e.g. does their job depend on this assessment? If so, they may be over-confident in order to appear knowledgeable). Motivational biases such as these can often be significantly reduced or entirely overcome by explaining that an honest assessment is required, not a promise.

Also, it may be possible to construct incentive structures encouraging honest assessments of information. Motivational biases can be manipulated because they are often under rational control.

Cognitive bias is more problematic because it emerges from incorrect processing of the information; in this sense it is not under conscious control. Cognitive biases are typically the result of using heuristics, the common-sense devices or rules of thumb derived from experience, used by people to make relatively quick decisions in uncertain situations. They are used because a full assessment of available information is difficult and/or time consuming or information is sparse. For example, when thinking about buying/selling shares from their portfolio, potential investors may have little real knowledge about what is going to happen to share prices in the future; given this limited information, they will adopt the heuristic of following the crowd, i.e. buying when the market is rising and selling when it is falling.

At least four types of heuristics that produce cognitive bias are commonly employed: availability, anchoring and adjustment, control and representativeness (Kahneman & Tversky 1982*b*; Tversky & Kahneman 1974).

Availability is the heuristic of assessing an event's probability by the ease with which occurrences of the event are brought to mind. This often works quite well, but can be biased by the prominence of certain events rather than representing their frequency. For example, headline news of airplane crashes will be brought to mind more readily than that of bicycle crashes, even though the latter are far more frequent. Similarly, the availability heuristic may cause geologists to recall the most interesting, attractive or complex field examples rather than those that are most often encountered, thus biasing their future interpretations.

Anchoring and adjustment form a single heuristic that involves making an initial estimate of a probability, called an 'anchor', and then revising or adjusting it up or down in the light of new information (Tversky & Kahneman 1974). This typically results in assessments that are biased towards the anchor value. For example, in deciding on an appropriate wage to demand in the face of an uncertain economic environment, workers will tend to anchor their demand around their existing wage.

The *control* heuristic is the tendency of people to act as though they can influence a situation over which they have no control. People value lottery tickets with numbers that they have chosen more highly than those with randomly selected numbers, even though the probability of a win is identical in both cases.

The *representativeness* heuristic is where people use the similarity between two events to estimate the probability of one from the other (Tversky & Kahneman 1982). Consider the following example:

> Linda is 31 years old, single, outspoken and very bright. She majored in philosophy. As a student, she was deeply concerned with issues of discrimination and social justice and also participated in anti-nuclear demonstrations. Please check off the most likely alternative:
>
> (1) Linda is a bank-teller.
> (2) Linda is a bank-teller and is active in the feminist movement.

If this problem were to be expressed in probabilistic/statistical terms (i.e. is 1 more likely than 1 and 2), most people with a basic knowledge of probability would realised that two events happening together is less probable than each event happening irrespective of whether the other occurred. However, when confronted with the details about Linda, most people find the second option more likely than the first, simply because the description appears to be more representative of a feminist stereotype and hence is more plausible. This is a *conjunction fallacy*; the former option is the most likely since the probability of the conjunction of two events can never be more probable than either event independently, i.e. a more detailed scenario is at best always equally (and usually less) probable than a simple scenario. In the same way that the probability of events compounded using logical *and* is often over-estimated, the probability of events compounded using logical *or* is often under-estimated (Bar-Hillel 1973).

Such biases can also create an unbounded probability problem: subjects tend to overestimate each probability in a set of exhaustive and mutually exclusive scenarios, so that the estimated sum of all probabilities is greater than one (Anderson 1998, p.15). Also, people do not correct their probability estimates when the set of exhaustive but mutually exclusive outcomes is augmented, again leading to an estimate of total probability in excess of one.

Other well-known biases introduced by the representativeness heuristic include the gambler's fallacy and base-rate neglect. The *gambler's fallacy* is the belief that, when a series of trials have all had the same outcome, then the opposite outcome is more likely to occur in the following trials, since random fluctuations seem more representative of the sample space.

Base-rate neglect is neglect of the relative frequency with which events occur. Consider the following example (the group was told that Dick came from a population of 30 engineers and 70 lawyers):

> Dick is a 30-year-old man. He is married with no children. A man of high ability and high motivation, he promises to be quite successful in his field. He is well liked by his colleagues.

This description provides no information about Dick's profession, but when subjects were asked to estimate the probability of Dick being an engineer, the median probability estimate was 50%, whereas the correct answer is 30%. Subjects ignored the base rate and judged the description as equally representative of an engineer and a lawyer (Tversky & Kahneman 1974, p. 1126).

An interesting and commonly used combination of the gambler's fallacy and base-rate neglect is called *probability matching*, a heuristic known to be used by humans and some other primates (Bliss *et al.* 1995). This is where a reaction from a given range is chosen in proportion to the probabilities of occurrence of various consequences. An example given by Lo (2001) was from World War Two. Bomber pilots were allowed to carry either a flak jacket or a parachute, but not both because of the extra weight. They knew that the probability of getting strafed by enemy guns (hence the flak jacket for protection from shrapnel) was three times that of being shot down (requiring a parachute). Pilots were observed to take flak jackets on three out of every four occasions and parachutes on the fourth occasion. This is not an optimal assessment of the probabilities. Pilots were more likely to have survived if they had taken a flak jacket 100% of the time because the probability of getting strafed by enemy guns was always more likely than the probability of being shot down; the flak jacket was always more likely to be of use.

Probability matching might occur in a geological context if people were asked to estimate the most likely type of geology at a set of subsurface locations in a reservoir, knowing that wells would be drilled at those locations and that their estimates would be checked. If all they knew about the geology in the reservoir was that it was either type 1 or type 2, and that type 1 was three times as likely to occur as type 2, then it is possible that, on average, they would posit geology type 1 as the most likely three times out of every four. Although a non-optimal prediction, this would be a natural tendency for anyone, including an expert, who was not intimately familiar with basic probability theory.

Other cognitive biases reflect emotional responses. For example, in most cases where elicitation has involved experts, the experts have individually been *over-confident* about their knowledge. Multiple experts undergoing the same elicitation procedure often produce barely overlapping or non-overlapping estimates of elicited parameter values. Even groups of experts are observed to display over-confidence in their consolidated results (see below). Over-confidence is especially a problem for extreme probabilities (close to 0% and 100%), which people tend to find hard to assess.

Other forms of emotional response affecting the heuristics employed include mood: people in a happy mood are more likely to use heuristics associated with top-down processing, i.e. to rely on pre-existing knowledge with little attention to precise details. By contrast, people in a sad mood are more likely to use bottom-up processing heuristics, i.e. to pay more attention to precise details than existing knowledge (Schwarz 2000, p. 434). Minsky (1997, p. 519) analyses some of the emotional constraints in the case of expert knowledge, arguing that the 'negative knowledge' associated with some emotional states may inhibit whole strategies of expert thought.

Of all of the biases described above, the most prevalent may be over-confidence and base-rate neglect (Baecher 1988). However, the frequency with which different heuristics are employed within the geosciences has never been assessed and would be an area of fruitful and useful future research.

Group bias

Until now we have assumed that experts are acting as atomistic agents. In reality, experts collect and confer, and, in so doing, generate and perpetuate more complex forms of bias associated with group interactions. Difficulties in acquiring probabilistic knowledge by individual experts are compounded because mistakes and misjudgements may be communicated to other experts. This process is made more complex by a source of individual bias that emerges from anchoring effects, i.e. if one individual's judgements are 'anchored' to those of another (Tversky & Kahneman 1974; Eichenberger 2001).

This implies that expert knowledge will not necessarily evolve along a predetermined, objective path but instead may exhibit path dependency and persistence. Traditionally, the evolution of expert knowledge has been analysed in philosophical terms in the context of Kuhn's theory of scientific revolution. More recently, the evolution of group knowledge has been explained using approaches from evolutionary biology and economic models of mimetic contagion (e.g. in stock markets). Each of these three approaches is briefly described below.

Kuhn's theory of scientific revolutions

Kuhn (1962) and Pajares (2001) explain how scientific thought emerges in the context of given paradigms, emerging from past scientific achievements. Students learn from experts and mentors and this moderates disagreement over fundamental theorems. This is *paradigm anchoring*, in which experts' beliefs are anchored to the existing dominant approach. Such paradigm-based research, however, forces thought within certain boundaries, and this may lead to herding of expert opinion towards prevailing hypotheses and approaches. Experts will be unprepared to overthrow an old paradigm unless there is a new paradigm to replace it, and it may take time for new paradigms to gain acceptability. In normal times 'mopping-up' exercises occur in which the anomalies that do not fit with the current paradigm are discarded (a form of cognitive dissonance may occur – cognitive dissonance being the process of rationalizing information that does not fit with preconceived notions of how the world works). Once a theoretical paradigm has become established, alternative approaches are resisted and paradigms will shift only when evidence and anomalies accumulate to such an extent that a scientific crisis develops. When this happens, a scientific revolution is precipitated.

The problem with this deterministic, descriptive account of how new theories emerge is that it does not illuminate the *processes* underlying the formation of expert opinion. By contrast, biological and economic research has focused more closely on underlying mechanisms, as explained below.

Analogies from evolutionary biology

Path dependency in the evolution of scientific beliefs can be described using biological analogies, for example, those based around the concept of a *meme*, the cultural equivalent of a gene (Dawkins 1976). Imitation is a distinguishing characteristic of human behaviour and a meme is a unit of imitation (Blackmore 1999). The discovery of 'mirror neurons' (neurons in the premotor areas of primate brains that are activated without conscious control and generate imitative behaviour in primates) has lent some scientific support to these biological explanations for imitative behaviour (Rizzolatti *et al.* 2002). This biological approach is also consistent with the use of neural networks for information processing (i.e. mathematical approaches that emulate adaptive learning processes observed in human brains).

Biological insights can be applied in the analysis of belief formation in a human context. Anderson (1998) asserts that successful memes survive (are remembered) and reproduce (are transmitted) effectively when they: (1) map effectively onto human cognitive structures, (2) incorporate a standardized decision structure, and (3) have been reinforced by dominant members of the scientific community. Lynch (1996, 2003) applies these insights in his analysis of the evolutionary replication of ideas and argues that 'thought contagion' affects a wide range of human behaviours and beliefs, including the analysis of stock market behaviour.

Economic analogies: herding and mimetic contagion

An interlocking literature assessing various possibilities for thought contagion has developed in economics, beginning with Keynes's analysis of uncertainty, rationality, subjective probabilities, herd behaviour and conventions (Keynes 1921, 1936, 1937). In Keynes's analysis, herding behaviours are linked back into an analysis of probabilistic judgement in a Bayesian setting. Differences in posterior judgements of probable outcomes may not reflect irrationality but instead may emerge as a result of differences in prior information. Rational economic agents may have an incentive to follow the crowd, and herding will result as a response to individuals' perceptions of their own ignorance. This herding will be rational if the individuals have reason to believe that other agents' judgements are based upon better information than their own; thus other people's judgements become a dataset in themselves. In this way, people will incorporate other people's opinions into their prior information set and their posterior judgements may exhibit herding tendencies. Shiller (2000, 2003) analyses these ideas in the context of feedback theories of endogenous opinion formation in which beliefs about the system determine the path of that system (e.g. as seen in stock markets). These ideas are also developed in Topol (1991), Schleifer (2000), Brunnermeier (2001) and Sornette (2003), among others.

Ideas about herding can be applied to the literature on the acquisition of expertise in an academic context by recognizing that divergent expert opinions reflect uncertainty rather than irrationality or misguided thought. The incorporation of the judgement of other experts into experts' prior information sets explains herding tendencies. However, while expert herding behaviour can be explained as a rational phenomenon, the existence of herding may still contribute to instability if the herd is led down the wrong path. Stable outcomes will only be achieved if the herd can be led along a path of increasing the stock of common (real) knowledge. In such cases, increases in the stock of reliable prior information will contribute to convergence in posterior probabilities.

If, however, the herd path fosters increasing noise within the system, then the process of opinion formation will become unstable. Further research is needed to assess the extent to which expert herds move in either stable or unstable directions. This can be done by assessing the extent to which herd leaders (experts) are selected on objective v. subjective grounds, and by assessing the extent to which herd leaders turn out to be right in the end. This would be another direction for fruitful and useful future research.

Expert elicitation techniques (methods of interrogating experts for information; see below) address these issues. It should be noted, however, that the implications for the social sciences v. the physical sciences may be different because the existence of an objective and immutable reality in many situations in the physical world contrasts with the endogenous, mutable nature of reality in the social, economic and cultural sphere. Feedback loops between belief and reality are therefore less likely to occur in the geosciences.

Elicitation theory

The preceding sections have outlined in general terms how and why people use new information to form probabilistic judgements. This section describes how these ideas are addressed when eliciting and processing expert knowledge and opinion. The key elements of elicitation methods are outlined below, highlighting various sources of individual and group bias and how these may be reduced.

How experts think

The first step in expert elicitation involves identifying experts. Wood & Ford (1993) outline four ways in which an expert's approach to problem solving differs from a novice's approach to problem solving: (1) expert knowledge is grounded in specific cases; (2) experts represent problems in terms of formal principles; (3) experts solve problems using known strategies; (4) experts rely less on declarative knowledge (the what) and more on procedural knowledge (the how).

Eliciting and documenting expert judgement

Biases in knowledge and judgement defined earlier will emerge for experts just as they emerge for ordinary people making everyday decisions. Suitable elicitation methods can sometimes correct the biases in expert opinion, and the problem tackled in the field of elicitation theory is to design the best way to interrogate experts or lay people in order to obtain accurate information about a subject in question.

There are no universally accepted protocols for probability elicitation and there is relatively little formal empirical evaluation of alternative approaches, although some examples and practical guidelines are outlined in Meyer & Booker (2003) and Meyer *et al.* (2003). There are, however, three common assessment protocols: the Stanford/SRI protocol, Morgan and Henrion's protocol and the Wallsten/EPA protocol (Morgan & Henrion 1990). Both the Stanford/SRI and Morgan and Henrion's protocols include five principal phases: (1) motivating the experts with the aims of the elicitation process, (2) structuring the uncertain quantities in an unambiguous way, (3) conditioning the expert's judgement to avoid cognitive biases, (4) encoding the probability distributions, and (5) verifying the consistency of the elicited distributions. The Wallsten/EPA protocol includes the preparation of a document that describes the objectives of the elicitation process, descriptions of cognitive heuristics and biases, and other relevant issues. The expert reads the document before the elicitation process, which is similar to a step in Morgan & Henrion's (1990) protocol.

Within each protocol the elicitor must decide exactly what problems to tackle or what questions to ask in order to maximize information about the topic of interest. Coupe & van der Gaag (1997) showed how a sensitivity analysis might sometimes be carried out in order to see which elicited probabilities would have most influence on the output of a Bayesian belief network. This issue is developed by Curtis & Wood (2004), who optimize the design of the elicitation process in real time. They take into account all information available, as it is elicited, and design the elicitation procedure using the experimental design method of Curtis *et al.* (2004). This must be an optimal strategy, although the details of any particular method must be tailored to particular tasks.

Editing expert judgement: the problem of calibration

If expert judgement is affected by the biases described earlier, then the results of elicitation from each individual expert will need to be calibrated against each other. This in turn requires some statistical model of the elicitation process. The main model proposed in the literature is that of Lindley *et al.* (1979), which requires the presence of an objective assessor to consolidate the results derived from subjective experts. It is not clear, however, why an assessor should be any more objective than the experts.

Other work that attempts to calibrate experts' judgements includes that of Lau & Leong (1999), who created a user-friendly Java interface for elicitation that includes graphical illustrations of possible biases and any inconsistencies in elicited probability estimates. The interface then enters into dialogue with the expert until consistency is achieved. This allows the experts themselves to try to compensate for their natural biases and inconsistencies without the need for a statistical model applied by an external elicitor. Other methods used to deliver the questions asked of the experts in graphical form were reviewed by Renooij (2001). While it is clear, however, that a careful choice of graphical representations of the questions and answers can reduce bias, it is likely that this can only provide partial compensation.

It should be noted that some of the heuristics described earlier perform extraordinarily well in some situations (Gigerenzer & Goldstein 1996; Juslin & Persson 2002). Gigerenzer & Goldstein (1996) examined a particular controlled task using a 'take the best' algorithm. This algorithm selects a best guess at the answer to a question from a set of possibilities by using the minimum number of heuristics from a ranked list such that a guess can be made. The 'take the best' algorithm worked as well as an algorithm that used full probabilistic information to make the guess, but at a fraction of the cost. The problem is that, in practical situations, it is not clear from the results alone whether or not the heuristics work well since there is no objective answer with which to compare them. The role of the elicitor is to try to reduce the use of heuristics unless there is no alternative. In the latter situation, the heuristic used should, at least, be explicitly understood by all involved so that the results can be treated as conditional on this heuristic being effective.

Group elicitation

As discussed previously, natural biases in human cognitive processes and biases caused by paradigm anchoring and scientific revolutions, mimetic contagion and herding effects result in individuals' cognitive biases being compounded in groups. If expert opinion evolves along a particular path just because others have started on that path then the link between subjective probabilities and underlying objective probability distributions may be completely broken. Expert opinion may be led further and further away from an objective grounding, and the evolution of knowledge becomes endogenous, determined by events within the system itself. Group biases may particularly affect the evolution of expert opinion in geoscientific contexts where the information available is relatively weak compared to that which would be necessary to constrain belief so as to be close to reality.

Suppose we wish to assess probabilities based on the estimates of several experts, each with different background knowledge. Few studies have addressed the issue of how best to combine such knowledge into a single probability distribution function (p.d.f.). Individual assessments will almost certainly differ, sometimes by orders of magnitude (see below). In order to reconcile these differences we can either combine the individual assessments into one or we can ask the experts to reach a consensus. The former approach assumes that nothing extra is gained by sharing knowledge and ideas among the experts. The second approach can be jeopardized by group interaction problems (the dominance of one expert over others, or the pressure for conformity). It remains unclear which method produces more accurate final probability estimates.

Philips (1999) studied a case where two groups of experts of varying relevant backgrounds assessed, both individually and in groups, the p.d.f. of very long-term corrosion rates of carbon steel drums containing nuclear waste, after the containers have been sealed in a concrete bunker underground. Philips's first important result was that the two approaches described above led to different probability assessments: the average of individual assessments of the p.d.f. did not match the group consensus p.d.f.

Secondly, individuals' median estimates in each group initially spanned three orders of magnitude. Individuals' p.d.f.s were assessed at three stages during the ensuing discussions, and it was found that this median spread decreased consistently during the formation of a joint consensus distribution (i.e. inter-expert discussion resulted consistently in some convergence between the individuals' views) in only one of the two groups. The observed convergence was, however, accompanied by an increase in the variance or spread of each individual's p.d.f. estimates. This is a typical feature of anchoring and adjustment – individuals simply increase the range of their initial p.d.f.s in order to encompass the range of the consensus p.d.f.

Thirdly, even after a consensus has been reached within each of the groups of experts, the resulting two groups' p.d.f.s differed in their median estimates by three orders of magnitude. This is the same magnitude of difference as was observed between individuals in each group. On further analysis, the reason turned out to be that each group agreed on different basic assumptions in the initial discussions (there was a threefold difference between the two groups' estimates of alkalinity on the outside of the steel drums, and one group considered this sufficient to accelerate the rate of corrosion). We can conclude that particular attention should be paid to surfacing initial assumptions during the elicitation process (but note that deciding on appropriate assumptions may in turn require another elicitation process!).

Geoscientific example

In relatively simple situations there have been apparently successful attempts to elicit uncertain information from experts within the geosciences. One such case is described by Rankey & Mitchell (2003). Six seismic interpreters with different levels and types of experience were asked to define the depth of the top of a roughly circular, buried Devonian pinnacle reef (a hydrocarbon reservoir) in the western Canadian Basin. They were provided with 3-D seismic data spanning the reef and, initially, with log data from two wells to which the seismic data had been matched in depth. After a first-pass interpretation they were provided with logs and depth matches from a further four wells, and also with literature on the geology of the area, after which they were allowed to revise their interpretation. Four of the six interpreters chose to do so, although all changes made were minor. To maintain consistency the base of the reef was picked prior to the experiment, considered 'known' for the purpose of this experiment, and given to all interpreters. The purpose of the exercise was to elicit information both on the location of the top of the reef and on its uncertainty.

Particular features to note about this elicitation process are:

- Interpreters were kept separate from one another during the exercise so that there were no group interaction effects.
- There would appear to be few motivational biases – none of the interpreters (to our knowledge) had a vested interest in any particular results.
- There may have been some anchoring after the first-pass interpretation: one of the interpreters who did not change his interpretation noted, 'I did ... not want to change any of my picks based on the additional well data – looks like I had it nailed.' There may have been an element of pride for some interpreters in appearing to get the first-pass interpretation correct.
- The basic assumptions intrinsic to the experiment, and made by all interpreters (e.g. the depth of the base of the reef, the assumed accuracy of the depth matches to well logs), themselves required an elicitation (interpretation) process. This was not detailed in the report by Rankey & Mitchell (2003), but it is clear that errors in such assumptions could have an impact on both the interpretations made in this experiment and, potentially, their uncertainty.

This exercise was agreed by all interpreters to be relatively simple: the seismic data was of good quality, the stratigraphy was simple and the well ties appeared to be good. One interpreter commented that there were few critical decision points, the main one being whether to pick a high or a low arrival on the seismic data through the reef's southern flank. No interpreters found any decisions difficult to make. The southern–central part of the reef is extensively dolomitized, the rest of the reef is limestone.

Most interpreters picked very similar top surfaces all over the reef, other than around the pinnacle margins. The largest variation was in the southern–central section. In fact, interpreters were picking a 'zero crossing' on the 3-D seismic data volume (a point at which the seismic data changes from positive to negative values) because this coincided with the reef crest on one of the wells. This zero crossing became less distinct and appeared to bifurcate into an upper and a lower zero crossing over the dolomitized area.

Subsequent modelling showed that the zero crossing was probably due to interference in the seismic waves ('tuning') between reflections from the limestone and from layers just above the reef (the reef crest was not the largest seismic impedance contrast in its vicinity). The difficulty in the southern–central section was probably due to different tuning effects caused by the switch from limestone to dolomite, causing the zero crossing to deviate from the reef top. Interpreters did not know the results of this modelling during this exercise.

Each interpreter was fairly confident of their own interpretation (their individual estimated uncertainty was low), yet the spread of interpretations around the southern–central margin was significant over an area of 200 m laterally. In this area, differences in gross pore volumes, predicted using seismic attributes around the interpreters' top-reef estimates, differed by up to 24%; predictions of gross rock volume varied by up to 13%. Interpreters had correctly identified this area as containing a significant decision point, but had assumed that they had made the correct decision. It therefore seems reasonable to conclude that the over-confidence bias occurred in most interpreters.

Ultimately there is no proof that the final uncertainty estimates obtained in this experiment were reasonable, as there is no report of a well being drilled in the more complex southern–central area to check whether the true top-reef depth lay within the bounds defined by the range of interpretations. However, this is one of the few geological interpretation cases within the hydrocarbon industry where a genuine attempt was made to hold a set of well-controlled trials in order to assess this uncertainty. In most cases a single individual interprets the seismic data, the interpretation being 'checked' by one other. In the above experiment this would have lead to an over-confident and biased interpretation; hence the results of this experiment at least demonstrated a successful improvement on more usual methods.

Conclusions

This paper explains how the judgements of experts can be biased by their use of heuristics to guide the formation of their opinions. We present research from cognitive psychology showing that such heuristics are used by both experts and lay people alike and often cause biases in individuals' perceived knowledge – commonly over-confidence and base-rate neglect. We show that models from economic and biological literature explain how, in conditions of uncertainty or asymmetric prior information, such biases can cause herding behaviour, potentially leading to instability in the stock of common, accepted knowledge. Elicitation theory is described that attempts to elicit robust information from experts by mediating the effects of such biases.

During the course of this discussion paper we identified several key directions in which future research is likely to be both useful and fruitful:

- measurement of the frequency with which different heuristics tend to be employed by experts within the geosciences (this will help to define the range of biases that may occur, and their likely prevalence);
- development of geoscience-specific methods to reduce biases originating from the use of such heuristics;
- creation of methods to detect herding behaviour amongst scientific experts;
- assessment of whether herd leaders are selected on objective or subjective grounds, or whether they are better described as emergent or self-created; also assessment (in retrospect) of the the frequency with which they turned out to be correct in their contested and uncontested opinions;
- research on how best to reconcile opinions from multiple experts to obtain the best probabilistic description of an unknown, objective reality (in cases where one exists).

Geological prior information is often partly opinion or judgement-based. When assembling prior information in order to augment it with new data, at the very least it is necessary to be aware of the heuristics and biases discussed herein in order to limit their propagation to future accepted knowledge. In itself this will help to reduce emergent group behaviour, such as herding, thereby increasing the probability that the stock of common knowledge will become stable and converge towards facts about the Earth as it really is.

References

ANDERSON, J. L. 1998. Embracing uncertainty: the interface of Bayesian statistics and cognitive psychology. *Conservation Ecology*, **2**, 2. [Online] World Wide Web Address: http://www.consecol.org/vol2/iss1/art2.

BAECHER, G. B. 1988. *Judgemental Probability in Geotechnical Risk Assessment*. Technical Report Prepared for The Office of the Chief, US Army Corps of Engineers. World Wide Web Address: http://www.ence.umd.edu/~gbaecher/papers.d/judge_prob.d/judge_prob.html.

BAR-HILLEL, M. 1973. On the subjective probability of compound events. *Organizational Behaviour and Human Performance*, **9**, 396–406.

BLACKMORE, S. 1999. *The Meme Machine*. Oxford University Press, Oxford.

BLISS, J. P., GILSON, R. D., & DEATON, J. E. 1995. Human probability matching behaviour in response to alarms of varying reliability. *Ergonomics*, **38**, 2300–2312.

BROCK, W. A. 1998. Chaos and complexity in economic and financial systems. *In:* VON FURSTENBURG, G. (ed.) *Acting Under Uncertainty: Multidisciplinary Approaches*. Kluwer Academic, Boston.

BRUNNERMEIER, M. K. 2001. *Asset Pricing Under Asymmetric Information*. Oxford University Press, Oxford.

BULMER, M. G. 1979. *Principles of Statistics*. Dover, New York.

CARNAP, R. 1950. *Logical Foundations of Probability*. Routledge & Kegan Paul, London.

COUPE, V. M. H., & VAN DER GAAG, L. C. 1997. Supporting probability elicitation by sensitivity analysis, *Knowledge Acquisition, Modeling and Management: Lecture Notes in Computer Science*, **1319**, 335–340.

COX, R. T. 1946. Probability, frequency and reasonable expectation. *American Journal of Physics* **14**, 1–13.

CURTIS, A. & WOOD, R. 2004. Optimal elicitation of probabilistic information from experts. *In:* CURTIS, A. & WOOD, R. (eds) *Geological Prior Information: Informing Science and Engineering*. Geological Society, London, Special Publications, **239**, 127–145.

CURTIS, A., MICHELINI, A., LESLIE, D. & LOMAX, A. (2004). A deterministic algorithm applied to experimental design applied to tomographic and microseismic monitoring surveys. *Geophysical Journal International*, **157**, 595–606.

DAWKINS, R. 1976. *The Selfish Gene*. Oxford University Press, Oxford.

EICHENBERGER, R. 2001. Economic incentives transform psychological anomalies. *In:* FREY, B. S. (ed.) *Inspiring Economics: Human Motivation in Political Economy*. Edward Elgar, Cheltenham, 21–36.

GIGERENZER, G. & GOLDSTEIN, D. G. 1996. Reasoning and fast and frugal way: models of bounded rationality. *Psychological Review*, **103**, 650–669.

GLEICK, J. 1997. *Chaos: Making a New Science*. Penguin Harmondsworth.

GOULD, P. 1970. Is *Statistix inferens* the geographical name for a wild goose? *Economic Geography*, **46**, 438–448.

JUSLIN, J. & PERSSON, M. 2002. Probabilities from Examplars (PROBEX): a "lazy" algorithm for probabilistic inference from generic knowledge. *Cognitive Science*, **26**, 563–607.

KAHNEMAN, D. & TVERSKY, A. 1982*a*. Subjective probability: a judgement of representativeness. *In:* KAHNEMAN, D., SLOVIC, P. & Tversky, A. (eds) *Judgement Under Uncertainty: Heuristics and Biases*. Cambridge University Press, Cambridge, 32–47.

KAHNEMAN, D. & TVERSKY, A. 1982*b*. On the psychology of prediction. *In:* KAHNEMAN, D., SLOVIC, P. & TVERSKY, A. (eds) *Judgement Under Uncertainty: Heuristics and Biases*. Cambridge University Press, Cambridge, 48–68. [Reprinted from *Psychological Review*, **80**, 237–251.]

KENNEDY, P. 1998. *A Guide to Econometrics*. 4th edition. Blackwell, Oxford.

KEYNES, J. M. 1921. *A Treatise on Probability*. Macmillan, London.

KEYNES, J. M. 1936. *The General Theory of Employment, Interest and Money*. Macmillan, London.

KEYNES, J. M. 1937. The general theory of employment. *Quarterly Journal of Economics*, **51**, 209–223.

KUHN, T. 1962. *The Structure of Scientific Revolutions*. University of Chicago Press, Chicago.

KYBURG, H. E. 1997. Expertise and context in uncertain inference. *In:* FELTOVICH, P. J., FORD, K. M. & HOFFMAN, R. R. (eds) *Expertise in Context: Human and Machine*. MIT Press and American Association for Artificial Intelligence, Cambridge, Massachusetts, 499–514.

LAU, A.-H. & LEONG, T.-Y. 1999. Probes: a framework for probability elicitation from experts. *American Medical Informatics Association Annual Symposium*, November 1999, 301–305.

LEE, P. M. 1997. *Bayesian Statistics: An Introduction*. 2nd edition. Arnold, London.

LINDLEY, D. V., TVERSKY, A. & BROWN, R. V. 1979. On the reconciliation of probability assessments (with discussion). *Journal of the Royal Statistical Society*, **142**, 146–180.

LO, A. W. 2001. Bubble, rubble, finance in trouble? [Edited luncheon address.] 3rd Annual Institute of Psychology and Markets Conference, 1 June 2001, New York City.

LYNCH, A. 1996. *Thought Contagion: How Belief Spreads Through Society*. Basic Books, New York.

LYNCH, A. 2003. An introduction to the evolutionary epidemiology of ideas. *The Biological Physicist*, **3**, 7–14.

MEYER, M. A. & BOOKER, J. M. 1991. *Eliciting and analyzing expert judgement: a practical guide*. Knowledge acquisition for knowledge-based system series, **5**. Academic Press, London.

MEYER, M. A., BUTTERFIELD, K. B., MURRAY, W. S., SMITH, R. E. & BOOKER, J. M. 2003. Guidelines for eliciting expert judgement as probabilities or fuzzy logic. *In:* ROSS, T., BOOKER, J. & PARKINSON, W. (eds) *Fuzzy Logic and Probability Applications: Bridging the Gap*. ASA-SIAM Series on Statistics and Applied Probability, Philadelphia, **7**.

MINSKY, M. 1997. Negative expertise. *In:* FELTOVICH, P. J., FORD, K. M. & HOFFMAN, R. R. (eds) *Expertise in Context: Human and Machine.* MIT Press and American Association for Artificial Intelligence, Cambridge, Massachusetts, 515–521.

MORGAN, M. G. & HENRION, M. 1990. *Uncertainty: A Guide to Dealing with Uncertainty in Quantitative Risk and Policy Analysis.* Cambridge University Press, Cambridge.

PAJARES, F. 2001. The structure of scientific revolutions. *Philosophers Web Magazine.* World Wide Web Address: http://www.philosophers.co.uk.

PHILIPS, L. D. 1999. Group elicitation of probability distributions: are many heads better than one? *Bayesian Research Conference; Decision Science and Technology, Reflections on the Contributions of Ward Edwards,* February 1999, Los Angeles. Kluwer Academic, Boston and London, 313–330.

PSHENICHNY, C. A. 2004. Classical logic and the problem of uncertainty. *In:* CURTIS, A. & WOOD, R. (eds) *Geological Prior Information: Informing Science and Engineering.* Geological Society, London, Special Publications, **239**, 111–126.

RANKEY, E. C. & MITCHELL, J. C. 2003. That's why it's called interpretation: impact of horizon uncertainty on seismic attribute analysis. *The Leading Edge,* **22**, 820–828.

RECKHOW, K. H. 2002. Bayesian approaches in ecological analysis and modeling. *In:* CANHAM, C. D., COLE, J. J. & LAUENROTH, W. K. (eds) *The Role of Models in Ecosystem Science,* Princeton University Press, Princeton.

RENOOIJ, S. 2001. Probability elicitation for belief networks: issues to consider. *The Knowledge Engineering Review,* **16**, 255–269.

RIZZOLATTI, G., CRAIGHERO, L. & FADIGA, L. 2002.The mirror system in humans. *In:* STAMENOV, M. I. & GALLESE, V. (eds) *Mirror Neurons and the Evolution of Brain and Language.* Advances in Consciousness Research, John Benjamins, Amsterdam, **42**, 37–59.

SHILLER, R. J. 2000. *Irrational Exuberance.* Princeton University Press, Princeton.

SHILLER, R. J. 2003. From efficient markets theory to behaviour finance. *Journal of Economic Perspectives,* **17**, 83–104.

SCHLEIFER, A. 2000. *Inefficient Markets: An Introduction to Behavioural Finance.* Oxford University Press, Oxford.

SCHWARZ, N. 2000. Emotion, cognition, and decision making. *Cognition and Emotion,* **14**, 433–440.

SKINNER, D. C. 1999. *Introduction to Decision Analysis.* Probabilistic Publishing, Delaware.

SMITH, M., SCALES, J. A. & FISCHER, T. 1992. Global search and genetic algorithms. *The Leading Edge,* **11**, 22–6.

SORNETTE, D. 2003. *Why Stock Markets Crash: Critical Events in Complex Financial Systems.* Princeton University Press, Princeton.

TOPOL, R. 1991. Bubbles and volatility of stock prices: effect of mimetic contagion. *Economic Journal,* **101**, 786–800.

TVERSKY, A. & KAHNEMAN, D. 1974. Judgement under uncertainty: heuristics and biases. *Science,* **185**, 1124–1131.

TVERSKY, A. & KAHNEMAN, D. 1982. Judgements of and by representativeness. *In:* KAHNEMAN, D., SLOVIC, P. & TVERSKY, A. (eds) *Judgement Under Uncertainty: Heuristics and Biases.* Cambridge University Press, Cambridge, 84–98.

WOOD, R. & CURTIS, A. 2004. Geological prior information and its application to geoscientific problems. *In:* CURTIS, A. & WOOD, R. (eds) *Geological Prior Information: Informing Science and Engineering.* Geological Society, London, Special Publications, **239**, 1–14.

WOOD, L. E. & FORD, J. M. 1993. Structuring interviews with experts during knowledge elicitation. *International Journal of Intelligent Systems,* **8**, 71–90.

Stratal patterns and lithofacies of an intact seismic-scale Carboniferous carbonate platform (Asturias, northwestern Spain): a virtual outcrop model

KLAAS VERWER[1], JEROEN A.M. KENTER[1], BEN MAATHUIS[2] & GIOVANNA DELLA PORTA[1]

[1]*Faculty of Earth and Life Sciences, Vrije Universiteit, De Boelelaan 1085, 1081 HV, Amsterdam, Netherlands (e-mail: klaas.verwer@falw.vu.nl)*

[2]*International Institute for Geo-Information Science and Earth Observation (ITC), PO Box 6, 7500 AA Enschede, Netherlands*

Abstract: Among the more challenging questions in geology are those concerning the anatomy of sedimentary bodies and related stratal surfaces. Though significant progress has been made on the interpretation of depositional environments, little systematic data are available on their dimensions and geometry. With the recent advances in computer power, software development and accuracy of affordable positioning equipment, it has now become possible to extract high-resolution quantitative data on the anatomy of sedimentary bodies. In Asturias, northwestern Spain, aerial photography provides continuous 2-D cross-sections of a seismic-scale, rotated to vertical, carbonate platform margin of the Early Carboniferous. Digital elevation models, orthorectified aerial photographic imagery and ground verification of stratal surfaces generated the elements that are required to reconstruct the true dimensions, angular relationships of bedding planes and the spatial distribution of facies units in this platform margin. Along with biostratigraphy this provides sufficient constraints to estimate rates of progradation, aggradation, growth and removal of sediments in the slope environment. Here we present a methodology to create outcrop models and integrate complementary types of data that provide new insights in sedimentology that were previously unattainable.

The anatomy of bedding planes (stratal patterns) and geometry of depositional bodies have been the focus of sedimentological studies since William Smith published the first geological map in 1815. Precise and accurate anatomical information on sedimentary bodies and their internal bedding geometries is essential for depositional models, sequence stratigraphy, prediction from seismic reflection data, flow models for hydrocarbon reservoirs and aquifers, remediation of contaminated aquifers, the reconstruction of basins and the general advancement of the field of sedimentology, to name just a few applications (Read 1985; Vail 1987; Sarg 1988). With the rapidly increasing demand for potable water and energy resources, the need for such data will increase in the future. To accommodate such demand, imaging techniques such as remote sensing, reflection seismology, ground-penetrating radar and time-domain electromagnetic surveying have significantly improved in resolution but lack sufficient ground truthing. The reason for this disparity is that, by definition, 'remote' sensing is used to acquire data on parameters that do not lend themselves readily to field verification. Indeed, the quantitative link between geological and physical observations is often absent. Though the need for the quantification of geometrical properties of sedimentary deposits is therefore clear and widely acknowledged (Kerans *et al.* 1995), little data are available in the open literature that provide high-resolution measurements of stratal anatomy and lithofacies boundaries. Sparse data are available on bedding properties (bed length and continuity) in ancient shallow-water carbonates and on clastic coastal and deep-water systems because much resides in confidential industry studies. Some of the few studies that currently exist include: the terminal Proterozoic Nama Group, a shallow carbonate ramp in Namibia (Adams *et al.* 2001, 2002, 2004); the Eocene Ainsa-II turbidite system (Jenette & Bellian 2003*a*, *b*; Løseth *et al.* 2003*a*, *b*); the Tanqua Karoo deep-water complex (Hodgetts *et al.* 2003); the Carboniferous Ross Formation, County Clare, western Ireland (Pringle *et al.* 2003); the Devonian of the Canning Basin

From: CURTIS, A. & WOOD, R. (eds) 2004. *Geological Prior Information: Informing Science and Engineering*. Geological Society, London, Special Publications, **239**, 29–41. 186239-171-8/04/$15.00

(Hasler *et al.* 2003), and a rotated carbonate platform margin in northern Spain (this paper and references cited).

One of the reasons for the apparent lack of such information has been the difficulty of capturing the 2-D and, where accessible, the 3-D geometry of stratal surfaces and boundaries between lithofacies units in a reliable and precise manner. Laser surveying and differential global-positioning systems (DGPS) are now refined, miniaturized, robust and portable, and the data are readily processed and integrated with relatively easy to use geographical information system (GIS) software packages. Since the costs of these systems have decreased sufficiently to allow academic groups to afford them, these tools can now be applied to sedimentological problems.

In Asturias, northwestern Spain, aerial photographs provide an excellent view over the architecture of a seismic-scale, rotated but intact carbonate platform (Bahamonde *et al.* 1997; Della Porta *et al.* 2002*a*, *b*, 2003, 2004; Kenter *et al.* 2002). Aerial photographs provide insights into general lithofacies distribution and architecture of the Carboniferous carbonate platform margin. A major problem in mapping geological features from aerial photos is that the image is distorted by the camera lens properties and relative position (flight path, height and camera orientation) with respect to the surface object, especially when the terrain has a highly variable relief. In addition, finite image resolution, layovers, foreshortenings, dilatations, shadowing and false crests contribute to the distortion of the imagery. A standard method of eliminating distortions is orthorectification of the aerial photographs (Greeve 1997). To derive a quantitative model of the stratal and lithofacies architecture of this steep-margined carbonate platform, ground control calibration measurements were made and stratal patterns were tracked using a real-time kinematic (RTK) DGPS.

The aim of this study is to provide a multi-layered digital geological model that incorporates: (1) a digital elevation model (DEM) with 2.5 m resolution; (2) high-altitude, low-resolution colour (*c.* 2 m) and low-altitude, high-resolution (*c.* 0.5 m) black-and-white orthorectified aerial photographs; (3) a lithofacies distribution map; and (4) stratal surfaces. This model can serve as quantitative analogue for subsurface carbonate depositional systems with similar origin and geometry known from, for example, the Pricaspian Basin (Kenter & Harris 2002; Harris & Kenter 2003). The digital model may allow quantitative and visual integration of all spatial and geological data and will significantly enhance the understanding of such systems.

Geological setting

A Carboniferous, rotated, but intact, seismic-scale carbonate platform is exposed in the Cantabrian Mountains, northwestern Spain. The Cantabrian Zone is one of the six zones into which the Iberian Massif was divided according to structural style and stratigraphy. The Cantabrian Zone is characterized by a set of imbricated thrust sheets deformed by thin-skinned tectonism (Julivert 1971). Within the Cantabrian Zone, five tectonostratigraphic provinces (Fig. 1) were distinguished and named, from the most internal to the most external, as Fold and Nappe, Central Asturian Coal Basin, Ponga Nappe, Picos de Europa and Pisuerga–Carrión Provinces (Julivert 1971). The Sierra de Cuera, located in the northeastern sector of the Ponga Nappe Province/Unit, was involved in thrust propagation during the Late Moscovian to Early Kasimovian (Marquinez 1989) and is the focus of this study.

During the Serpukhovian to Moscovian, the Cantabrian Zone was a marine foreland basin, where the subsiding proximal areas were occupied by thick siliciclastic wedges of turbiditic and deltaic successions separated by calcareous units (e.g. Central Asturian Coal Basin), while in the more distal and stable areas (e.g. Picos de Europa, Ponge Nappe) extensive carbonate platforms nucleated (Colmonero *et al.* 1993). These carbonate successions comprise the Bashkirian Valdeteja Formation (850 m thick), which is predominantly progradational, and the Moscovian Picos de Europa Formation (800 m thick), which is mostly aggradational. They overlie the Serpukhovian Barcaliente Formation, which consists of 350 m-thick, dark, finely laminated mudstone that represents an extensive and stable shelf that served as the foundation for the Upper Carboniferous platforms (Bahamonde *et al.* 1997).

The Upper Carboniferous (Lower Bashkirian to Lower Moscovian; Della Porta *et al.* 2002*a*) Sierra de Cuera outcrop corresponds to a south-verging thrust sheet tilted in the Early Kasimovian (Marquinez 1989). In Sierra de Cuera, no evident separation between the Valdeteja and the Picos de Europa Formations has been recorded (Bahamonde *et al.* 1997).

During the Early Bashkirian, a low-angle carbonate ramp nucleated on the Barcaliente Formation. Initial vertical aggradation was followed by horizontal progradation. During

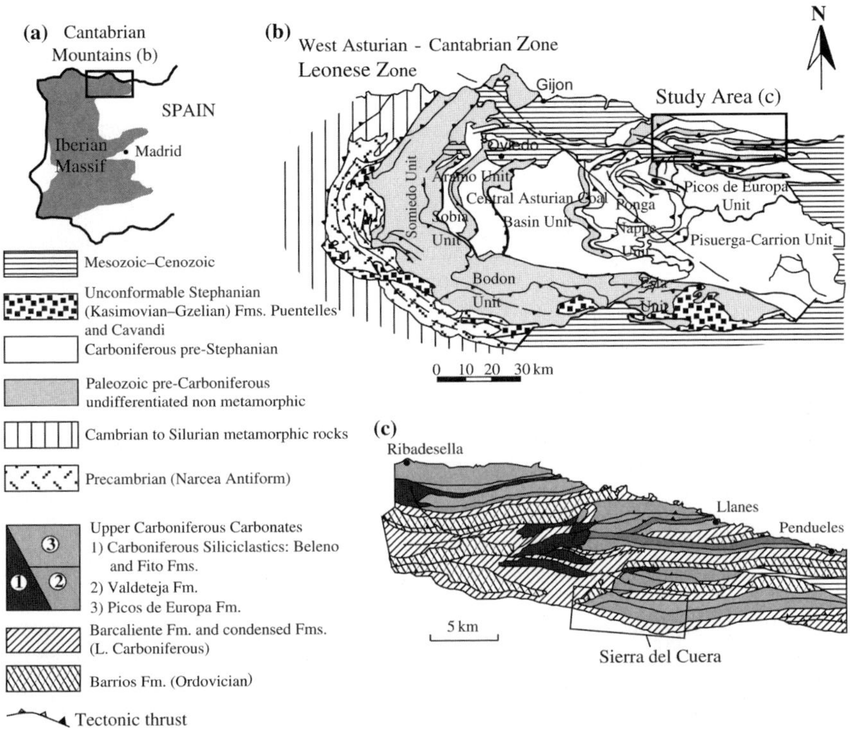

Fig. 1. (**a**) Location of the Cantabrian Mountains in the northeastern sector of the Hercynian Iberian Massif. (**b**) Schematic tectonic map of the Cantabrian Mountains area with major tectonic units (modified after Julivert 1971) and an overview of the study area. (**c**) Study area with location of Sierra de Cuera platform outcrop.

the Bashkirian, dominant progradational phases, alternating with several aggradational phases, produced steep clinoforms. At the transition between Bashkirian and Moscovian, the flat-topped shallow-water platform (1 km thick, 10 km wide) developed and grew by alternation of mostly aggrading and minor prograding phases during the Early Moscovian (Kenter *et al.* 2002).

Due to the nearly vertical orientation of bedding planes after tectonic tilting (70° to 90°), aerial photographs provide excellent images of cross-sections of the depositional system. Structural deformation, mapped to investigate the potential distortion on the platform architecture (Fig. 2), consists at present of strike-slip fractures and faults with minor displacement (up to a few tens of metres) as confirmed by biostratigraphic analysis and offsets of marker beds across fault zones. The major trends are oriented NE–SE, NW–SE, ENE–WSW and WNW–ESE. Rotation of stratal patterns introduced by strike-slip faults can be up to 2° determined on the basis of the aggrading inner platform strata, which are respectively horizontal and rotated on the opposite blocks of the fault.

Lithofacies-stratal pattern zones

The following five lithofacies-stratal pattern zones were observed: inner platform, outer platform, upper slope, lower slope and toe-of-slope to basin. More extensive documentation on the lithofacies the reader may refer to Bahamonde *et al.* (1997); Della Porta *et al.* (2002*a*, *b*, 2003, 2004) and Kenter *et al.* (2002).

Inner platform

The inner platform deposits contain shoaling-upward cycles with a transgressive interval of coated grainstones with oncoids, followed by normal marine algal-skeletal packstone to wackestone that forms massive banks, bioclastic grainstone to packstone, and, near the top, restricted lagoonal peloidal packstone to grainstone with calcispheres. These cycles have a thickness of 2.5–25 m and can be traced from the

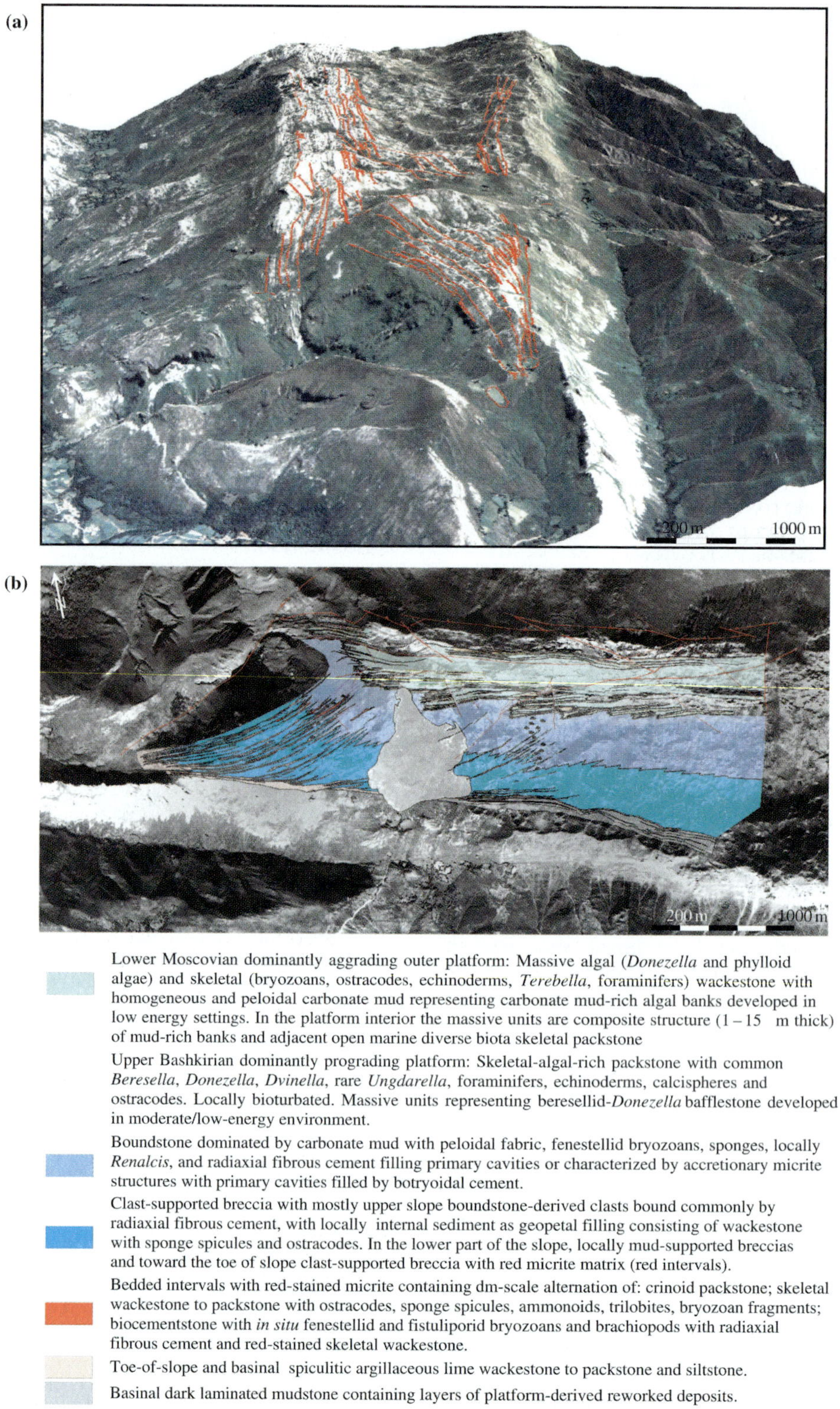

Fig. 2. (**a**) Drape image looking eastwards across the Sierra de Cuera virtual outcrop model, with traced stratal patterns (red lines). Horizontal resolution of the image is *c.* 2 m. (**b**) Orthorectified aerial photograph showing lithofacies map. Horizontal resolution of the image is 0.5 m.

platform break into the platform interior for a distance of at least 6 km.

Outer platform

From the platform interior towards the platform break there is a 1 km-wide zone that represents a lateral facies change from inner to outer platform. The mud-rich banks of the inner platform grade laterally into massive units of boundstone facies that alternate with thick crinoid packstone intervals and thin beds of ooid-coated skeletal grainstones. Generally, in the progradational setting, ooid grainstone shoals dominate, whereas, during aggradation, boundstone prevails in the outer platform.

Upper slope

Two distinctly different microbial boundstone margins are recognized in the field: (1) low-angle slopes and ramps, deposited during the nucleation phase of the platform, of nearly pure micritic limestone; and (2) steep (25° to 45°) slopes where microbial boundstone dominates the uppermost 300 m. The boundstones alternate thin-bedded crinoid–bryozoan grainstone and skeletal wackestone with red-stained micritic matrix. The boundstone facies appears to form massive units, several metres thick, that do not exhibit a mound-shape or depositional relief on the sea floor. This geometry might be the result of boundstone nucleation and growth on an inclined sea floor dipping at least 30°.

Lower slope

Below 300 m palaeowater depth, clast-supported breccia dominates the slope. Clasts are mostly derived from the upper-slope microbial boundstone zone.

Toe-of-slope to basin

Finally, below 600–700 m, argillaceous lime mudstone beds interfinger with grainstone to wackestone intervals of mostly platform-top derived skeletal and peloidal grains and thick intervals of upper-slope derived breccia.

Methodology

Field procedures

Since the US government relaxed the constraints on high-resolution GPS (so-called 'selective availability'), this system has become available in the public domain, enabling navigation to an accuracy of up to 10 m with normal handheld consumer GPS units. Even before this, techniques to improve this resolution to a millimetre- to centimetre scale existed in academia and industry. This technique is called differential GPS (DGPS) (Leica Geosystems AG 1999). One DGPS technique uses correction data from a second satellite system (e.g. Omnistar) or a local radio base station to achieve a resolution up to centimetres. A reference receiver is always positioned at a point with fixed or known coordinates. The other receivers are free to roam around and are known as rover receivers. The baseline(s) (i.e. the length of the 3-D vector between a pair of stations for which simultaneous GPS data has been collected and processed with differential techniques) between the reference and rover receiver(s) are calculated. For high-precision positioning, it is possible to achieve centimetre resolution using real-time kinematic technique (RTK). In RTK, a reference station of known, fixed position uses a radio link to broadcast GPS correction data to the associated rover units. The corrections allow for rover units to obtain a millimetre- to centimetre-scale position accuracy relative to the reference receiver.

In this study the following systems were used:

(1) A Trimble DGPS from Geometius Surveying & GIS Solutions (in summer 2001); this used reference data from a French base station and achieved a horizontal resolution of *c.* 30 cm post-processing.

(2) An SR530 DGPS from Leica Geosystem (in summer and autumn 2002). RTK reference data was used from an own set-up base station connected with the rovers via a radio modem. This resulted in centimetre precision. In case of loss of radio link between the reference and rover units, the receiver switched to Omnistar reference data to maintain decimetre-scale positional accuracy. Raw reference and rover data were post-processed by Geometius/Trimble in 2001, using PathFinder software, and by the Vrije Universiteit in 2002, using Leica SKI-Pro.

Stratal patterns corresponding to lithofacies boundaries are expressed in the weathering profile by the alternation of prominent ridges and recessive intervals. Stratal patterns identified in the field that were visible on the aerial photographs were walked out. Simultaneously, the stratal patterns were mapped with the DGPS. Both DGPSs allowed points and lines in the field to be coded according to attributes such as lithofacies along with their spatial position. Comments and codes were stored in the system and marked on aerial photographs.

Digital elevation model

With the increase in availability and advance of computer manipulation of digital terrain data, or DEMs, in the late twentieth century, it has become possible to quantify and portray land surfaces over large areas (Maune 2001). DEM data are being used widely to model geodynamic and surface processes, rates and physiographical effects, mainly in hydrogeology and geomorphology (Rice-Snow & Russell 1999; Whipple & Tucker 1999; Montgomery *et al.* 2001; Azor *et al.* 2002; Montgomery & Brandon 2002). In this study, a DEM and associated aerial imagery are used, in concert with GPS data, to quantify the geometry of a rotated Carboniferous carbonate platform and to obtain accurate measurements of the angular relationship among stratal surfaces and between these and the spatial distribution of lithofacies in a dip section; this is information that cannot be derived from uncorrected imagery or in the field.

Successful methods of obtaining and applying topographical data to geological problems have included: laser total station surveys (e.g. Bracco Gartner & Schlager 1997; Heimsath *et al.* 1997; McCormick *et al.* 2000; McCormick & Kaufman 2001; Jenette & Bellian 2003*a*, *b*; Kerans *et al.* 2003*a*, *b*); GPS total station surveys (e.g. Santos *et al.* 2000), aerial photo digitization (e.g. Dietrich *et al.* 1995); airborne laser altimetry surveys (e.g. Roering *et al.* 1999); and satellite imagery (e.g. Duncan *et al.* 1998). With the advances in software development in the last couple of years it has become possible to create DEMs with a relatively inexpensive and user-friendly approach using satellite and aerial photo imagery.

A DEM is created by comparing two images and identifying ground points that appear in both images. A set of ground control points (GCPs) are identifiable features on the images whose absolute 3-D ground coordinates have been measured. GCPs are essential for creating a DEM. Stereoscopic digital aerial photograph images combined with DGPS data generated a DEM of the Sierra de Cuera by using ERDAS IMAGINE OrthoBase Pro and ARCGIS software. The horizontal resolution of the created DEM is 2.5 m.

Aerial photograph orthorectification

An orthophoto is an aerial photograph that has the geometric properties of a map. Therefore, orthophotographs can be used as maps to make measurements and establish accurate geographic locations of features. The unprocessed aerial images display strong distortions (e.g. layovers, foreshortenings, dilatations, shadowing, false crests). A normal (i.e. unrectified) aerial photograph does not show features in their correct locations because of displacements caused by the tilt of the camera, terrain relief, sensor attitude/orientation errors and internal sensor errors. In contrast, an orthophotograph is both directionally and geometrically correct. The images do not display relief displacement and thus the measurements obtained from it are accurate. Aerial photographs were scanned using a photogrammetric scanner at a resolution of 25 μm. The corresponding image resolution is approximately 1.5 m for the colour aerial photographs and 0.5 m for the black-and-white photos. Orthorectification was performed in ERDAS IMAGINE OrthoBase Pro (Erdas Inc. 2001).

Lithofacies mapping

Mapping of the lithofacies distribution within the five key zones used digitized orthorectified aerial photography in combination with field sections, in GIS software. This resulting map was georeferenced (i.e. correct location information is attributed to the digital map International Ellipsoid 1950 UTM Zone 30 North) and placed over black-and-white orthorectified aerial photographs (Fig. 2b).

Virtual outcrop model

The digital geological model consists of superposing the different datasets outlined above into 'layers'. The base layer in this model is the DEM. The second layer is an orthorectified high-altitude colour aerial photograph. This 14.5 × 12.5 km area provides an overview of the Sierra de Cuera area. The third layer is composed of a mosaic of orthorectified low-altitude black-and-white aerial photographs. This mosaic covers *c.* 6 × 6.5 km. With an image resolution of 0.5 m, it is possible to resolve geological features. The fourth layer contains the lithofacies map with stratal surfaces and overlies the previous layers. Digitizing the different lithofacies in GIS software provides true 3-D coordinates of different elements of the platform and direct spatial information. Layers 2–4 are draped onto the basal DEM and can be viewed in pseudo 3-D in GIS software (e.g. ESRI ArcScene or ERDAS IMAGINE Virtual

GIS). Finally, the DGPS-mapped stratal pattern data is added to the model (Fig. 2a).

By integrating all the layers, it is possible to quantitatively analyse different layers, and thus different types of data, at the same time. For instance, it is possible to: calculate morphology v. lithofacies; analyse angles of deposition of the sediment; analyse the rotation due to thrusting; determine length, frequency and distribution of stratal patterns, faults and facies; 'map' estimated rates of production of boundstone; and generate measurements of palaeowater depth, lateral progradation, vertical aggradation and downward shifts of the platform break due to sea-level changes, as well as other spatial statistics.

Figure 3 displays a tentative reconstruction of the original prograding platform strata, corrected for the fault displacement. During the dominantly prograding phase II (Upper Bashkirian), the platform stratal pattern is horizontal-parallel 0.5–1.8 km landwards of the platform break. These strata are also nearly parallel to the Moscovian (phases III and IV) platform interior. Outer platform stratal surfaces exhibit a roll-over, i.e. from horizontal they gradually increase in basinwards inclination towards the platform break with dips of 9°–12° to 2°–6°. After each prograding increment, the initial dip of the outermost platform strata is 9°–12°, it decreases to 4°–6° during the following 50–100 m of vertical aggradation and tends towards horizontal within 100–300 m upsection. Figure 3 shows the divergence of the outermost portion of the stratal surfaces and their thinning landwards. Stratal surfaces either terminate in onlap or thin and flatten to horizontal parallel surfaces.

In the prograding slope (phase II), the upper 100–200 m lack stratal surfaces because they consist of massive boundstone. In the intermediate portion the clinoforms have inclinations between 20° and 28° and a topographical relief of 650–750 m. The shape of these clinoforms is slightly sigmoidal. At the toe-of-slope, stratal surfaces gradually flatten to horizontal. Interfingering occurs at very low angles, separating breccia beds and muddy spiculitic basinal facies. In phase III, the platform is horizontally bedded with continuous surfaces in the inner and outer platform settings. The platform break is a sharp transition from horizontal platform strata into steep clinoforms of upper slope strata. Clinoforms have an average depositional relief of 850 m and declivities up to 45°. Strata are nearly planar for the uppermost 250–300 m, concave for the remainder of the slope and flatten to horizontal in the toe-of-slope to basin. The declivities and shape of Sierra de Cuera clinoforms are controlled by the characteristics of the composing lithofacies as demonstrated for ancient and modern carbonate slopes (Kenter 1990).

The relative percentages per area of slope lithofacies during the Late Bashkirian and the latest Bashkirian–Vereian time intervals have been calculated (Table 1).Boundstones dominate the uppermost slope between depths of 0–150 m, in particular in the Upper Bashkirian, with 40% and 33% respectively, while the portion of redeposited layers increases from 0.1% in phase II to 12% in phase III. The lower part of the

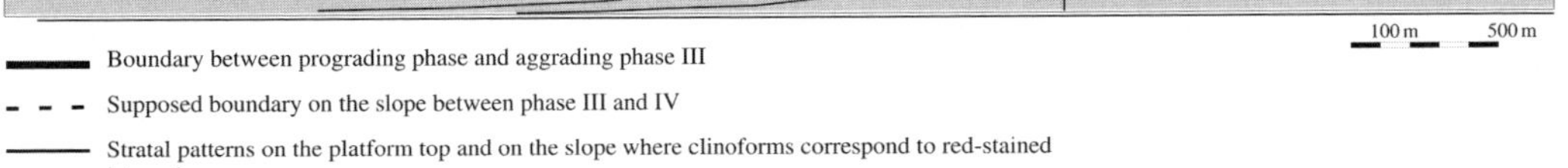

Boundary between prograding phase and aggrading phase III

Supposed boundary on the slope between phase III and IV

Stratal patterns on the platform top and on the slope where clinoforms correspond to red-stained intervals

Fig. 3. Model of platform architecture and stratal geometries, tentatively corrected from the fault displacements, during prograding phase II, aggrading phase III and aggrading to slightly prograding phase IV. Measurements of lateral progradation, vertical aggradation and downwards shift of platform break are reported in metres with the arrow pointing to the growth direction. The traced clinoforms correspond to the red-stained intervals as tracked with DGPS on the outcrop (modified from Della Porta *et al.* 2004).

Table 1. *Relative percentages of slope lithofacies area*

Relative percentage of lithofacies area	Late Bashkirian progradation	Latest Bashkirian to Vereian aggradation
Whole slope		
Boundstone	40%	33%
Breccia	50%	52%
Redeposited and 'red-stained' layers	10%	15%
Uppermost slope (0–150 m)		
Boundstone	99.5%	87%
Breccia	0.4%	1%
Redeposited and 'red-stained' layers	0.1%	12%
Lower part upper slope (150–300 m)		
Boundstone	83%	34%
Breccia	13%	50%
Redeposited and 'red-stained' layers	4%	16%
Lower slope (300–500 m)		
Boundstone	10%	–
Breccia	80%	92%
Redeposited and 'red-stained' layers	10%	8%
Lower slope to toe-of-slope (500 m to 700–850 m)		
Boundstone	10%	–
Breccia	79%	72%
Redeposited and 'red-stained' layers	21%	28%

upper slope (150–300 m) is boundstone-dominated during the Late Bashkirian (83% boundstone). Red-stained layers and redeposited lithofacies become dominant in this portion of the slope (16% and 50% respectively v. 34% of boundstone) at the Bashkirian–Moscovian boundary. The upper part of the lower slope (300–500 m) is breccia-dominated (80–92%). Layers of platform- and slope-derived material increase in the lowermost part of the slope (500 m to 700–850 m).

Next, it is possible to calculate progradation/aggradation ratios (Table 2),which are useful descriptors of platform evolution when used in the context of other key parameters, such as carbonate factory size, production rates, shelf-to-basin bathymetry and antecedent topography (Kerans & Tinker 1999).

Platform architecture evolution consisted of westward progradation for more than 10 km with development of the steep clinoforms in the Bashkirian (phase I). During the upper Bashkirian (phase II), progradation alternated with several minor aggradational phases. During phase II, the platform prograded basinward for a total of 2910 m and aggraded in the inner part by 165 m (uncorrected for compaction or missing rocks). Using the duration of 7 Ma for the Late Bashkirian (Harland *et al.* 1990), estimates of rates of lateral progradation are 415 m/Ma (equivalent to μm/yr). Menning *et al.* (2000) state a duration for the Westphalian A/Langsettian of 3 Ma, which is correlated with the Bashkirian in Harland *et al.* (1990). The progradation rate obtained with a time duration of 3 Ma is 970 m/Ma. This range of 415–970 m/Ma comprises all possible variation and errors introduced by the uncertainty in time of the Upper Bashkirian, by errors in the location of the chronostratigraphic boundaries based on biostratigraphy and by the location of the lower limit of the prograding phase II already within the Upper Bashkirian. The shift to dominant aggradation (Vereian; phase III) occurred near the Bashkirian–Moscovian transition. Resulting aggradation rates are 288 m v. 160 m of lateral progradation. The aggradation rate was approximately 108–114 m/Ma v. a progradation rate of 76–80 m/Ma (Vereian time duration of 2.1 Ma according to Harland *et al.* 1990, and of 2 Ma according to Menning *et al.* 2001) (Della Porta *et al.* 2004).

Accumulation rates calculated on mapped cross-sections measuring the number of pixels of each area expressed in m^2/Ma (Table 1) yield values for the Upper Bashkirian of 838 182 m^2/Ma (using 3.0 Ma, according to Menning *et al.* 2001) to 359 221 m^2/Ma (using 7.0 Ma, accord-

Table 2. *Progradation/aggradation ratios*

	Late Bashkirian		Vereian
	(3 Ma interval)	(7 Ma interval)	(2.1 Ma interval)
Progradation rates	970 m/Ma	415 m/Ma	76–80 m/Ma
Aggradation rates	55 m/Ma (interior); 140 m/Ma (break)	23 m/Ma (interior); 60 m/Ma (break)	108–114 m/Ma –
Accumulation rates measured in cross-section			
Whole depositional system	838 182 m^2/Ma	359 221 m^2/Ma	637 011 m^2/Ma
Platform	152 156 m^2/Ma	65 210 m^2/Ma	431 354 m^2/Ma
Slope	699 010 m^2/Ma	299 575 m^2/Ma	214 060 m^2/Ma
Boundstone (*in situ*)	386 533 m^2/Ma	165 657 m^2/Ma	699 944 m^2/Ma
Boundstone + breccias	682 790 m^2/Ma	295 624 m^2/Ma	194 165 m^2/Ma
Redeposited packstone–grainstone	12 484 m^2/Ma	5350 m^2/Ma	12 051 m^2/Ma

ing to Harland *et al.* 1990), and 637 011 m^2/Ma for the Vereian for the whole depositional system. It emerges that, during progradation, the greatest contribution of sediment supply is produced by the slope boundstone. During aggradation, the flat-topped platform was responsible for the main sediment supply to the slope. The slope, and in particular the boundstone, had higher production rates (at least two to five times greater) during progradation than aggradation. One reason for this is the increased tectonic subsidence, which created greater accommodation space and relative sea-level rises with greater amplitude (Della Porta *et al.* 2003). The presented estimates are conservative because sedimentation rates decrease with increasing time of observation, i.e. sedimentation rate is inversely proportional to the square root of the change in time (Schlager 1999). Therefore, accumulation rates evaluated for 1–10 Ma range would increase by a factor of 1000–3000 when scaled to a year basis, and by a factor of 30–1000 when scaled to 1 ka.

In general, progradational styles building out for several kilometres are mostly recorded for low-angle carbonate platform slope systems with a dominant bank-top sediment source (Schlager 1981; McNeill *et al.* 2001). The Sierra de Cuera platform and, for example, the Tengiz platform (Pricaspian Basin, Kazakhstan) both have a highly productive microbial cement boundstone factory, extending from the platform break to nearly 300 m depth, and a lower slope dominated by (mega)breccias and grain flow deposits derived from the margin and slope itself (Harris & Kenter 2003). The broad depth range of microbial cement boundstone increases the potential for production during both lowstands and highstands of sea level and thereby facilitates progradation. Rapid *in situ* lithification of the boundstone provides stability to the steep slopes, but also leads to readjustment through shearing and avalanching. What controls the microbial cement boundstone formation remains debatable (Kenter & Harris 2002), but its presence is a key factor in the progradational geometry of these margins.

Application to reservoir analogues

Age equivalent and facies-similar slope deposits are observed in the subsurface of the Pricaspian Basin in Kazakhstan (Tengiz, Korolev, Kashagan) (Kenter & Harris 2002). The geometries of the different sedimentary bodies determine the mechanical and petrophysical properties and influence flow characteristics, reservoir properties and eventually development and production strategies of several giant carbonate reservoirs. Since the reservoir zone is difficult to image, because there is a thick salt cap and cores provide only 1-D information, the adoption of the Asturias quantitative outcrop model of lithofacies distribution and stratal anatomy provided crucial input to constrain similar zones in the subsurface. Figure 4 depicts a schematic cross-section through Tengiz, showing the major depositional sequences and lithofacies types (modified after Weber *et al.* 2003). Although vertically exaggerated in scale, the key Upper Visean to Serpukhovian lithofacies zones in the eastern flank show, from shallow to deep (or margin to basin), *in situ* microbial boundstone, boundstone breccia and lower- to toe-of-slope alternating carbonate sand, breccia tongues and basinal spiculitic lime mudstone. Initially, core observations were sparse and scattered, and a hypothetical model was built on information

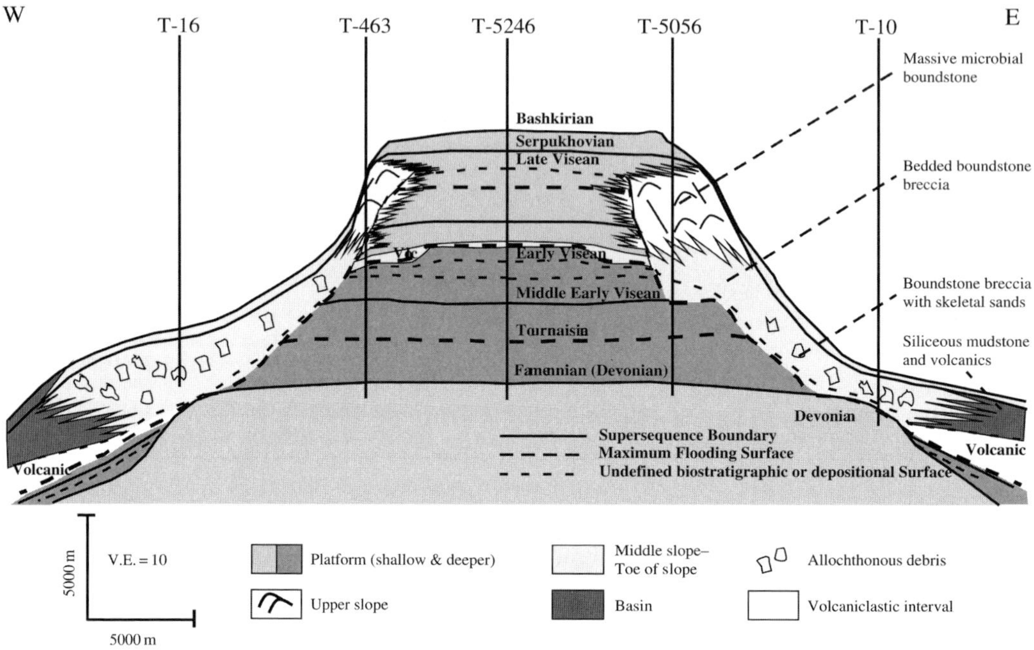

Fig. 4. Schematic cross-section through Tengiz showing the major depositional sequences, lithofacies types and approximate positions of some key wells. Thin black lines, supersequence boundaries; thick dashed lines, flooding surfaces; thin dashed lines, higher order (smaller-scale) sequence boundaries. Vertical exaggeration (V.E.) ten times horizontal. Modified after Weber *et al.* (2003).

from Asturias, where spatial information on slope lithofacies type and distribution from Asturias (this paper and references cited) was transposed to the Tengiz cross-section. The resulting facies map was compared and contrasted with the seismic data and used for mapping seismic facies as well for determining the position of the boundary between slope and platform. This provided information for the reservoir model since the microbial boundstone forms a zone with low matrix permeability but high fracture flow. Spatial modelling (Gocad) demonstrated that upper slope boundstone makes up approximately 25% of the platform volume in the Upper Visean to Bashkirian reservoir unit (Weber *et al.* 2003). Recent wells and continuous core penetrations like T-5056, T-463 and others validated the predictive value of the initial model based on Asturias.

Studies of analogues are valuable because they constrain interpretations and lend predictability to unravelling facies patterns in reservoirs. These patterns help to understand the lateral continuity of stratification, variation within layers, heterogeneity, and performance of reservoir examples. Though the results are not fully representative for the subsurface, given the depositional and diagenetic diversity encountered in carbonate environments (so it is not possible to exactly mirror the two platforms quantitatively), the model does help us to understand the lithofacies character at the platform margin and of the platform interior sediments. In this, outcrop studies provide the only data on the possible distribution of reservoir facies and the specific spatial distribution of the different platform elements in more than 1-D.

Conclusions

The mapping (tracking) of key stratal surfaces and facies boundaries with DGPS allows the quantitative measurement of platform geometry and facies architecture. This particular outcrop study not only contributes to the understanding of reservoir geometry in the subsurface of the Pricaspian Basin by providing a realistic image of the stratigraphic architecture, but also provides the required geometrical framework to calculate accretion and net production of deep-water microbial boundstone controlling the platform style. To date, construction of 3-D models from outcrops has been limited by outcrop character and the ability to capture scaled images of the outcrop that can be interpreted in 3-D in a manner similar to 2-D or 3-D seismic data. Recent advances made in DGPS equipment and in the development of GIS and remote-sensing software create the opportunity for the construction of high-resolution 3-D models with a high degree of spatial integrity.

The resulting seismic-scale model combines the high resolution of the orthorectified aerial photographs with a sub-decimetre accuracy of mapped stratal patterns. The ability to make direct comparisons of reservoirs and analogue data within this type of carbonate system greatly improves the value of outcrop information. It not only provides the information to reliably estimate the spatial distribution of reservoir properties in equivalent geologic settings, it also advances more academic research themes, such as the production models of boundstone deposits, ranges and modes of platform elements, measurements of lateral progradation, vertical aggradation, and downward shifts of the platform break in this type of margin. This will enhance the understanding of carbonate production through geological time and allow for a more precise comparison between the different carbonate factories. In addition, forward modelling of depositional systems may finally be provided with long-needed quantitative information on anatomy and rates.

Tengizchevroil (Atyrau, Kazakhstan) and TotalFinaElf are acknowledged for financial support. In particular, D. Fischer (Tengizchevroil) is thanked for stimulating discussions on the applicability of the Asturias slope model as an analogue for the slope environment in fields in the Pricaspian Basin. We thank D. McCormick and V.P. Wright for valuable reviews of the manuscript. We thank editor R.A. Wood for her helpful comments. J. Bahamonde is acknowledged for field assistance.

References

ADAMS, E. W., SCHROEDER, S., GROTZINGER, J. P., MCCORMICK, D. S., AMTHOR, J. E., AL-HASHIMI, R., AL-JAAIDI, O., AL-SIYABI, H. & SMITH, D. W. 2001. Digital mapping of the geometry and stratal patterns of microbial reefs, terminal Neoproterozoic Nama Group, Namibia. *Geological Society of America Annual Meeting, 1–11 November 2001, Boston*, 256.

ADAMS, E. W., GROTZINGER, J. P., SCHROEDER, S., AL-SIYABI, H., AMTHOR, J. E. & MCCORMICK, D. S. 2002. Shape, size and distribution of microbial reefs in a carbonate-ramp setting (Nama Group, Namibia). *16th International Sedimentological Congress, 8–12 July 2002, Johannesburg*, 326.

ADAMS, E. W., SCHROEDER, S., GROTZINGER, J. P. & MCCORMICK, D. (2004). Digital reconstruction and stratigraphic evolution of a microbial-dominated, isolated carbonate platform (terminal Proterozoic, Nama Group, Namibia). *Journal of Sedimentary Research*, **74**, 479–497.

AZOR, A., KELLER, E. A. & YEATS, R. S. 2002. Geomorphic indicators of active fold growth – South Mountain–Oak Ridge Anticline, Ventura Basin, southern California. *Bulletin of the Geological Society of America*, **114**, 745–753.

BAHAMONDE, J. R., COLMENERO, J. R. & VERA, C. 1997. Growth and demise of Late Carboniferous carbonate platforms in the eastern Cantabrian Zone, Asturias, northwestern Spain. *Sedimentary Geology*, **110**, 99–122.

BRACCO GARTNER, G. L. & SCHLAGER, W. 1997. Quantification of topography using a laser-transit system. *Journal of Sedimentary Research*, **67**, 987–989.

COLMENERO, J. R., AGUEDA, J. A., BAHAMONDE, J. R., BARBA, F. J., BARBA, P., FERNANDEZ, L. P. & SALVADOR C. I. 1993. Evolución de la cuenca de antepaís namuriense y westfaliense de la Zona Cantábrica, NW de España. *12th International Carboniferous–Permian Congress, 22–27 September 1991, Buenos Aires, Comptes Rendues*, **2**, 175–190.

DELLA PORTA, G., KENTER, J. A. M. & BAHAMONDE, J. R. 2002*a*. Microfacies and palaeoenvironment of *Donezella* accumulations across an Upper Carboniferous high-rising carbonate platform (Asturias, NW Spain). *Facies*, **46**, 159–168.

DELLA PORTA, G., KENTER, J. A. M., IMMENHAUSER, A. & BAHAMONDER, J. R. 2002*b*. Lithofacies character and architecture across a Pennsylvanian inner-platform transect (Sierra de Cuera, Asturias, Spain). *Journal of Sedimentary Research*, **72**, 898–916.

DELLA PORTA, G., KENTER, J. A. M., BAHAMONDE, J. R., IMMENHAUSER, A. & VILLA, E. 2003. Microbial boundstone dominated steep carbonate platform slope (Upper Carboniferous, N Spain): microfacies, facies distribution and stratal geometry. *Facies*, **49**, 175–208.

DELLA PORTA, G., KENTER, J. A. M. & BAHAMONDE, J. R. (2004). Depositional facies and stratal geometry of prograding and aggrading slope-to-platform transitions in a high-relief Upper Carboniferous carbonate platform (Cantabrian Mountains, N Spain). *Sedimentology*, **51**, 267–296.

DIETRICH, W. E., REISS, R., HSU, M.-L. & MONTGOMERY, D. R. 1995. A process-based model for colluvial soil depth and shallow landsliding using digital elevation data. *Hydrological Processes*, **9**, 383–400.

DUNCAN, C. C., KLEIN, A. J., MASEK, J. G. & ISAAKS, B. L. 1998. Comparison of Late Pleistocene and modern glacier extents in central Nepal based on digital elevation data and satellite imagery. *Quaternary Research*, **49**, 241–254.

ERDAS Inc. 2001. *IMAGINE OrthoBase Pro User's Guide ERDAS IMAGINE V8.5*. Erdas Inc., Atlanta.

GREEVE, C. 1997. *Digital Photogrammetry: An Addendum to the Manual of Photogrammetry*. American Society of Photogrammetry and Remote Sensing, Bethesda.

HARRIS, P. M. & KENTER, J. A. M. 2003. Steep microbial-dominated platform margins: examples and implications. *American Association of Petroleum Geologists Annual Meeting, 11–14 May 2003, Salt Lake City*, A71.

HARLAND, P., HINE, A. C., VARGO, G. A., ELROD, J. A. & JAAP, W. C. 1990. *A Geologic Time Scale 1989*. Cambridge University Press, Cambridge.

HASLER, C.-A., ADAMS, E. W., WOOD, R. A. & DICKSON, J. A. D. 2003. Fine-scale forward modelling: example from a Devonian reef of the Canning Basin. *Analogue and Numerical Forward Modelling of Sedimentary Systems: From Understanding to Prediction, 9–11 October 2003, Utrecht*, p. 40.

HEIMSATH, A. M., DIETRICH, W. E., NISHIIZUMI, K. & FINKEL, R. C. 1997. The soil production function and landscape equilibrium. *Nature*, **388**, 358–361.

HODGETTS, D., VIK, E., HODGSON, D., DRINKWATER, N. & FLINT, S. 2003. 3D architecture of a deepwater turbidite depositional system from outcrop and wells. (2) Construction, visualization and analysis of reservoir models. *American Association of Petroleum Geologists International Conference & Exhibition, 21–24 September 2003, Barcelona*, A41.

JENNETTE, D. & BELLIAN, J. 2003*a*. 3-D digital characterization and visualization of the solitary channel complex, Tabernas Basin, southern Spain. *American Association of Petroleum Geologists International Conference & Exhibition, 21–24 September 2003, Barcelona*, A46.

JENNETTE, D. C. & BELLIAN, J. A. 2003*b*. Application of laser-scanned outcrop data to build models of deepwater reservoirs: examples from the Tabernas and Ainsa Basins, northern and southern Spain. *American Association of Petroleum Geologists Annual Meeting, 11–14 May 2003, Salt Lake City*, A85.

JULIVERT, M. 1971. Décollement tectonics in the Hercynian cordillera of northwest Spain. *American Journal of Science*, **270**, 1–29.

KENTER, J. A. M. 1990. Carbonate platform flanks: slope angle and sediment fabric. *Sedimentology*, **37**, 777–794.

KENTER, J. A. M. & HARRIS, P. M. 2002. Prograding steep and high-relief carbonate platform margins. *American Association of Petroleum Geologists Annual Convention, 10–13 March 2003, Houston*, A92.

KENTER, J. A. M., VAN HOEFLAKEN, F., BAHAMONDE, J. R., BRACCO GARTNER, G. L., KEIM, L. & BESEMS, R. E. 2002. Anatomy and lithofacies of an intact and seismic-scale Carboniferous carbonate platform (Asturias, NW Spain): analogues of hydrocarbon reservoirs in the Pricaspian basin (Kazakhstan). *In:* ZEMPOLICH, W. G. & COOK, H. E. (eds) *Paleozoic Carbonates of the Commonwealth of Independent States (CIS): Subsurface Reservoirs and Outcrop Analogs*. SEPM (Society for Sedimentary Geology), Tulsa, Special Publications, **74**, 181–204.

KERANS, C. & TINKER, S. W. 1999. Extrinsic stratigraphic controls on development of the Capitan Reef Complex. *In:* SALLER, A. H., HARRIS, P. M., K. P. L. & MAZZULLO, S. J. (eds) *Geologic Framework of the Capitan Reef*. SEPM (Society for Sedimentary Geology), Tulsa, Special Publications, **65**, 15–39.

KERANS, C., LUCIA, J. & WANG, F. 1995. The role of outcrop data in performance prediction of nonfractured carbonate fields. *American Association of Petroleum Geologists International Conference & Exhibition, 24–24 September 2003, Barcelona*, A49.

KERANS, C., BELLIAN, J. & PLAYTON, T. 2003*a*. 3D modeling of deepwater carbonate outcrops using laser technology (Lidar). *American Association of Petroleum Geologists Annual Meeting, 11–14 May 2003, Salt Lake City*, A91.

KERANS, C., BELLIAN, J. & PLAYTON, T. 2003*b*. 3D mapping and visualization of carbonate slope and basin floor reservoir strata: bringing the outcrop to the desktop. *American Association of Petroleum Geologists International Meeting, 10–13 September 1995, Nice*, 1226.

LEICA GEOSYSTEMS AG. 1999. *GPS System 500, Getting Started with Real-Time Surveys*. Leica Geosystems AG, Heerbrugg, Switzerland.

LØSETH, T. M., THURMOND, J., SØEGAARD, K., RIVENÆS, J. C. & MARTINSEN, O. J. 2003*a*. Building reservoir model using virtual outcrops: a fully quantitative approach. *American Association of Petroleum Geologists Annual Meeting, 11–14 May 2003, Salt Lake City*, A110.

LØSETH, T. M., THURMOND, J., SØEGAARD, K., RIVENÆS, J. C. & MARTINSEN, O. 2003*b*. Visualization and utilization of 3D outcrop data. *American Association of Petroleum Geologists International Conference & Exhibition, 21–24 September 2003, Barcelona*, A56.

MARQUINEZ, J. 1989. *Mapa Geológico de La Región del Cuera y Picos de Europa*. Trabajos de Geologia, Universidad de Oviedo.

MAUNE, D. F. 2001. *Digital Elevation Model Technologies and Applications: The DEM User Manual*. American Society for Photogrammetry & Remote Sensing, Bethesda.

MCCORMICK, D. S. & KAUFMAN, P. S. 2001. *Outcrop Characterization with Digital Geologic Mapping*. American Association of Petroleum Geologists, Bulletin **85**.

MCCORMICK, D. S., THURMOND, J. B., GROTZINGER, J. P. & FLEMING, R. J. 2000. *Creating a Three-Dimensional Model of Clinoforms in the Upper San Andreas Formation, Last Chance Canyon, New Mexico*. American Association of Petroleum Geologists, Bulletin **84**.

MCNEILL, D. F., EBERLI, G. P., LIDZ, B. H., SWART, P. K. & KENTER, J. A. M. 2001. Chronostratigraphy of a prograded platform margin: a record of dynamic slope sedimentation, western Great Bahama Bank. *In:* GINSBURG, R. N. (ed.) *Subsurface Geology of a Prograding Carbonate Platform Margin, Great Bahama Bank: Results of the Bahamas Drilling Project*. SEPM (Society for Sedimentary Geology), Tulsa Special Publications, **70**, 101–134.

MENNING, M., WEYER, D., DROZDZEWSKI, G., VAN AMERON, H. W. J. & WENDT, I. 2000. A Carboniferous time scale 2000: discussion and use parameters as time indicators from central and western Europe. *Geologische Jahrbuch*, **A156**, 3–44.

MENNING, M., WEYER, D., WENDT, I., RILEY, N. J. & DAVYDOV, V. I. 2001. Discussion on high-precision 40Ar/39Ar spectrum dating on sanidine from the Donets Basins, Ukraine: evidence for correlation problems in the Upper Carboniferous. *Journal of the Geological Society*, **158**, 733–736.

MONTGOMERY, D. R. 2001. Slope distributions, threshold hillslopes, and steady-state topography. *American Journal of Science*, **301**, 432–454.

MONTGOMERY, D. R. & BRANDON, M. T. 2002. Topographic controls on erosion rates in tectonically active mountain ranges. *Earth and Planetary Science Letters*, **201**, 481–489.

PRINGLE, J. K., CLARK, J. D., WESTERMAN, A. R. & GARDINER, A. R. 2003. Using GPR to image 3D turbidite channel architecture in the Carboniferous Ross Formation, County Clare, western Ireland. *In:* BRISTOW, C. S. & JOL, H. (eds) *GPR in Sediments*. Geological Society, London, Special Publications, **211**, 309–320.

READ, J. F. 1985. Carbonate Platform Facies Models. American Association of Petroleum Geologists, Bulletins, **69**.

RICE-SNOW, S. & RUSSELL, J. 1999. Long-range persistence of elevation and relief values along the Continental Divide in the conterminous U.S. *4th International Conference on GeoComputation, Fredericksburg*, 25–28 July 1999, p. 43.

ROERING, J. J., KIRCHNER, J. W. & DIETRICH, W. E. 1999. Evidence for non-linear, diffusive sediment transport on hillslopes and implications for landscape morphology. *Water Resource Research*, **35**, 853–870.

SANTOS, M. L. M., GUENAT, C., BOUZELBOUDJEN, M. & GOLAY, F. 2000. Three-dimensional GIS cartography applied to the study of the spatial variation of soil horizons in a Swiss floodplain. *Geoderma*, **97**, 351–366.

SARG, J. F. 1988. Carbonate sequence stratigraphy. *In:* WILGUS, C. K., HASTINGS, B. S., KENDALL, C. G. St.C., POSAMENTIER, H. W., ROSS, C. A. & VAN WAGONER, J. C. (eds) *Sea Level Changes: An Integrated Approach*. SEPM (Society for Sedimentary Geology), Tulsa, Special Publications, **42**, 155–181.

SCHLAGER, W. 1981. The paradox of drowned reefs and carbonate platforms. *Bulletin of the Geological Society of America*, **92**, 197–211.

SCHLAGER, W. 1999. Scaling of sedimentation rates and drowning of reefs and carbonate platforms. *Geology*, **27**, 193–186.

VAIL, P. R. 1987. Seismic stratigraphy interpretation using sequence stratigraphy. *In:* BALLY, A. W. (ed.) *Atlas of Seismic Stratigraphy*. American Association of Petroleum Geologists, Studies in Geology, **27**, 1–10.

WEBER, L. J., FRANCIS, B. P., HARRIS, P. M. & CLARK, M. (2003) Stratigraphy, lithofacies, and reservoir distribution, Tengiz Field, Kazakhstan. *In:* AHR, W. M., HARRIS, P. M., MORGAN, W. A. & SOMERVILLE, I. D. (eds) *Permo-Carboniferous Platforms and Reefs*. SEPM (Society for Sedimentary Geology) Special Publications, **78**, 351–394.

WHIPPLE, K. X. & TUCKER, G. E. 1999. Dynamics of the stream-power river incision model: implications for height limits of mountain ranges, landscape response timescales, and reseach needs. *Journal of Geophysical Research*, **104**, 17 661–17 674.

Digital field data acquisition: towards increased quantification of uncertainty during geological mapping

RICHARD R. JONES[1,2], KENNETH J. W. MCCAFFREY[3], ROBERT W. WILSON[3] & ROBERT E. HOLDSWORTH[3]

[1]*CognIT a.s, Meltzergate 4, N-0257 Oslo, Norway*

[2]*Currently at: e-Science Research Institute and Geospatial Research Ltd, Department of Earth Sciences, University of Durham, DH1 3LE, UK*

[3]*Department of Earth Sciences, University of Durham, DH1 3LE, UK (e-mail: r.r.jones@durham.ac.uk)*

Abstract: Traditional methods of geological mapping were developed within the inherent constraints imposed by paper-based publishing. These methods are still dominant in the earth sciences, despite recent advances in digital technology in a range of fields, including global-positioning systems, geographical information systems (GIS), 3-D computer visualization, portable computer devices, knowledge engineering and artificial intelligence. Digital geological mapping has the potential to overcome some serious limitations of paper-based maps. Although geological maps are usually highly interpretive, traditional maps show little of the raw field data collected or the reasoning used during interpretation. In geological mapping, interpretation typically relies on the prior experience and prior knowledge of the mapper, but this input is rarely published explicitly with the final printed map. Digital mapping techniques open up new possibilities for publishing maps digitally in a GIS format, together with spatially referenced raw field data, field photographs, explanation of the interpretation process and background information relevant to the map area. Having field data in a digital form allows the use of interpolation methods based on fuzzy logic to quantify some types of uncertainty associated with subsurface interpretation, and the use of this uncertainty to evaluate the validity of competing interpretations.

Although the methodology of geological mapping has remained largely unchanged since William Smith's pioneering work 200 years ago (Smith 1801, 1815), recent developments in digital technology have the potential to revolutionize the way in which geological field data is gathered, stored, processed, displayed and distributed. The digitalization of field mapping is occurring through advances in global-positioning systems (GPS), geographical information systems (GIS), highly portable hand-held personal digital assistants (PDAs), high-powered 3-D computer graphics, and satellite communication equipment. In this paper we discuss the fundamental importance of this new technology in helping to remove some of the inherent limitations of traditional methods of geological mapping.

Prior information

In common usage the concept of 'prior information' conveys a variety of meanings, largely dependent upon context, ranging from philosophical to mathematical. The aim of this paper is not to provide an exhaustive discussion on the semantics of prior information, but rather to focus pragmatically on specific issues that have direct relevance to the process of geological mapping.

Relationship between information and knowledge

In relation to geosciences it is convenient for us to consider 'prior information' in a broad way that also encompasses related concepts such as 'data' and 'knowledge'. During the last decade many business sectors, including the hydrocarbon and mining industries have focused heavily on maximizing re-use of corporate knowledge and expertise. This highly profit-driven process has given rise to very pragmatic research in areas of applied knowledge engineering. Workers in these fields (e.g. van der Spek & Spijkervet 1997;

From: CURTIS, A. & WOOD, R. (eds) 2004. *Geological Prior Information: Informing Science and Engineering.* Geological Society, London, Special Publications, **239**, 43–56. 186239-171-8/04/$15.00 © The Geological Society of London.

Table 1. *The knowledge hierarchy with examples from geological mapping*

	Level	Example
5	Expertise (= ability to apply knowledge effectively to perform tasks accurately and efficiently within existing constraints)	The finished geological map! A geological summary of the mapped area Scientific paper describing new fossil found in area
4	Knowledge (= application of information to be able to make decisions, solve problems, or perform tasks)	Boundary between rock types A and B is in valley Rock A is granite Mineral P is biotite
3	Information (= data in a specified context, data with meaning)	Rock A is observed at location GR 234 789 Rock A contains minerals P, Q, R Mineral P has colour that is black
2	Data (= inputs that can be represented explicitly in symbolic form)	Colour is black Lustre is shiny GPS location at time T is 123456 from satellite 1 GPS location at time T is 123678 from satellite 2
1	Inputs (= sensory signals, machine measurements)	Geologist receives visual sensory signals GPS unit receives satellite signals

Liebowitz & Beckman 1998) generally view 'information' as just one part of a knowledge hierarchy that ranges from basic data input to sophisticated interpretation by an expert (Table 1).

Interdependence of data, information and knowledge

In traditional views concerning the way in which scientific progress occurs, heavy emphasis is placed on the role of induction in the derivation of scientific theories based on observed phenomena. Because little consideration was generally given to the fact that sensory inputs might be misleading, objective 'scientific' observations could be treated as facts. This view of scientific methodology was largely dominant until the early twentieth century and still remains influential in much of geoscience education today:

$$\text{observation} = \text{fact} \xrightarrow{\text{induction}} \text{scientific theory} \xrightarrow{\text{deduction}} \text{prediction.}$$

The inductivist view of scientific methodology has been refuted in several ways, and individual scientific theories are now more generally viewed as belonging to larger knowledge structures (the 'paradigm' of Kuhn 1962, or the 'research programme' of Lakatos 1974). Of central importance in the rejection of induction is that observations cannot be made independently of prevalent theory and that the formation of individual theories builds upon prior scientific knowledge. Thus there is a solid philosophical foundation for stating that prior information (*sensu lato*) is always an influential factor during scientific research, including geological mapping and other disciplines within the Earth sciences. Put more simply, this discussion merely emphasises something that most geologists will take for granted: the end product of a geological mapping process is a non-unique, subjective interpretation that is markedly influenced by the previous background, prior experience and expertise of the geologist, and it is an interpretation made within the context of current geological understanding.

Maps created using traditional methods of production and distribution often emphasise the knowledge and expertise (i.e. levels 4 and 5 in Table 1) of the mapper, while raw data and information (i.e. levels 2 and 3 in Table 1) gathered during the mapping process usually remain unpublished in field maps and notebooks. This situation is particularly prevalent with regional maps typically produced by national surveys. Digital methods of production and distribution allow the possibility of presenting not only the final interpretation but also additional data/information to the user, so that individual observations become reproducible, assumptions or inferences made can be questioned and overall interpretations more readily tested.

Accessibility of knowledge

Researchers who have studied the transfer of knowledge between workers in modern industry have recognized that there is great variation in the accessibility of an individual's knowledge and expertise (Nonaka 1994; Nonaka & Takeuchi 1995). While some knowledge often exists in an *explicit* format that is readily available and that can be communicated and understood by other colleagues, there is usually also a large amount of *implicit* knowledge that is not (yet) in a format that is easily accessible to others (Table 2). Implicit knowledge exists within the mind of a person, but can be made readily available to others (i.e. made explicit) through query and discussion, or through a conscious decision to document what one knows about a subject. In contrast, a large amount of the knowledge possessed by people is usually *tacit* (Polanyi 1958, 1966). Tacit knowledge also exists within the mind of a person, but is generally in a form that is not easily accessible to other people, either because the owners are not aware that they possess the knowledge, or because they are unable to express it in a useful, understandable format. An example is the knowledge of how to ride a bicycle: although you may know how to cycle without having to think, it is very difficult to give an explicit description of how to perform the task to people who have never tried it themselves. Similarly, the way in which a skilled map-reader navigates by converting the symbols shown on a map into a mental 3-D picture of the terrain is a process that is done automatically at a subconscious level and is difficult to describe explicitly in a meaningful way. A further example is the way in which a geologist interprets structural symbols on a map to form an image of the 3-D architecture that the map depicts. Traditionally the transfer of tacit knowledge of this type is achieved by repeated learning through 'trial and error', often in a teacher/pupil or master/apprentice situation. More recently, specialized knowledge elicitation techniques have been devised that extract tacit knowledge and represent it in more explicit forms that can be manipulated by a computer (e.g. Boose 1986; Kidd 1987; Ford & Sterman 1998; White & Sleeman 1999).

Workers who are considered to be 'experts' within a particular domain typically possess high levels of tacit knowledge. This is often particularly acute within geological mapping, where the quality of output will be greatly dependent upon the skill and expertise of the mappers, although their proficiency is rarely immediately obvious simply by looking at the final result.

Conditional probability

Ideally, when faced with uncertainty associated with interpretation of individual field observations, the geologist should attempt to make further field observations in order to resolve any outstanding issues. In practice, this process will almost always be restricted by limitations imposed through lack of exposure and limited resources (manpower, time, money), and the geologist is almost always obliged to supplement direct observation with a mixture of insight, guile, guesswork and 'gut feeling'. These aspects reflect the geologist's personal bias and represent a prior belief which, if it can be expressed explicitly, can be used to predict the likelihood of a particular interpretation being true. This forms the basis of Bayesian statistics, in which the probability that hypothesis H is true, given the occurrence of event E (in context I) is given by:

$$P(H|EI) = (P(H|I)*P(E|HI))/P(E|I).$$

Thus Bayesian probability can provide a mathematical framework for expressing the uncertainty associated with geological interpretation and offers a possible way of extending and improving traditional methods of geological mapping.

Table 2. *Different levels of accessibility of knowledge*

Knowledge accessibility	General examples	Examples from mapping
Explicit	Scientific papers published in journals Textbooks Instruction manuals	Detailed outcrop map ('green-line' mapping)
Implicit	Unpublished observations Working hypotheses Undocumented troubleshooting fixes	Observations not recorded explicitly on final map General geological theory that has influenced specific interpretation
Tacit	Instant recognition of minerals based on 'look and feel' rather than explicit physical tests Interpretation of seismic sections	Expertise of the mapper The mapper's preconceived bias, insight, gut-feeling and intuition

Traditional geological mapping

The ability to produce accurate field maps and to record associated observational data in a notebook lies at the core of Earth science activities (e.g. Barnes 1981; McClay 1987) and forms the basis by which geological maps are constructed. While a geological map is a 2-D representation of the distribution of rock formations in an area, it also conveys, through symbols and graphics, the 3-D geometry and form of the rocks and structures in the area. The gathering of field data occurs in a broad range of natural environments and is typically carried out by individuals or small teams of geoscientists working on foot or using various forms of transport; consequently any equipment or techniques used to gather information must be highly mobile and easy to maintain. For these reasons, it perhaps not surprising that field mapping has, since its inception, remained a paper-based activity using maps, field notebooks and compass-clinometers.

The scientific aims of a study and the time available for fieldwork – which is often dictated by funding – will determine the type of geological mapping to be carried out. Barnes (1981) identifies four main types of geological mapping activities:

(1) *Reconnaissance mapping* typically covers a large area and is carried out in order to find out as much as possible about a poorly known region over a short period of field study. Significant amounts of work may be done using remote sensing techniques or interpretation of aerial photographs.

(2) *Regional mapping* typically results in geological maps at 1 : 50 000 scale recorded on an accurate topographical base map. Such mapping is generally the result of systematic programmes of field-based data gathering, fully supported by photogeological interpretation and integration of other subsurface geological or geophysical datasets.

(3) *Detailed mapping* generally refers to maps made at 1 : 10 000 or larger scales and in many cases these are produced to document key geological relationships in detail. Many require the field geologist to produce their own base map simultaneously, using planetable-, chain- or cairn-mapping techniques (see Barnes 1981 and references therein).

(4) *Specialized mapping* refers to maps which are constructed for specific purposes and do not necessarily include all aspects of the observed geology. These include mine plans and maps showing geotechnical, geophysical or geochemical data.

The following information is particularly critical to all field-based data gathering and observational activities: accurate location; geological context; the spatial/temporal relationship to other data or observations gathered at that or other location(s). In addition, all field maps and notebooks must be legible and readable by another geologist, and they must clearly distinguish between observed facts and inferences drawn from those facts (Ramsay & Huber 1987). Generally speaking, observations and data are gathered at a series of localities, the positions of which are marked by hand onto a topographic or aerial photographic base map, with all data measurements and observations being recorded simultaneously in a field notebook. Ideally, the extent of visible outcrop ('green-line mapping') and location of mappable geological boundaries (including some indication of how well constrained these boundaries are in terms of the available exposure, using solid, dashed or pecked lines) will be added to the base map by the field geologist as he or she moves between localities. Most geologists are encouraged to interpret their observations and measurements as they map and to modify these interpretations as more information is acquired. Thus field mapping is an iterative exercise in which both data gathering and interpretation occur simultaneously. This also means, however, that field-based data gathering presents a number of very significant challenges viewed from a perspective of prior information. In particular, we highlight four main problems:

(1) Field mapping involves extensive use of tacit knowledge, in which the *a priori* assumptions made when making interpretations, or even when gathering data, are either not stated or may not even be considered; thus 'facts' and 'inferences' are often not clearly separated in many studies.

(2) The workflow from field mapping to published map is generally a complex process involving data collection, interpretation, data reduction and final map drafting. The map is an abstraction at one specific scale of a large amount of data collected at the outcrop scale. Therefore the vast majority of 'inputs', 'data' and 'information' are typically excluded from the final result as they generally cannot all

be incorporated into the final paper map. Most maps are therefore dominated by interpretations ('knowledge' + 'expertise'). In many cases, some of the interpretation is made in locations far removed from the field either before or after the actual data gathering was carried out.

(3) All paper maps are inherently limited in terms of what they show by their scale. In many cases this means that they lack precise spatial accuracy, meaning that *reproducible* observations or measurements are often difficult or impossible.

(4) The final map generally shows little expression of the uncertainties involved in its production, and where uncertainty is depicted it is primitive, *ad hoc* and qualitative. Thus traditional geological mapping remains a highly interpretative, subjective art form in which uncertainty is difficult to quantify in any statistically meaningful way.

Digital geological mapping

GIS has evolved from its early use as a computer mapping system and is now defined as 'an information management system for organizing, visualizing and analysing spatially-orientated data' (Coburn & Yarus 2000, p.1). Since GIS became commercially available in the 1980s, GIS products are now used in a large number of applications that deal with spatial data, including social and economic planning, marketing, facilities management and environmental and resource assessment (Rhind 1992; Longley *et al.* 2001). Bonham-Carter (2000) describes the core GIS activities in a geoscience project as being: (1) data organization, (2) data visualization, (3) data search, (4) data combination, (5) data analysis, and (6) data prediction and decision support. The combination of these capabilities and the ability to handle large databases (up to a terabyte) indicate the power of the GIS approach for handling spatial data and its attraction for geoscience users such as the petroleum and mining industries. A generalized work flow for digital geological mapping is shown in Figure 1.

In its original guise, GIS largely dealt with 2-D data that was mapped onto the earth's surface (Rhind 1992). It was recognized that in order to deal with volumetric spatial information or 3-D geometries from subsurface data, a 3-D GIS or a 'GSIS' (geo-scientific information system) was required, and such systems have now been developed for commercial purposes (Turner 1992, 2000). Gocad® is one example of a powerful software system that is capable of displaying and analysing complex 3-D geological subsurface architectures (Mallet 1992).

Digital geological mapping (DGM) is a methodology by which a geologist collects GPS-located field data in a digital format. The method has been adapted from digital mapping and surveying techniques that are now widely used in construction, engineering and environmental industries. Early pioneers that have customized these techniques for use in Earth science fieldwork include Struik *et al.* 1991; Schetselaar 1995; Brodaric 1997; Briner *et al.* 1999; Bryant *et al.* 2000; Pundt & Brinkkotter-Runde 2000; Maerten *et al.* 2001 and Xu *et al.* 2001. Digital acquisition is gradually becoming more commonplace, particularly in North America, although adoption by European national surveys has been slow (Jackson & Asch 2002) and teaching of digital data capture is still not widespread in Europe.

The DGM we describe here is a mapping system that would be suitable for most geological purposes. The system involves the integration of three key technological components: (1) a GPS receiver usually capable of obtaining differential correction data that enable sub-metre positional accuracy; (2) a PDA or other digital data-logger, and (3) mobile GIS software. Mobile GIS is a specialized version of PDA software that can exchange information with more general-purpose desktop GIS. When used in 3-D mode we suggest that DGM provides an onshore equivalent to a wider definition of digital mapping as used by the petroleum industry to make 3-D structural interpretations in subsurface data.

In a GIS, information is usually displayed as a series of layers that can be superimposed, with each comprising a single type of data. Typically this may comprise features or objects that have distinct shape and size, or field data that vary continuously over a surface (Longley *et al.* 2001) as summarized in Table 3. The advantage of a GIS-based mapping system is that any number of different types of data may be geo-referenced and included as a separate layer in the database. These can then be displayed and analysed in conjunction with newly acquired field data (Fig. 2). Examples of data that may be included are field photographs, regional geophysical maps, aerial photography, satellite imagery, topographic data, previously digitized geological information, sample catalogues, geochronological data, geochemical data etc. In this way the GIS graphical user interface represents a single entry point to a wide range of spatially related relevant data that can be easily accessed in seconds. By comparison, such disparate types of data would traditionally be spread widely

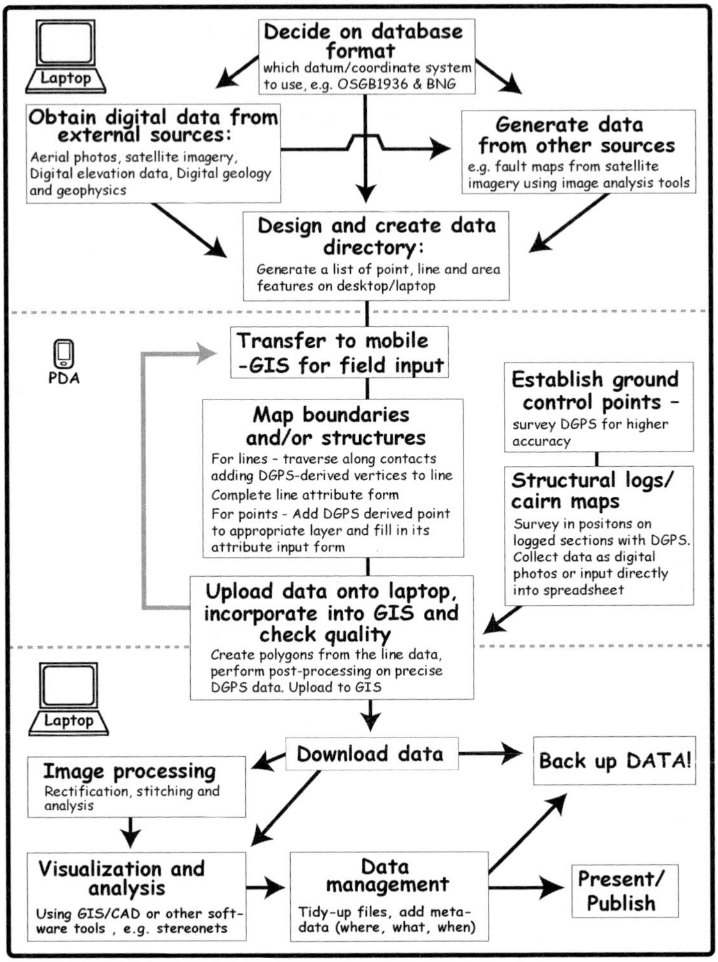

Fig. 1. Generalized workflow for digital geological mapping.

between field notebook, paper maps, isolated files on a computer, boxes of photographic slides or prints, library journals and loose papers.

Other advantages of DGM over traditional mapping include: (1) improved time efficiency, especially regarding data management, analysis and output; (2) because digitally captured data has high spatial precision, a significant reduction in uncertainty regarding location errors; (3) easier visualization of 3-D geometries because, when

Table 3. *Data types in a typical GIS system*

GIS data	Type	Geological data	Specific example
Point	Object	Locations of structural stations, anywhere a measurement is made or a sample is taken	Bedding strike & dip Gravity measurement Geochemical sample
Line	Object	Boundaries between areas or linear objects	Contact between rock units Fault trace
Polygon	Object	Areas of rock units	Formation extent Area of igneous intrusion
Raster	Field	Data sampled on a matrix of equally sized squares	Elevation (digital elevation model) Landsat image

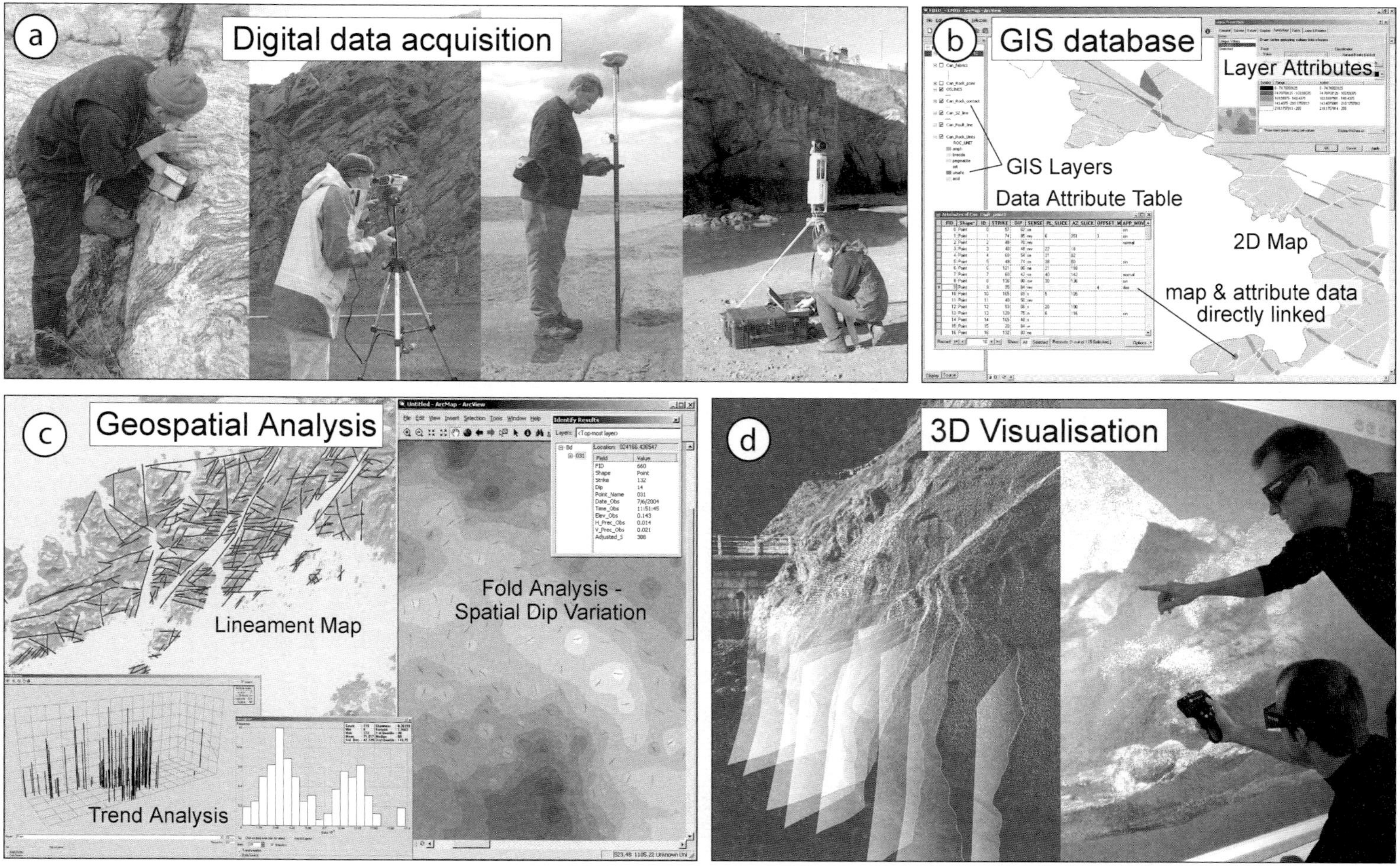

Fig. 2. Examples of digital data capture (Northumberland coast, England), 3-D visualization (West Orkney Basin, Scotland), and GIS analysis and map display from GIS database (Achmelvich Bay, NW Scotland).

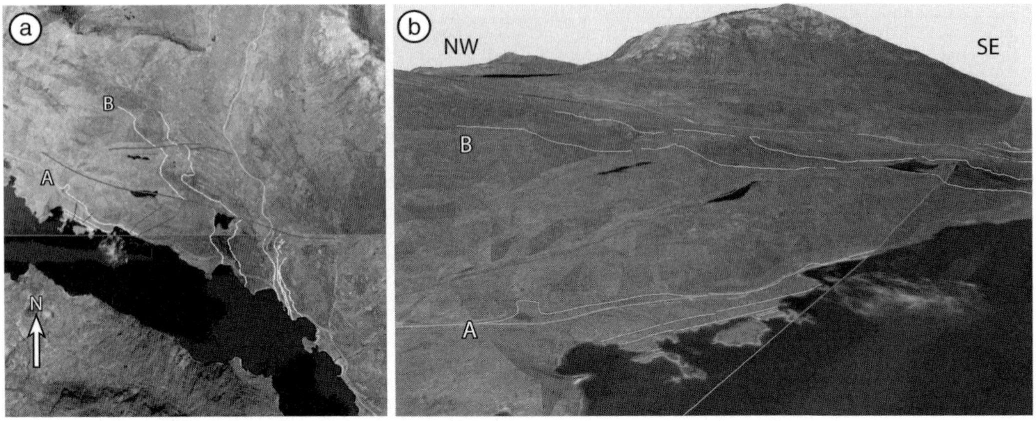

Fig. 3. Digital geological mapping of part of the foreland to the Moine Thrust at Loch Assynt, northern Scotland: (**a**) map view using GIS, showing geological boundaries (upper GIS layers) and aerial photograph (lowest GIS layer); (**b**) oblique perspective (2.5-D view) of the same data, from the SW, showing that boundary A is subhorizontal and boundary B dips to the east. Display has × 1.5 vertical exaggeration.

elevation data is acquired, DGM is inherently 2.5-D or 3-D (see below), (McCaffrey *et al.* 2003). At present the disadvantages of DGM include the relatively high cost of robust equipment, the potential for data loss in the event of equipment failure and the reluctance of more conservative geologists to explore the increased possibilities offered by new technology.

2-D, 3-D and 2.5-D data

In DGM, data are collected in either 2-D or 3-D mode. If the objective is purely to produce a map of the region then point, line and polygon data may be stored using only x and y coordinates (i.e. longitude and latitude). Real-time differential GPSs regularly give precision to approximately 1 m in the horizontal plane and survey systems that post-process positional data can attain centimetre-scale accuracy. The positional precision and accuracy that may be achieved using GPS is dependent on variations in the input satellite configuration (an error summarized by the dilution of precision statistic calculated continuously by GPS receivers). Topography or buildings can limit the number of input satellites available to a GPS receiver and thus accurate positioning near a cliff or in a deep valley may be difficult to achieve.

Geological information gathered by traditional geological mapping has been displayed on 2-D representations such as geological maps, but this format has disadvantages as described above. While GIS software products are often used to produce traditional geological maps, the systems also allow more flexible methods of visualization that can be easily tailored to individual requirements. For example, on-screen data can be viewed at different scales using the zoom and pan functions with different combinations of data layers visible as required.

GIS data may be overlain, or ‘draped’, onto a digital elevation model, in the form of a surface fitted to a raster map of elevation values, in order to produce a display that has been referred to as a ‘2.5-D’ representation (Longley *et al.* 2001). These data may then be displayed using a 3-D viewer that allows rotation to different vantage points as well as zoom and pan. One particularly useful geological application of 2.5-D displays is to study the relationship of geological formations and structure to topography (Fig. 3; McCaffrey *et al.* 2003).

For accurate 3-D reconstructions of geological architectures, the z coordinate (i.e. elevation) for all positions is essential. Most mobile GIS applications allow this to be incorporated into the data table. Despite GPS having poorer resolution in the z direction, in good conditions a differential GPS can give a vertical precision of approximately 1 m. Alternatively, 2-D data may be converted to 3-D by locating the positions onto a digital elevation model; however the resolution is limited by the horizontal spacing of the grid nodes (typically 20–50 m) and the precision of the values at each node (typically $\pm$ 3–10 m).

DGM and prior information

DGM is capable of incorporating prior information for the following reasons:

Table 4. *Examples of different types of uncertainty arising during geological mapping*

Level	Type of uncertainty	Examples
Data acquisition	Positional	How sure am I of my current location? How reliable is my base map? What is the precision of my GPS measurements? Is the borehole straight, or has it deviated without me knowing?
	Measurement	What is the precision of my clinometer? What is the accuracy of my dip/strike readings?
	Scale-dependent variability	How much does the dip and strike vary over the scale of the outcrop? Is my reading representative of the surrounding area?
	Observational	Is this rock best described as a granite? Is this fossil the brachiopod *Pentamerus*? Is that a stretching lineation or an intersection lineation?
	Temporal	How reliable is this way-up criteria? Is the relative age of these structures identified correctly?
	Sampling bias	Is my data biased by the natural predominance of sub-horizontal exposures? Has my sampling been skewed by me focusing only on the zones of high strain?
Primary interpretation	Correlation	Is this limestone the same unit as the limestone at the last outcrop? Is it valid to correlate the S2 fabric here with the S2 fabric observed on the other side of the area?
	Interpolation	How likely is it that all the ground between these two outcrops of slate also consists of slate? How much control do I have over the geometry of this fold trace?
	Inference from topography	Is there really a fault running along the unexposed valley floor? Does this sharp change in slope correspond to a lithological boundary?
Compound interpretation	Finished 2-D map Geological cross-section 3-D structural model	How can I quantify the uncertainty associated with this sophisticated interpretive model that I have slowly built up through a long iterative process of data collection and individual primary interpretations?

- The powerful data-handling capability and scaling functionality of GIS means that the *a priori* framework, newly collected data and interpretation can all be maintained in a single digital model.
- Several types of data can be stored together, all tied to their geospatial position within the GIS model, e.g. attribute data, metadata, photographs, sketches, ideas, notes, video, speech etc.
- Having data in digital format is the starting point for quantification of uncertainty (as discussed below).

In order to improve the inclusion of prior information in DGM workflows, more work needs to be done to integrate the various steps involved in the process of data acquisition, interpretation and final model. User-friendly data-gathering methods need to be developed to make it possible for geologists to capture information in ways that are more intuitive (e.g. using more rapid ways of data entry for PDAs, such as speech recognition), rather than ways dictated by the non-flexibility of existing hardware and software.

Quantification of uncertainty in geological mapping

Irrespective of whether a mapper uses traditional or digital methods, the acquisition and interpretation of field data inevitably involves a wide range of different types of uncertainty (Table 4). Uncertainties accrue with the onset of data acquisition, and they accumulate and propagate throughout the overall interpretation process. Some sources of uncertainty can readily be expressed in terms of a quantitative assessment of the precision of measurement for a piece of equipment (e.g. error tolerance of GPS or clinometer measurements). Other uncertainties can be reduced and quantified by repeating observations so that a measure of variance can be ascertained (e.g. variation of dip and strike at a single exposure). However, much uncertainty typically associated with geological interpretation is less easy to quantify. This is uncertainty that is associated not with accuracy of individual measurements but rather with the non-uniqueness of multiple solutions, each of which seems to be a viable interpretation to the problem, based on available data.

Interpolation

Geological mapping tends to produce sparse datasets. This is usually because the amount of exposure is limited, but even when there are high levels of exposure it is generally impractical to study all exposed rock in detail. Therefore one of the most important aspects of creating a geological map involves the interpolation of data to fill the areas between the intermittent data points actually measured. Interpolation below (and above) the surface of the Earth is, of course, also central to producing 2-D cross-sections and 3-D models. Most GIS systems have in-built analysis tools for the interpolation of geospatial point data across a topographical surface. These include deterministic methods (based on curve fitting of mathematical functions) and more advanced geostatistical methods ('kriging') that combine statistical analysis with mathematical curve fitting (Fig. 2c) and which also provide a statistical measure of uncertainty across the whole surface. The most basic approaches to kriging incorporate simplistic probability distributions, and the values of uncertainty derived are simply based on the sparseness of data. Kriging has a tendency to smooth out sharp variations between adjacent data points, so care is needed when showing spatial variation in parameters that might change abruptly across discontinuities such as faults.

Geostatistical methods of interpolation rely upon the basic assumption of spatial autocorrelation; i.e. points that are spatially near to each another tend to be more similar than those further away. This assumption is often acceptable for many types of geoscientific data (e.g. elevation values, mineral concentrations, pollution levels etc.), which can generally be mapped and interpolated using standard GIS functionality.

While geostatistical kriging methods usually work well for interpolation of geospatial point data distributed on the earth's surface, they are generally less well suited to interpretation of subsurface structure unless a reasonable amount of data is available at depth (e.g. dense borehole data, seismic grid, data from mine workings or high topographical relief). Although geological surfaces (bedding, foliation, fractures) tend to have high spatial autocorrelation, for subsurface interpretation based on outcrop data alone the availability of data is typically much too sparse in the *z* direction to allow meaningful interpolation at depth. Structural measurements at outcrop include vector data (strike, dip, plunge, azimuth) that describe the 3-D orientation of geological surfaces, as well as crucial supporting information concerning structural polarity (younging, facing, vergence) and temporal relationships (relative age of beds, structures, cross-cutting relationships, multiple generations of structures). Therefore, structural measurements typically encapsulate important additional information that should be used as a prior input to the interpolation process (Fig. 4). Simply using a predefined mathematical curve (cf. deterministic interpolation methods) or statistical probability distribution (cf. kriging) disregards the extra structural information gathered by the geologist and is therefore much less likely to produce a realistic interpolation.

Qualitative uncertainties

Many types of uncertainty are difficult to express quantitatively and are more suited to a *qualitative* evaluation by the geologist. Although this may be abhorrent to inductivists, who believe that science consists only of quantitative, objective measurement, a subjective statement such as 'This rock looks sheared and I am reasonably confident that it is a mylonite' is a more useful and representative observation statement than having to make a binary choice between 'This *is* a mylonite' and 'This is *not* a mylonite'. An obvious strategy to tackle this situation would be for field geologists to record specifically an estimate of confidence with every observation as a matter of routine. However, most geologists will perceive this data as superfluous and gathering it as an additional, unnecessary burden, because the general lack of methodology that has developed within geological mapping allows such information to be used in a systematic way.

The potential now exists for this situation to change, following recent advances in mathematics and computer technology, especially

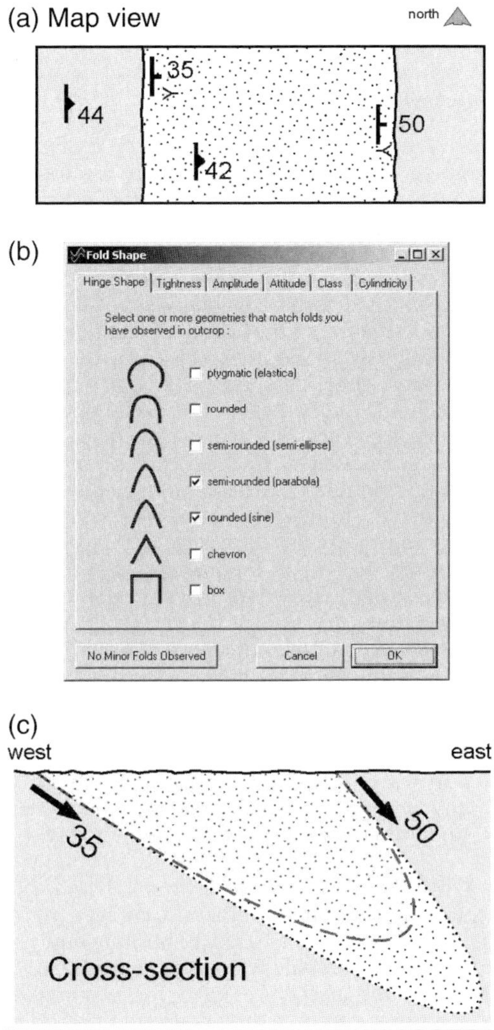

Fig. 4. Use of structural field observations as prior input for subsurface interpolation: (**a**) example sketch map showing dip and strike of bedding and cleavage with younging data; (**b**) data entry dialogue box for field observations of minor fold hinge geometry, which are used to influence subsurface interpolation; (**c**) cross-section showing example of subsurface interpolation. A range of large-scale fold geometries are consistent with evidence from minor folds observed at the surface. The dotted fold profile traces a fold surface that is 33% longer than the dashed profile and forms a fold in the stippled rocks with 70% additional area.

within several branches of artificial intelligence (AI). In particular, developments in fuzzy logic provide a formal framework in which relative terms (such as '*quite* sheared', '*very* fractured') can be transformed to discrete numerical values

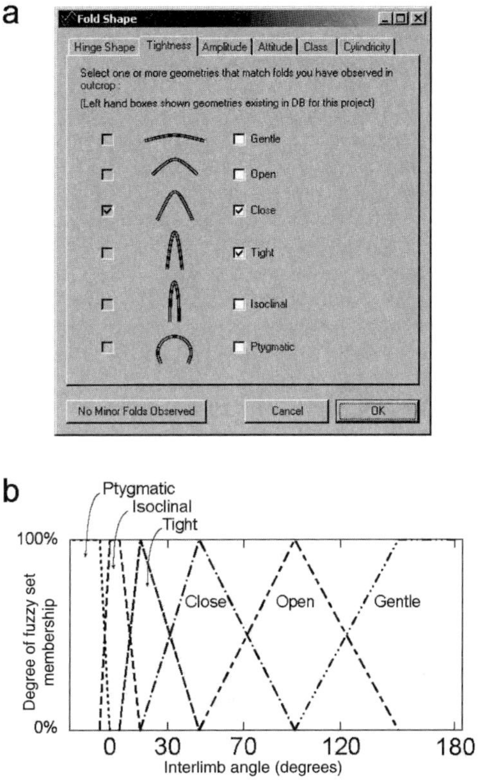

Fig. 5. Further use of structural field observations as prior input for subsurface interpolation: (**a**) data entry dialogue box in which the user can specify the typical inter-limb angles of minor folds seen in outcrop; (**b**) fuzzy set definition showing mapping of inter-limb angle to fuzzy descriptors ('open', 'tight' etc.). The range of inter-limb values indicated by the geologist is used to constrain the range of realistic subsurface interpolations and also to test for internal inconsistencies between field data and other user inputs.

and represented within binary computer code. Figure 5 shows an example of the use of fuzzy descriptors to represent the tightness of inter-limb angles in folds. The value(s) of fold tightness chosen by the user as representative of minor folds observed in outcrop are combined with other user input for fold data and used to constrain subsurface interpolation. Checks are also made to identify potential areas where the data might not be internally consistent using a fuzzy rule base with rules such as:

> If hinge shape is 'chevron' or 'box' and interlimb angle is 'close' then conflict risk is moderate.

If hinge shape is 'chevron' or 'box' and 'interlimb_angle' is less than or equal to 'tight' then conflict risk is significant.

Other branches of AI (e.g. Bayesian networks, neural networks, genetic algorithms, constraint satisfaction techniques) also have potential for finding solutions for complicated non-linear models involving very many variables. All these methodologies have a proven track record of finding good solutions to real problems with a much shorter amount of computer processing than traditional approaches (e.g. Buttenfield & Mark 1991; Braunschweig & Day 1995; Hamburger 1995; Murnion 1996; Jones *et al.* 1998; Nordlund 1999; Ouenes 2000; Peddle 2002; Luo & Dimitrakopoulos 2003).

Future trends in geological mapping

For the last two decades the growth in information technology (IT) has generally been so great that geologists have struggled to keep abreast of technological advances. Within geological mapping, earth scientists have been slow to improve workflows and methods of interpretation that exploit the newly developing technologies. There is no indication that the current rate of growth within IT is set to diminish, and the following trends are likely to provide increased opportunities for geoscientists to improve the process of geological mapping:

- Portable equipment will continue to become lighter, cheaper, more robust, more powerful, more intuitive and user-friendly, and more integrated with the user.
- An increased number of 2-D and 3-D analytical tools will be incorporated into existing GIS software to provide a single integrated tool for geological mapping and interpretation.
- Interpretation tools will propagate information about uncertainties through the modelling process so that various interpretations can be tested in parallel and an indication of overall uncertainty can be given for each interpretation.
- Satellite communications technology combined with GRID facilities will bring supercomputing power to field geologists (a 'PersonalGRID'). This will increase the possibilities for ongoing iterative interpretation of field data while still in the field.
- Speech recognition software combined with semantic-based search technology can help to encourage the geologist to verbalize (i.e. make explicit) more of the decision-making processes involved in mapping.

Conclusions

Although traditional processes of geological mapping have a proven track record established over a period of 200 years, there are nevertheless important methodological shortcomings seen from a scientific perspective:

- Paper-based published maps generally show only a fraction of the field data that have been collected and used as the basis for the map's creation. Other data remain hidden in the field notebook and in the mind of the mapper.
- Published maps rarely make reference to the reasoning used during interpretation of the basic field data. Reasoning typically relies heavily on prior information and knowledge that represents the experience of the geologist before the onset of the mapping project.
- Paper-based maps are by necessity published at a fixed scale. The skill of the cartographer is to present as much relevant information as possible while maintaining legibility at the chosen scale, but inevitably there is a loss of precision, especially with respect to geospatial positioning.
- With traditional maps there are generally only very limited possibilities for expressing any uncertainty concerning the given interpretation, and these are not quantitative.

While the above limitations have always been acceptable as long as there were no viable alternatives to paper-based publishing, today's information technology makes it possible to store all the necessary data for a mapping project on a compact disk that costs just a few pence. Digital-mapping techniques have the potential to improve the scientific validity of the mapping process in the following ways:

- Collected field data can be stored and distributed together with the interpreted map in a single digital model within a GIS. In future it should be possible to capture and store an even wider range of multi-media datatypes in a seamless way (including metadata, photographs, sketches, ideas and notes, video, speech etc.), all tied to the appropriate geospatial position within the GIS model.
- Prior inputs used as the basis for interpretations can be stated explicitly within the same GIS mode. Although this alone does not involve the quantification of uncertainty associated with interpretation, it does represent a radical improvement to geological

mapping practice as it shows the user of the map not only the mapper's interpretation but also the raw data upon which the interpretation is based.

- Traditional paper-based maps can only display a finite amount of information for a given scale of map without loss of legibility. The inherent scalability of a GIS model makes it possible to store a huge amount of data for any geospatial locality, so that the amount of data available to the map user through the user interface is not restricted in the same way.
- As progressively more analytical tools are incorporated into GIS software, geospatially referenced data can be analysed and interpreted within a single software environment.
- Many types of uncertainty that arise during the mapping process can be either quantified or estimated qualitatively in a way that can be represented digitally (using AI techniques).

Digital geological mapping is still in its infancy. Future work should concentrate on the following challenges:

- The digital workflow should be continuously improved to make it more intuitive and quicker to capture field data in digital format.
- Further integration of analysis tools and GIS is needed.
- Methodologies should be further developed that use the uncertainties associated with individual data or interpretations as the input to produce an overall estimate of uncertainty associated with a given model. Alternative interpretations can be modelled simultaneously, with cumulative uncertainty calculated for each model.
- Efforts should be made to combine portable GIS with GRID technology in order to provide the field geologist with powerful analysis tools while in the field. This can increase the ability of the geologist to analyse their data while still on the outcrop, thereby helping to optimize strategy for further data collection.

Thanks to N. Drinkwater (Schlumberger) and D. Jeanette (Bureau of Mineral Resources, Texas) for constructive reviews. N. Holliman (e-Science) and S. Waggott (Halcrow) provided essential technical expertise. J. Imber, N. de Paola, P. Clegg and other members of Durham RRG and B. Bremdal at CognIT contributed with useful debate. Immeasurable inspiration has been gathered from the pioneering efforts of O. McNoleg (McNoleg 1996, 1998). E. McAllister and others at Leeds RDR made early progress with DGM. Much of this work was part of the Ocean Margins Link project (NER/T/S/2000/01018), funded by Statoil UK Ltd, British Petroleum and the National Environment Research Council.

References

BARNES, J. W. 1981. *Basic Geological Mapping*. Geological Society, London, Handbook Series, Open University Press, Maidenhead.

BONHAM-CARTER, G. 2000. An overview of GIS in the geosciences. *In:* COBURN, T. C. & YARUS, J. M. (eds) *Geographic Information Systems in Petroleum Exploration and Development*. American Association of Petroleum Geologists, Computer Applications in Geology, **4**, 17–26.

BOOSE, J. H. 1986. *Expertise Transfer for Expert System Design*. Elsevier, Amsterdam.

BRAUNSCHWEIG, B. & DAY, R. 1995. An overview of AI techniques and of their use in the Petroleum Industry. *In:* BRAUNSCHWEIG, B. & DAY, R. (eds) *Artificial Intelligence in the Petroleum Industry: Symbolic and Computational Applications*. Editions Technip, Paris, France, 3–56.

BRINER, A. P., KRONENBERG, H., MAZUREK, M., HORN, H., ENGI, M. & PETERS, T. 1999. FieldBook and GeoDatabase: tools for field data acquisition and analysis. *Computers & Geosciences*, **25**, 1101–1111.

BRODARIC, B. 1997. Field data capture and manipulation using GSC FIELDLOG v3.0. *In:* SOLLER, D. R. (ed.) *Proceedings of a Workshop on Digital Mapping Techniques: Methods for Geologic Map Capture, Management and Publication*. US Geological Survey Open File Report 97–269, 77–81.

BRYANT, I., CARR, D., CIRILLI, P., DRINKWATER, N., MCCORMICK, D., TILKE, P. & THURMOND, J. 2000. Use of 3D digital analogues as templates in reservoir modelling. *Petroleum Geoscience*, **6**, 195–201.

BUTTENFIELD, B. P. & MARK, D. M. 1991. Expert systems in cartographic design. *In:* TAYLOR, D. R. F. (ed.) *Geographic Information Systems: The Microcomputer and Modern Cartography*. Pergamon Press, Oxford, 129–150.

COBURN, T. C. & YARUS, J. M. 2000. *Geographic Information Systems in Petroleum Exploration and Development*. American Association of Petroleum Geologists, Computer Applications in Geology, **4**.

FORD, D. & STERMAN, J. 1998. Expert knowledge elicitation for improving mental and formal models. *System Dynamics Review*, **14**, 309–340.

HAMBURGER, J. 1995. GeoCos 3D: A diagnosis system, based on the FROG formalism, to help geologists set up valid interpretations. *In:* BRAUNSCHWEIG, B. & DAY, R. (eds) *Artificial Intelligence in the Petroleum Industry: Symbolic and Computational Applications*. Editions Technip, Paris, France, 57–85.

JACKSON, I. & ASCH, K. 2002. The status of digital geological mapping in Europe: the results of a census of the digital mapping coverage, approaches and standards of 29 European geological survey organisations in the year 2000. *Computers & Geosciences*, **28**, 783–788.

JONES, R. R., LARSEN, S. J. & HANSEN, G. I. 1998. An expert system for building virtual copies of real world environments. *In:* CANTU, F. J., SOTO, R., LIEBOWITZ, J. & SUCAR, E. (eds) *Expert Systems: Application of Advanced Information Technologies*. Cognizant, New York, 248–255.

KIDD, A. L. (ed.) 1987. *Knowledge Elicitation for Expert Systems: A Practical Handbook*. Plenum Press, New York.

KUHN, T. S. 1962. *The Structure of Scientific Revolutions*. University of Chicago Press, Chicago.

LAKATOS, I. 1974. Falsification and the methodology of scientific research programmes. *In:* LAKATOS, I. & MUSGRAVE, A. (eds) *Criticism and the Growth of Knowledge*. Cambridge University Press, Cambridge, 91–196.

LIEBOWITZ, J. & BECKMAN, T. 1998. *Knowledge Organizations*. CRC Press, Florida.

LONGLEY, P. A., GOODCHILD, M. F., MAGUIRE, D. J. & RHIND, D. W. 2001. *Geographic Information Systems and Science*. Wiley & Sons Ltd, Chichester.

LUO, X. & DIMITRAKOPOULOS, R. 2003. Data-driven fuzzy analysis in quantitative resource assessment. *Computers & Geosciences*, **29**, 3–13.

MAERTEN, L., POLLARD, D. D. & MAERTEN, F. 2001. Digital mapping of three dimensional structures of the Chimney Rock fault system, central Utah. *Journal of Structural Geology*, **23**, 585–592.

MALLET, J. L. 1992. GOCAD: A computer aided design program for geological applications. *In:* TURNER, A. K. (ed.) *Three-Dimensional Modeling with Geoscientific Information Systems*. NATO ASI Series C, Mathematical and Physical Sciences, Kluwer Academic, Dordrecht, **354**, 123–141.

MCCAFFREY, K. J. W, HOLDSWORTH, R. E., CLEGG, P., JONES, R. R. & WILSON, R. 2003. *Using Digital Mapping and 3D Visualization to Enhance and Modernise Undergraduate Fieldwork Experience*. Planet Magazine, (Special edition), **5**.

MCCLAY, K. R. 1987. *The Mapping of Geological Structures*. Geological Society, London, Handbook Series, Open University Press, Maidenhead.

MCNOLEG, O. 1996. The integration of GIS, remote sensing, expert systems and adaptive co-kriging for environmental habitat modelling of the Highland Haggis using object oriented, fuzzy logic and neural network techniques. *Computers & Geosciences*, **22**, 585–588.

MCNOLEG, O. 1998. Professor Oleg McNoleg's guide to the successful use of geographical information systems (Ten ways to say nothing with GIS). *International Journal of Geographical Information Science*, **12**, 429–430.

MURNION, S. 1996. Spatial analysis using unsupervised neural networks. *Computers & Geosciences*, **22**, 1027–1031.

NONAKA, I. 1994. A dynamic theory of organizational knowledge creation. *Organization Science*, **5**, 14–37.

NONAKA, I. & TAKEUCHI, H. 1995. *The Knowledge-Creating Company: How Japanese Companies Create the Dynamics of Innovation*. Oxford University Press, Oxford.

NORDLUND, U. 1999. FUZZIM: forward stratigraphic modeling made simple. *Computers & Geosciences*, **25**, 449–456.

OUENES, A. 2000. Practical application of fuzzy logic and neural networks to fractured reservoir characterization. *Computers & Geosciences*, **26**, 953–962.

PEDDLE, D. R. 2002. Optimisation of multisource data analysis: An example using evidential reasoning for GIS data classification. *Computers & Geosciences*, **28**, 45–52.

POLANYI, M. 1958. *Personal Knowledge: Towards a Post-Critical Philosophy*. Routledge & Kegan Paul, London.

POLANYI, M. 1966. *The Tacit Dimension*. Routledge & Kegan Paul, London.

PUNDT, H. & BRINKKOTTER-RUNDE, K. 2000. Visualization of spatial data for field based GIS. *Computers & Geosciences*, **26**, 51–56.

RAMSAY, J. G. & HUBER, M. I. 1987. *The Techniques of Modern Structural Geology. Vol. 2. Folds and Fractures*. Academic Press, London.

RHIND, D. W. 1992. Spatial data handling in the Geosciences. *In:* TURNER, A. K. (ed.) *Three-Dimensional Modeling with Geoscientific Information Systems*. NATO ASI Series C, Mathematical and Physical Sciences, Kluwer Academic, Dordrecht, **354**, 13–27.

SCHETSELAAR, E. M. 1995. Computerized field-data capture and GIS analysis for generation of cross sections in 3-D perspective views. *Computers & Geosciences*, **21**, 687–701.

SMITH, W. 1801. *General Map of Strata in England and Wales*.

SMITH, W. 1815. *A Delineation of the Strata of England and Wales, with Part of Scotland*. J. Carey, London.

STRUIK, L. C., ATRENS, A. & HAYNES, A. 1991. Hand-held computer as a field notebook and its integration with the Ontario Geological Survey's 'FIELDLOG' program. *Geological Survey of Canada, Current Research, Part A*, 279–284.

TURNER, A. K. 1992. *Three-Dimensional Modeling with Geoscientific Information Systems*. NATO ASI Series C, Mathematical and Physical Sciences, Kluwer Academic, Dordrecht, **354**.

TURNER, A. K. 2000. Geoscientific modeling: past, present and future. *In:* COBURN, T. C. & YARUS, J. M. (eds) *Geographic Information Systems in Petroleum Exploration and Development*. American Association of Petroleum Geologists, Computer Applications in Geology, **4**, 27–36.

WHITE, S. & SLEEMAN, D. 1999. A constraint-based approach to the description of competence. *In:* FENSEL, D. & STUDER R. (eds) *Knowledge Acquisition, Modeling and Management: Lecture Notes in Computer Science*, **1621**, 291–308.

VAN DER SPEK, R. & SPIJKERVET, A. 1997. *Knowledge Management: Dealing Intelligently with Knowledge Management Network*. CIBIT, Utrecht.

XU, X., BATTACHARYA, J. A., DAVIES, R. K. & AIKEN, C. L. V. 2000. Digital geologic mapping of the Ferron sandstone, Muddy Creek, Utah, with GPS and reflectorless laser rangefinders. *GPS Solutions*, **5**, 15–23.

Three-dimensional geological models from outcrop data using digital data collection techniques: an example from the Tanqua Karoo depocentre, South Africa

D. HODGETTS[1,2], N. J. DRINKWATER[3], J. HODGSON[1], J. KAVANAGH[1], S. S. FLINT[1], K. J. KEOGH[1,4] & J. A. HOWELL[1,5]

[1]*Stratigraphy Group, Department of Earth and Ocean Science, University of Liverpool, Brownlow Street, Liverpool L69 3BX, UK (e-mail: david.hodgetts@man.ac.uk)*

[2]*Now at Department of Earth Science, University of Manchester, Oxford Road, Manchester M13 9PL, UK*

[3]*Schlumberger Cambridge Research, High Cross, Madingley Road, Cambridge CB3 0EL, UK*

[4]*Statoil ASA, TEK F&T UTV TOS, Forushagen, 4035 Stavanger, Norway*

[5]*Now at Department of Geology, University of Bergen, Allegaten 41, N-5007 Bergen, Norway*

Abstract: Recent technological advances have made the collection of digital geological data from outcrops a realistic and efficient proposition. The world-class exposures of Permian basin-floor turbidite fans of the Tanqua depocentre, Karoo Basin, South Africa have been the focus of one such study. These outcrops are faulted at a subseismic scale (displacements of up to 40 m), with continuous exposures of up to 40 km in depositional dip and 20 km strike directions. Digital data collection has been undertaken using a variety of methods: differential global-positioning systems (DGPS) mapping, surveying using laser total station and laser rangefinders, ground- and helicopter-based digital photography and photogrammetry, and digital sedimentary outcrop logging as well as geophysical data from boreholes. These data have then been integrated into several 3-D geological models of the study area, built using a subsurface reservoir-modelling system. The integrated dataset provides insights into the stratigraphic evolution of a deep-water fan complex by allowing true 3-D analysis and interpretation of data collected in the field. The improved understanding of these deep-water fan systems will improve existing models of offshore analogues by enhancing understanding of geometries and trends not resolvable from existing offshore data and by identifying potential problematic areas for fluid flow. Initial results from the application of this approach have been successfully applied to the conditioning of stochastic geological models of a subsurface deep-water reservoir from the North Sea.

Currently there is a noticeable increase in success in the discovery of deep-water reservoirs in deep and ultra deep-water depths (Vergara *et al.* 2001). Such prospects are associated with difficult operating environments where appraisal and development wells can cost tens of millions of pounds; accurate prediction of essential recovery parameters (e.g. subsurface reservoir geometry, pay thickness etc.) is critical and an understanding of the uncertainty in these factors is commonly a prerequisite to successful field development. One problem of this frontier exploration is that major economic decisions on projected field development are currently fast-tracked with an emphasis on minimum amounts of drilling, the absence of well workovers and an increased use of subsea facilities (Dromgoole *et al.* 2000). The critical decisions to implement these facilities are being based on commonly very limited amounts of data: typically a single, often non-cored discovery well combined with a 3-D seismic survey.

In addition to such new prospects, existing fields with deep-water reservoirs which are nearing or passing their production plateaus are often the focus for significant improved oil recovery plans to extend the overall field life. One accepted key factor is that production histories of existing fields have turned out to be significantly different from those initially predicted (Dromgoole *et al.* 2000). In both exploration and development

From: CURTIS, A. & WOOD, R. (eds) 2004. *Geological Prior Information: Informing Science and Engineering*. Geological Society, London, Special Publications, **239**, 57–75. 186239-171-8/04/$15.00

categories therefore, it is necessary to have a good understanding of reservoir architecture that can be used to predict reservoir production strategies (Kolla *et al.* 2001).

In response to this, this paper demonstrates that, by full and detailed characterization of high-quality onshore analogues, improvements in the understanding of depositional systems and geometries may be developed that will have a direct impact on the characterization and modelling of both prospective and existing subsurface fields. To this end, this paper documents an extensive field-surveying exercise that was undertaken to map in 3-D the spectacularly well-exposed turbidites of the Skoorsteenberg Formation, Tanqua Karoo Basin, South Africa. This was part of a European Union-funded research project, known as NOMAD (**No**vel **M**odelled **A**nalogue **D**ata for more efficient exploitation of deep-water hydrocarbon reservoirs). This high-resolution surveying has been undertaken over three field seasons (20 weeks in total, involving field teams of four to eight geologists) using high-resolution differential global-positioning system (DGPS) equipment, digital photography and photogrammetry, and traditional field geology techniques. The data have been collected and maintained in their true 3-D framework from outset to the final model, and we believe this is the first such dataset collected on this scale with such a high level of data integration. This paper is intended to fully document and illustrate the complete workflow involved from collecting these field data, building the high-resolution geological models and then applying the results and lessons learnt to a particular subsurface-modelling issue for a North Sea deep-water oil reservoir.

Geology and stratigraphy of the Tanqua depocentre

The Tanqua depocentre is located in the south-western corner of the Karoo Basin, South Africa (Fig. 1), which developed as one of several major retro-arc foreland basins (others include the Paranà, Beacon and Bowen Basins) inboard of a fold-thrust belt (the Cape Fold Belt). An oroclinal bend in the Cape Fold Belt bounds the Tanqua depocentre to the west (the N–S-trending Cederberge branch) and to the south (the E–W-trending Swartberg Branch) (Fig. 1).

The Karoo Basin sediments unconformably overly Precambrian rocks of the Namaqua–Natal Belt to the north and Ordovician to Early Carboniferous rocks of the Cape Supergroup to the south (Tankard *et al.* 1982). Basal Karoo Supergroup glaciogenic strata of the Dwyka Group are overlain by the Prince Albert Formation (180 m of shale and cherty shale beds, with an age of 288 ± 3 Ma; Bangert *et al.* 1999), the Whitehill Formation (black, carbonaceous shales with pelagic organisms) and the Collingham Formation (fine-grained sheet turbidites and intercalated ashes; 270 ± 1 Ma; Turner 1999), all indicative of post-glacial long-term sea-level rise.

Within the Tanqua depocentre, the Collingham Formation is overlain by a thick succession of basinal shales (the Tierberg Formation) succeeded by the Skoorsteenberg Formation, a 400 m-thick succession of sand-rich submarine fan systems with alternating siltstone and shale intervals (Bouma & Wickens 1991; Wickens 1994). Overlying submarine slope and prodeltaic (Kookfontein Formation), shoreface (Koedoesburg Formation) and fluvial (Abrahamskraal Formation, Beaufort Group) successions mark the overall progradation of the sedimentary system to the north and east during the Mid- to Late Permian.

The Skoorsteenberg Formation is very well exposed over 45 km in a S–N direction (broadly down dip) and over 15–20 km in an E–W direction (broadly across strike), and previous researchers have identified five sand-rich submarine fan systems (Fans 1–5; Bouma & Wickens 1991; Wickens 1994; Wickens & Bouma 2000; Fig. 2). As part of the NOMAD project, the recent drilling and full coring of seven wells in locations identified as being strategically important in developing an accurate and extensive 3-D geological and reservoir model has resulted in significant additional data. A total thickness of 1247 m of core was recovered and has been logged at a 0.5-cm scale. In addition, some of the wells were logged using downhole geophysical and geochemical techniques such as image, density, neutron and gamma-ray logs to provide an additional tool for comparison and contrast with data of a resolution and scale of sampling typically utilized in the oil industry.

Model area

The study area contains five deep-water fan systems, each up to 50 m thick, separated stratigraphically by siltstone- and hemipelagic shale-prone intervals (Wickens 1994; Johnson *et al.* 2001; Fig. 2). The area is thrust faulted at a predominantly subseismic scale (with faults of up to 40 m displacement). Apart from this compressional signature, the study area is characterized by excellent, continuous exposures in both depositional strike and dip directions.

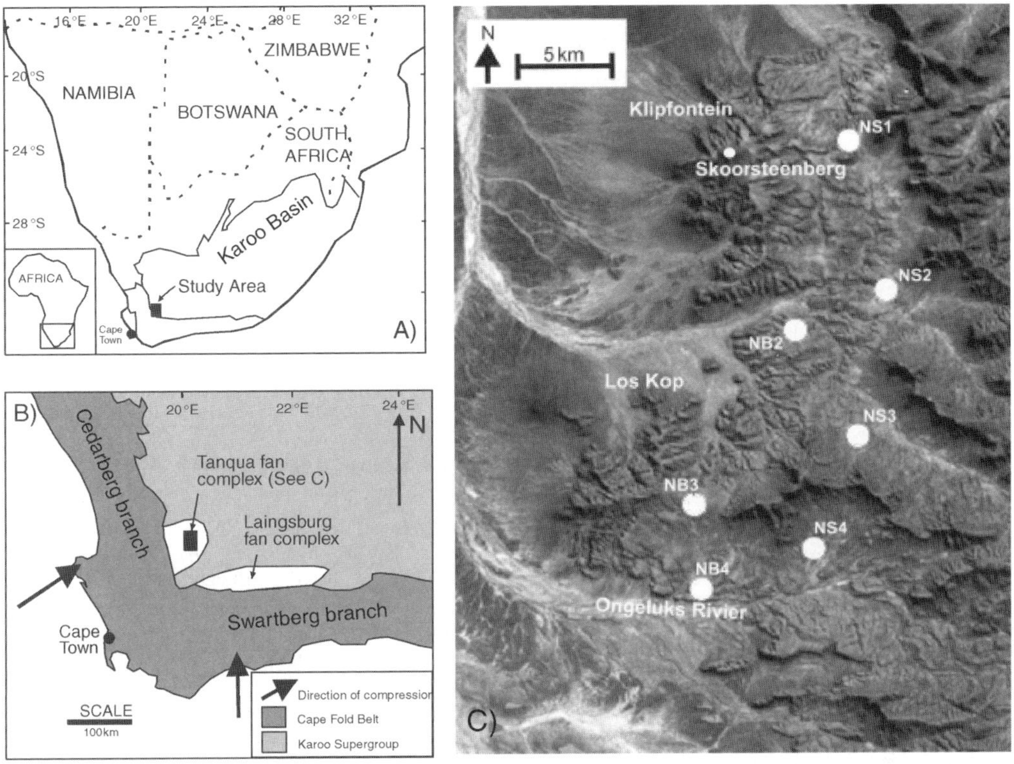

Fig. 1. (**a**) and (**b**) Location of the study area in South Africa in relation to the limbs of the Cape Fold Belt (adapted from Wickens 1994). (**c**) Landsat image of the study area (see (b) for location) with important localities and locations of boreholes. Study area is approximately 20 × 40 km. The seven well locations are marked; NB, 4 in (10 cm) diameter wells; NS, 3 in (7.5 cm) diameter wells.

The reservoir model area itself measures some 20 × 40 km in total, with the sizes of the individual fan-system models varying according to fan extent and exposure quality. Models of three of the five fan systems have been built with full 3-D high-resolution facies realizations.

Digital field data collection methodologies

Differential global-positioning system (DGPS) mapping. The system used for the 3-D surveying of the five turbidite fan systems was the Trimble RTK (real-time kinematic) DGPS, which utilizes a base station to broadcast RTK corrections to rover units which, in turn, are used to collect the 3-D positional data. The rover units use a survey controller to operate the DGPS and to collect the data. Each data point collected has a unique point ID associated with it, and a non-unique feature code that describes the information that data point represents, e.g. bedding plane, fault etc. Associated with each feature code is a library of attributes that allows information to be attached to the DGPS-mapped data points in the field. These attributes may include dip and azimuth information, palaeocurrent directions and fault displacement data (Tables 1 & 2). With this particular system, the attribute library has to be defined at the start of the database project associated with that particular data collection phase; any modifications to the attribute library mean that a new project has to be started, though the old projects and data will still be maintained. It is therefore absolutely essential to give careful thought to the design of the attribute library prior to the commencement of significant data measurement.

Extensive DGPS mapping of the fan-system bounding surfaces was used to define an envelope or stratigraphic framework into which the facies model would be populated. Successful construction of the structural framework required both the mapping of top and base markers for each of the five fan systems and

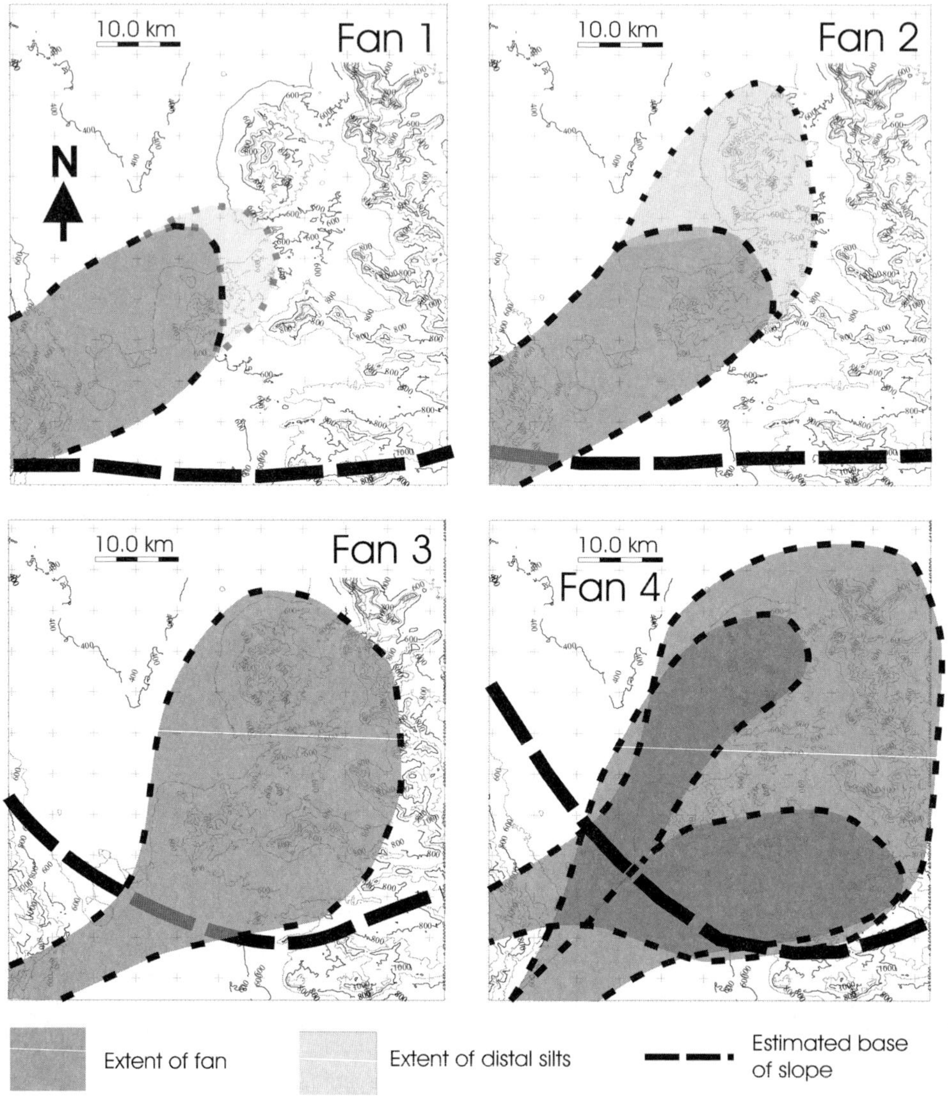

Fig. 2. The model study area with borehole locations and the projected extents of the four oldest deep-water fan systems. The outline positions of the fan systems have been significantly refined from the borehole data collected during the NOMAD project. Thick dashed line, estimated position of the base-of-slope. The estimated extents of distal siltstone sheets beyond the sandstone pinchout are indicated for Fans 1 and 2.

Table 1. *Example of feature codes used for the NOMAD data collection*

Feature code	Description
BASE FAN	Marker for base of fan system
TOP FAN	Marker for top of fan system
FAULT	Fault data
BEDDING	Bedding-plane data
PALAEOCURRENT	Measured palaeocurrent
GCP	Ground control point for photogrammetry
BASE STATION	Location of the base station

Table 2. *Example attribute library used for e.g. a base fan-system marker during the NOMAD field-data acquisition*

Attribute	Variable type (Range)	Description
Fan-system number	Integer	Fan number
Bedding		
Dip	Floating point (0–90)	Dip of bedding plane
Azimuth	Floating point (0–360)	Azimuth of bedding
Palaeocurrent		
Azimuth	Floating point (0–360)	Palaeocurrent direction
Type	String	Type of palaeocurrent (from ripple or groove etc.)
Confidence	Integer	Measure of confidence of reading

high-resolution mapping of the structural information of positions and displacement directions of the faults (Fig. 3).

The successful identification and hence mapping of the base and tops of the fan systems could often be problematic. Due to the nature of these depositional systems, the top and base of a fan are rarely represented by a sharp contact. The base is usually represented by a gradual transition from thin interbeds through to thick-bedded sandstone. In these cases a rule was used to define the base of fan as the lowest 20 cm sandstone bed (the 'generic base' marker). This is a diachronous surface that represents the onset of major sand deposition at that time and in that position. This marker was also identified and recorded during sedimentary logging at outcrop to ensure that the DGPS and logging data could be tied together during the model-building phase. In addition, in order to reduce error introduced by the interpretation of such boundaries during field mapping through the use of a variety of field geologists, an initial, pre-mapping reconnaissance of particular stratigraphic features was conducted throughout the model area. Ideally, each member of the field team would then consistently identify and measure the same surface. This training

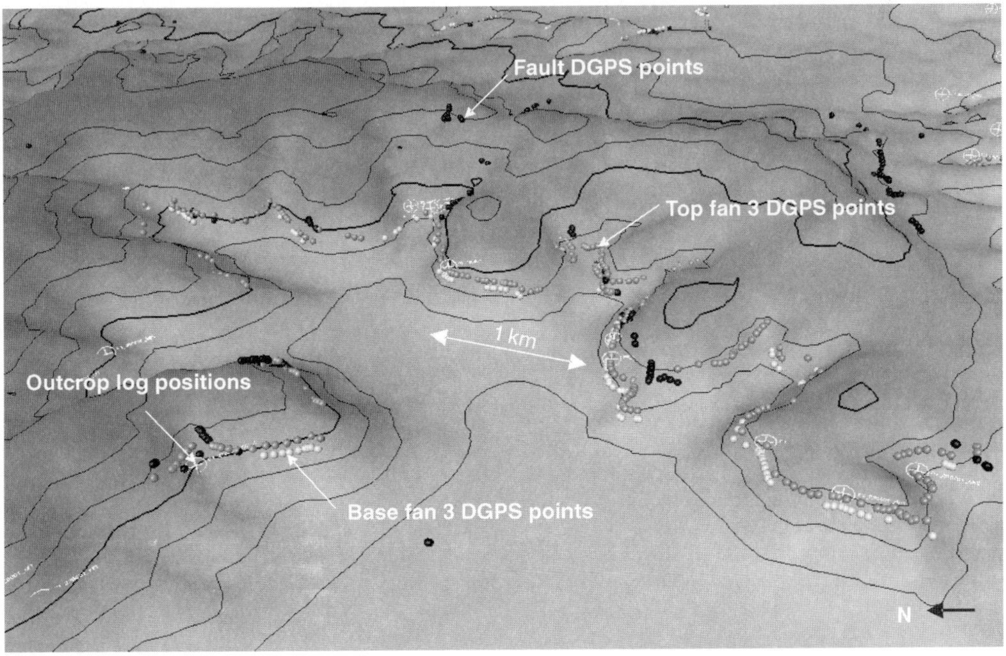

Fig. 3. DGPS-surveyed dataset for part of the NOMAD field area. The background image is a Landsat image draped over a 20 × 20 m horizontal-resolution digital elevation model (DEM). Datapoints represent the true 3-D position of the mapped marker horizon (black, faults; light grey, base of fan-system 3; dark grey, top of Fan system 3). Tops of measured outcrop logs are marked using 'circle and cross' motif.

approach was similarly adopted during the sedimentary-logging process so that consistency of facies interpretations was being standardized. Each geologist was also encouraged to record data using standardized logging sheets, again to reduce 'user error' and improve consistency of interpretation. In an ideal situation, all data, including the outcrop logs, would have been recorded with DGPS surveying by the same operator but, due to limited availability of equipment, manpower and field time, this was not always possible. An example of a GPS point dataset is shown in Figure 3.

Advantages of DGPS mapping:

- single-person operation;
- speed of mapping;
- ability to cover large areas;
- long range (5 km from base station).

Disadvantages of DGPS mapping:

- cumbersome rover units;
- satellite coverage problems when close to the outcrop; this can usually be overcome by choosing the time of day when satellite coverage is most appropriate for the area being surveyed, although there are some areas (back of valleys etc.) which may never have adequate satellite coverage.

Total-station surveying. For more detailed mapping of small-scale study areas (up to 2×2 km), total-station surveying techniques were used. The total station is a laser-based surveying system using a manned optical-surveying station, with one or more people on the outcrop equipped with survey poles and prisms (Fig. 4). Accuracy is at the sub-3 cm scale in 3-D and all data are recorded digitally in the survey controller. Feature codes and attribute libraries similar to those used in the mapping were adopted for these studies but, in addition to just mapping the top and base fan generic markers, key internal stratigraphic surfaces were surveyed in detail, e.g. erosion surfaces, key depositional surfaces and amalgamation horizons. These internal surfaces were first identified at the field-logging stage and then subsequently surveyed, again maintaining the integration between survey data and field log data.

Advantages:

- less cumbersome while on the outcrop than the rover units; ideal in less accessible areas;
- does not need satellite coverage while on the cliff face; only needed for set-up of the total station, which is usually away from the outcrop, and therefore away from satellite shadow.

Disadvantages:

- requires more than one person to operate, so personal radios are essential;
- can only survey areas in line of sight of total station, and hence there is usually some downtime while the unit is repositioned;
- requires human access, so is not suitable for all locations; the systems can be used in reflectorless mode, although range is severely limited.

Digital photography and photogrammetry. Outcrop photographs are a useful and common tool used to illustrate geological structures and to aid correlations between measured sections on a large scale. The effect of perspective inherent in all photographs is, however, a well-recognized common limitation. Digital photogrammetric techniques may be used to remove these effects and extract 3-D information from the photographs.

Photogrammetry has been used in many fields (e.g. see Warner 1995; Oka 1998; Reid & Harrison 2000; among others) and is defined (Slama 1980) as

> . . . the art, science, and technology of obtaining reliable information about physical objects and the environment through processes of recording, measuring, and interpreting photographic images and patterns of electromagnetic radiant energy and other phenomena.

It allows detailed 3-D information to be extracted from stereo pairs of digital images by knowing information about the camera and position of the camera relative to the objects being studied. Photogrammetry is applicable from both aerial (in this case from a helicopter) and land-based photography.

The main uses for this kind of data collection come in two main areas: those sections of outcrop with extensive exposure, where detailed data on thickness variations of beds may need to be collected (more than can be normally achieved by widely-spaced sedimentary logs) and/or areas which are otherwise inaccessible to direct measurement by field geologists (Fig. 5). Both visualization and interpretation within the NOMAD project have been carried out

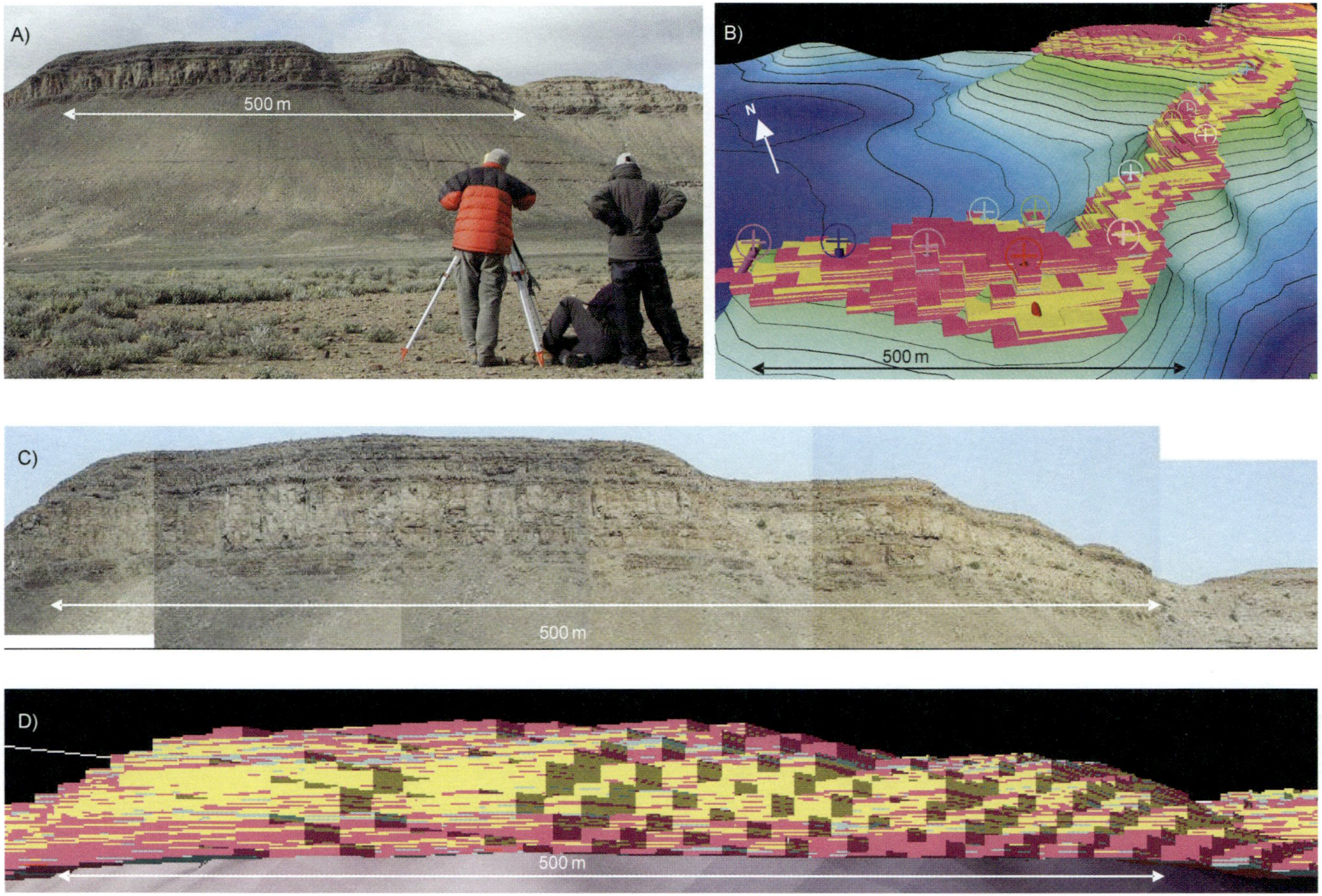

Fig. 4. (**a**) An example of total-station mapping set-up in the field (see text for description of total station use). (**b**) Example of a small-scale study model from the Los Kop area in the Tanqua Basin, Fan system 3. Model size is 2600 × 2000 m with a grid resolution of 10 × 10 × 0.25 m per cell. Model cells are colour-coded in relation to facies association types (pink, structured sandstone; yellow, structureless sandstone). Positions of measured outcrop sections are marked by 'circle and cross' motif. (**c**) Close-up view of the surveyed section of the outcrop in (a) and (b) for comparison with the final reservoir model. (**d**) Close-up of the reservoir model viewed from the same position as the photograph in (c) (see text for further explanation).

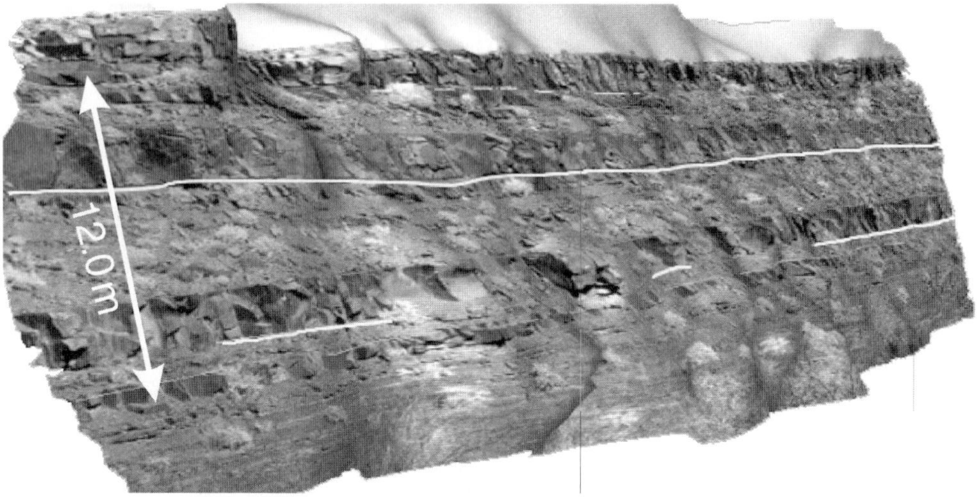

Fig. 5. Photogrammetric model of an outcrop from Fan system 4 showing digital photographic images projected onto a digital elevation model. The model is approximately 50 m in length. Key stratigraphic surfaces are marked on the surface of the model as white lines.

using Schlumberger's Petrel subsurface-modelling software with photogrammetric rectification carried out using Erdas IMAGINE photogrammetric software.

Advantages:

- good for collecting lots of high-resolution geometrical data quickly;
- covers extensive areas of outcrop efficiently;
- relatively inexpensive in terms of equipment if using photographs taken from land-based systems;
- can be used to collect data from areas that are otherwise inaccessible.

Disadvantages:

- long data-processing time to get good results;
- may need expensive helicopter coverage for some areas;
- results dependent on the resolution of the images;
- currently there are only limited types of software designed to work effectively with this kind of data.

Outcrop sedimentary logging. Outcrop sedimentary logging is one area that is recognized as lagging behind in terms of digital data collection in geological mapping exercises. This is due to the highly qualitative, largely subjective, usually paper-based nature of outcrop logging carried out in the past, as many field practitioners have felt that highly quantitative data collection was not necessary.

In order to rectify this situation, Microsoft Windows-based software has been developed at the University of Liverpool over the past three years to enable the effective collection of sedimentological log data in a digital, quantitative form without loss of information (often the perceived problem with this kind of approach in the past). This software is designed to run on a laptop computer and allows not only log description, but also correlation of logs (which should, as in the NOMAD project, be done in the field during the data collection phase) and a variety of log analyses (e.g. bed thickness, facies proportions, power spectral density, palaeocurrent derivation etc.). The primary aim of the use and development of this software was to maintain data in a 3-D framework (as a check for consistency) right from the outset of the data collection phase of NOMAD. This has been achieved with map, cross-section and 3-D views of the data collected, with all data georeferenced with GPS data. Data may be exported in a variety of formats, via copy and paste for graphical (computer-aided design, or CAD) output, or via several ASCII formats for import into proprietary software. During NOMAD, data was still collected by the field geology team using field notebooks and then processed at a later stage, primarily because of the recognized difficulty of using laptop computers in hostile field environments for extended periods. To counteract this difficulty, a version of this software for handheld computers is currently under development, using an integrated GPS module for data georeferencing.

Outcrop-log georeferencing was conducted with handheld Garmin GPS 12 units. These simple 12-channel receivers with no differential correction can give accuracies (when using the averaging function) of between 3 m and 15 m horizontally depending on conditions, though 3.5–7 m is a more usual range of accuracy (these errors may be further reduced using WAAS or EGNOS corrections available on more up-to-date handsets). In terms of vertical resolution these receivers do not give meaningful results, in many cases giving the top of a cliff section as being below the point measured at the base. As a result, the marker for the 'generic' fan base of any individual fan unit identified in the log was corrected vertically to the elevation of that point of the surface gridded from the DGPS data collected using rovers, ensuring that both datasets were now in the correct spatial location relative to each other. The small horizontal errors in the handheld GPS data were not considered to be as problematic because this error would be much smaller than the horizontal cell size (100 × 100 m) used in the construction of the final geological model. With reference to the NOMAD project, over 350 new sedimentary outcrop logs were recorded by the field team, in addition to those collected by previous research efforts (Johnson *et al.* 2001; van der Werffe & Johnson 2003). Within NOMAD, efforts were made to identify these earlier logs in the field, based on their stated grid references, and they were subsequently georeferenced into the final geological models using the Garmin GPS 12 handheld units described above.

Borehole/Geophysical log data. Seven boreholes have been drilled as part of the NOMAD project with full core recovery. Four of these boreholes are small diameter (3 in/7.5 cm wide) with natural gamma-ray logs only, but the three larger diameter wells (4 in/10 cm wide) have been geophysically logged with several of Schlumberger's logging tools, namely: Platform Express (including spectral gamma-ray sonic, and neutron-density logs.), Formation Micro Imaging (FMI) based on resistivity and the new Elemental Capture Spectroscopy Sonde (ECS), a geochemical logging tool. Tool choice was dictated by one of the strategic aims of the project – in this case, lithological characterization. As these tools were being used in non-porous, fully cemented subcrops, tools designed to measure a petrophysical response (such as the nuclear magnetic resonance, or NMR, tools) were not applied during NOMAD.

One of the key objectives of the NOMAD drilling and logging campaign was to characterize length-scale data of geological facies associations away from the wellbores; i.e. what is the range of length/width/thickness distributions of, for example, channel lag facies identified in subsurface image data? This is a common problem in deep-water subsurface reservoir characterization and has implications for the determination of stochastic object geometries during reservoir modelling. To help assess this issue, the location of all the wide-diameter boreholes (Fig. 1, NB2–4) was chosen with this objective in mind. The slim-diameter wells were located using a more 'exploration-based' approach, further away from outcrop locations, with the objective of using the results to help populate the final geomodel (Fig. 1, NS 1–4). Both narrow and wide cores were slabbed, polished and logged at a centimetre-scale with the sedimentary-logging software mentioned previously. Data from the core and wireline logs formed key inputs in the conditioning of the geological models of the fan systems.

Data integration and model building

Model building has been undertaken in Schlumberger's Petrel, a PC-based reservoir-modelling package. Data from all the above sources have been digitized and then integrated into 3-D geomodels of each of the turbidite fan systems.

Fault modelling from outcrop data. The first stage of building a reservoir model is to define the structural model, i.e. the fault framework. Traditionally in 3-D modelling, fault data comes from interpretations of seismic data and is in the form of fault sticks (a series of data points down the fault plane along the intersection of the in-line of the cross-line). During the NOMAD outcrop data collection exercise, commonly only a single point defined the fault, sometimes with an associated dip and azimuth. Note that this is a measurement at one exposed point on the fault plane and may not represent the true overall strike and dip of the fault plane; commonly, single-point fault measurements were associated with no other data. If dip azimuth data did exist, a fault stick was calculated from this; if not then the following method was used to calculate the dip of the fault. When surveying fault traces in the field, several points were recorded along the fault plane and the fault plane was traced out as far as confidently possible. This series of points was then used to calculate the dip and azimuth of the fault plane. Using three points, which fully describe a plane, the plane equation was used to calculate the dip and azimuth values (Fig. 6). Topographical, Landsat and aerial images were also implemented to help extend the fault model by tracing lineations associated with known fault positions. In this way all the faults that affect the data in the model area were accounted for. Once the fault model had been defined (with appropriate fault linkages) a 'pillar grid' was created in Petrel- this is the 3-D framework into which surfaces and 3-D grids are built.

Horizon and zone modelling. 'Horizons' are surfaces within the model that define the overall stratigraphic structure, while 'zones' are the intervals within the model that represent genetically related units. These should be defined in such a way as to make the model-building process simpler by subdividing the model into geologically meaningful intervals, obeying key geological relationships (e.g. unconformities, onlap/downlap etc.). In the NOMAD models the zone definition was based on digitally mapped, fan-wide intervals of silt- and shale-prone turbidites that mark times of reduced sand supply to the basin floor.

Horizon gridding presented several problems in the NOMAD dataset, mainly because of the uneven distribution of data across the model (i.e. a high density around the outcrop belt and a low density elsewhere). Even though the outcrop provided good 3-D coverage there are still areas where extrapolation/interpolation was required across a large area. Further problems arose from fault blocks containing few or no data points within them. Extra control points and contours needed to be added to ensure a geologically valid surface that matched data from both outcrop and boreholes.

The interpreted points for the top and base of each zone, identified through correlation principles within each outcrop and core log, were used to create a series of point datasets for each zone, which contain information on the thickness of each zone for every outcrop log or well. This dataset was then used to grid an isopach surface for each zone. It was found to be much simpler to build each zone from isopachs rather than from a horizon, as only the fan geometry had to be handled, and not the structure imposed on it by deformation as well. These isopachs were built up from the basal surface in turn and corrected back to the original well picks to ensure a good tie between the 3-D model and the log data. Errors in the isopachs were summed through the zone-building process so the correction process had to be done, even though the isopachs are derived from the original well picks. Without post-corrections, errors in zone thickness will be in the order of tens of centimetres.

Facies-association models. The original description of the NOMAD outcrop and core log dataset resulted in the determination of 14 separate facies types. To simplify the model-building process, this facies classification was 'lumped' into a coarser, six-strong facies association classification scheme; subtly different facies types were grouped together to derive this. The full facies-association models recognized structureless sands, structured sands, channel lags, hemipelagics, siltstone heterolithics and 'background' (equivalent to no-exposure recorded during outcrop logging).

Using this framework, fully 3-D models of three of the five fan systems have been constructed, using Petrel. The full 3-D facies-association models are a combination of outcrop data (constraints) and conceptual understanding/trends (controls), which were used to populate the volume with objects of appropriate dimensions and geometry. The conceptual trends were based on trends seen in outcrop observations, correlation panels and from *a priori* knowledge of other systems, be they modern or ancient. The trend maps created from these data and observations defined many features of the model; for example:

- Per zone:
  Flow direction (Fig. 7): conditioned by palaeocurrent information; controls the orientation of the objects used to populate the model.
- Per facies association:
  Object size (Fig. 8b): derived from outcrop observation and conceptual trends.
  Body insertion probability (Fig. 8a): defines the horizontal distribution of objects, based on trends seen in outcrop data, correlation panels and conceptual trends.

Vertical trends also play a significant part in the population of the facies-association models. In this case the vertical trends are calculated using a data analysis tool in Petrel and represent vertical probability distributions of the different facies associations for each zone within the model (Fig. 7c).

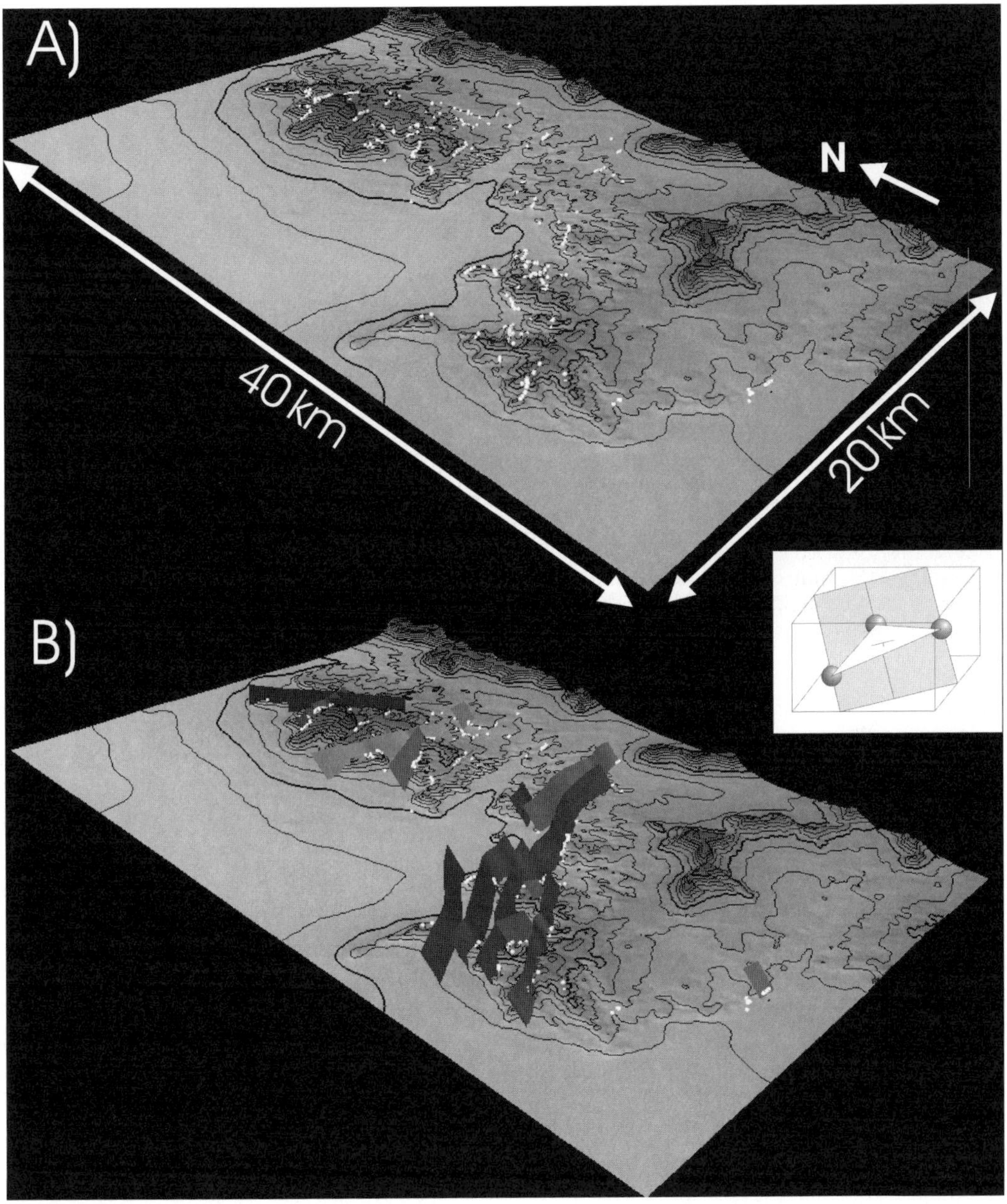

Fig. 6. Calculation of fault sticks from GPS-mapped data. (**a**) The DGPS-mapped data points (white dots) showing the location of measurement of known fault planes. Inset: Calculation of fault-plane orientation from three data points that completely describe that plane. (**b**) The final fault model with fault planes and linkages derived from the DGPS-mapped data.

Object geometry and size was much more problematical. In outcrop it is rare to be able to obtain more than a simple 'apparent' width from measured objects, so again conceptual models were utilized along with knowledge of the outcrop. Object shapes chosen for this study were:

- Flat-based, convex-topped ellipses (erosional remnants of structured sands, siltstone interbeds of hemipelagic shales);
- Flat-topped, round-based ellipses (scour forms);
- Fan shapes (also scours widening down depositional dip);

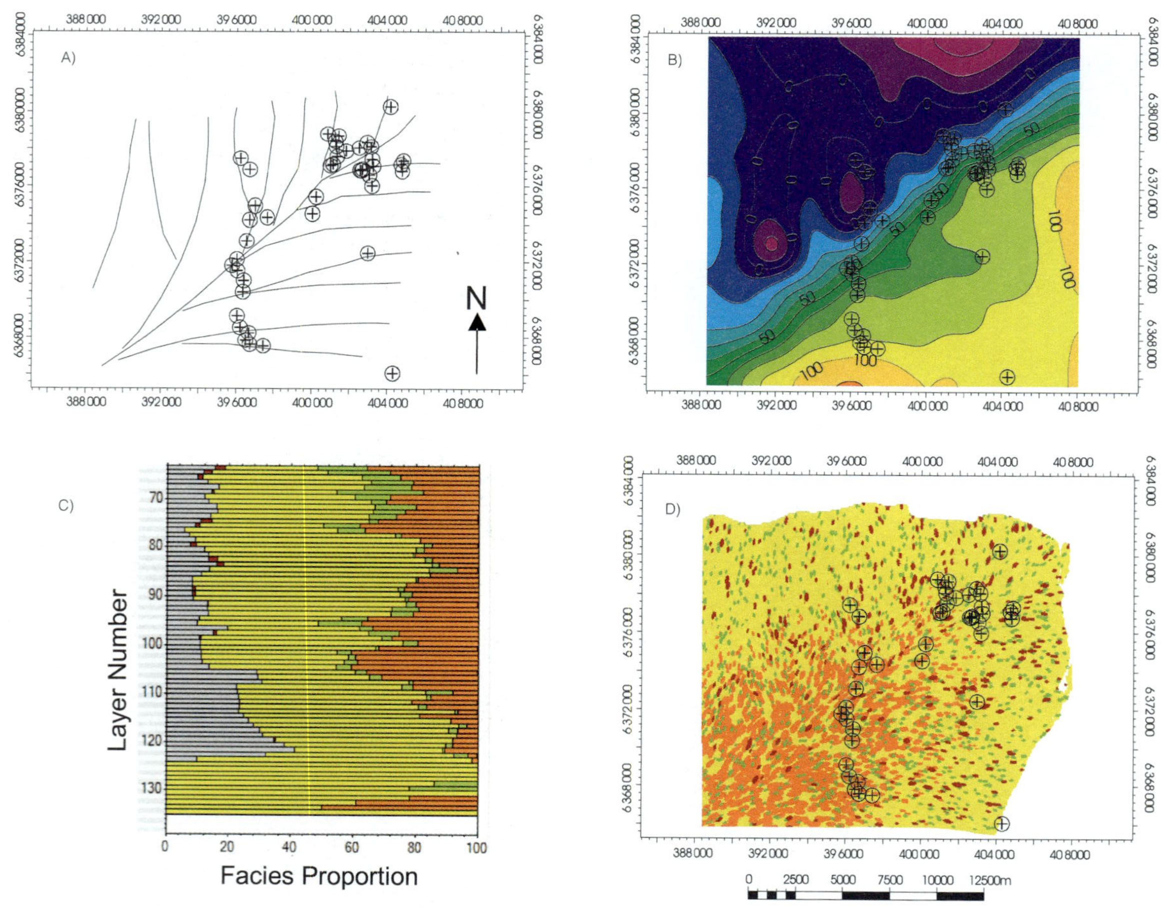

Fig. 7. (**a**) Flow-direction map combining measured palaeocurrent data (measurement location marked with 'circle and cross' motif) and conceptual flow model. (**b**) Converted flow-direction gridded map for use in object conditioning (red and blue, N–NE flow directions; green and yellow, NE–W flow directions). (**c**) Vertical facies association proportion for each layer in a zone (layers are 0.25 m thick). This vertical trend is used as an additional input for conditioning in 3-D facies-association modelling. Key for both (c) and (d): yellow, structureless sand; orange, structured sand; green, siltstone interbeds; brown, hemipelagics; grey, measure of no exposure in the original outcrop log data. (**d**) The final reservoir model, resulting from the combination these the areal and vertical trends.

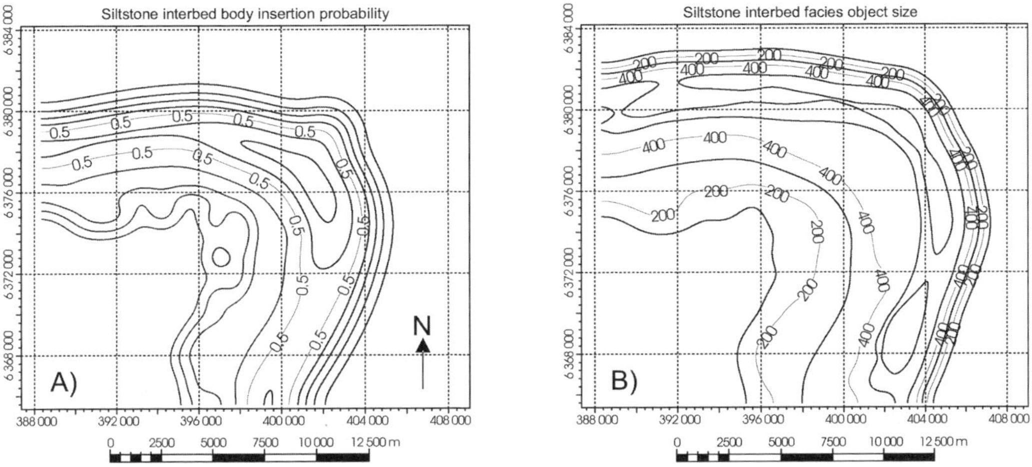

Fig. 8. (**a**) Insertion probability and (**b**) object size distribution maps for 'siltstone interbeds' objects in one of the more sand-prone zones of Fan system 2. Probability of siltstone interbeds occurring is lower in the more axial (high-energy) parts of the fan and increases towards the margins. Object size is also smaller in the axial parts of the model where the preservation potential is lower, but increases towards the margins where there is less erosion from subsequent events and therefore a higher preservation potential. These trend maps are used in conjunction with the flow-direction maps and vertical trends (Fig. 7b and c) in the conditioning of the stochastic model.

- Channel forms (increasing width and decreasing thickness down depositional dip as the channel becomes less confined.

In the NOMAD high-resolution geological models, however, the object thickness is set deterministically to be the model vertical resolu-

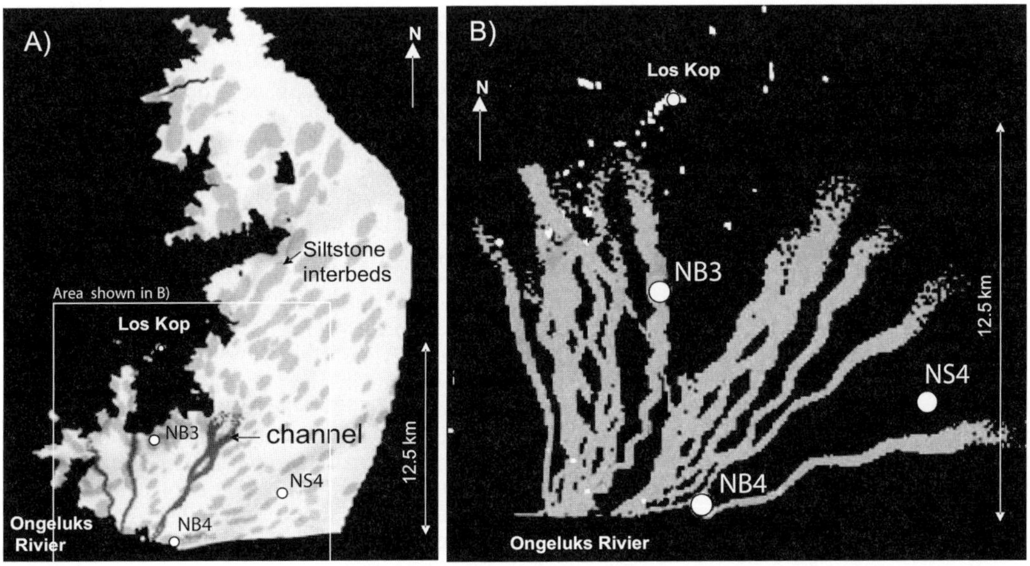

Fig. 9. Fan 3 integrated geomodel. (**a**) Deterministically mapped channels in a single sand-rich zone. Grey ellipses are 'structured sandstone' objects (representing erosional remnants of larger structured sand elements) set within a structureless sand facies association. Model is cut away to show outcrop and subsurface elements only. (**b**) All Fan 3 channels, with three of the boreholes marked. Other facies are not displayed in this view (see text for explanation of fan termination).

tion (i.e. 0.25 m). This has been used because most objects (e.g. beds or bedsets) in the measured outcrop dataset are 25 cm or less, and those objects greater than 25 cm tend to be composite objects built from several sub-objects representing individual events (e.g. individual flow events compositing to form an amalgamated bedset). Appropriate vertical and aerial-map trends will thus ensure a geologically valid distribution.

In the Tanqua area, turbidite channels do not follow the same patterns as their fluvial counterparts. Typically, in a fluvial channel the thickness of the channel tends to vary proportionally to the width, and this is how most reservoir-modelling systems are designed to populate volumes with channel bodies. In the Tanqua setting, channel thickness varies inversely to the channel width; as the channel widens and becomes less confined it becomes thinner. One other key piece of observational data that needs to be honoured is that, for many examples within the fan, channels become transitional to a sheet and so the channel concave-down geometry *sensu stricto* will cease. In order to simulate this, geometry maps for channel thickness and width have been created based on channel widths from known outcrop examples at different locations down the depositional dip profile. These are then used as conditioning for all the channels in the model (Fig. 9).

One other issue is that the modelling system does not allow channels to terminate within the model during the object population phase. Channels may be given start points and allowed to flow out of the model, and this was considered as an option, but defining start points would have reduced the potential for multiple stochastic realizations. Channel termination has therefore been achieved by creating a parameter of 'distance from feeder systems' within the fan system, then populating the channels into a separate parameter. Cells that maintain a 'channel facies' parameter at greater than a given distance from the source point are then replaced with a 'background' facies value using a direct assignment process within the modelling software. The channel objects in the truncated channel parameter are then merged with the main facies realization to result in the final facies association model (see Fig. 9).

Application of results and use of prior knowledge

Digital surveying has resulted in a major improvement of our understanding of the Tanqua fan deposits, in terms of sedimentology, structure and stratigraphy, and an enhanced appreciation of facies-association distributions on local and regional scales. Integration of this data has also provided significant new insights into lateral distribution and extent of the fans, especially when combined with the exceptional core and log results from the borehole data. The original mapping of this system (Wickens 1994) combined with later studies (Johnson *et al.* 2001; van der Werffe & Johnson 2003) used traditional paper-based mapping techniques and this was successful in erecting the initial stratigraphic and sedimentologic framework of the system. The important advantages and improvements on this original work gained by using a digital-surveying approach have been numerous:

(1) Initially, a fundamentally better understanding and inherently more accurate and data-rich description of fan-system and sand-body geometry has been obtained from the digital mapping in the field. This is mainly the result of the data being highly quantitative and it having been collected and mapped onto an accurate 3-D framework. The more traditional methods, i.e. collecting logs, constructing 2-D architectural panels from fence diagrams or from 2-D correlated photopanels and then estimating object sizes from those, results in errors being introduced from incorrect projection of the logs onto the 2-D plane, or from failure to adequately correct measurements in relation to the palaeoflow direction (to give true rather than apparent object size). Understanding of object size, shape and orientation is a key factor in reducing errors in reservoir-model building, as these directly impact on the flow simulation of the model.

(2) In addition, careful analysis of the palaeocurrents in their correct 3-D position, both aerially within the fan-system envelope and stratigraphically within the correct zone, provides significant insights into the spatio–temporal evolution of the flow regime within the fan. Knowledge of the flow regime has allowed the orientation of objects in a deterministic and, more importantly, predictive sense. This is achieved by the quantitative analysis of palaeocurrents and the integration of that data with the observed facies distributions. As a result of this integration, the current facies associations models are conditioned to a full palaeoflow reconstruction on a per zone basis; in the example of Fan 4, this is a 23-strong zonation, with a vertical resolution of 0.5–3.5 m per zone.

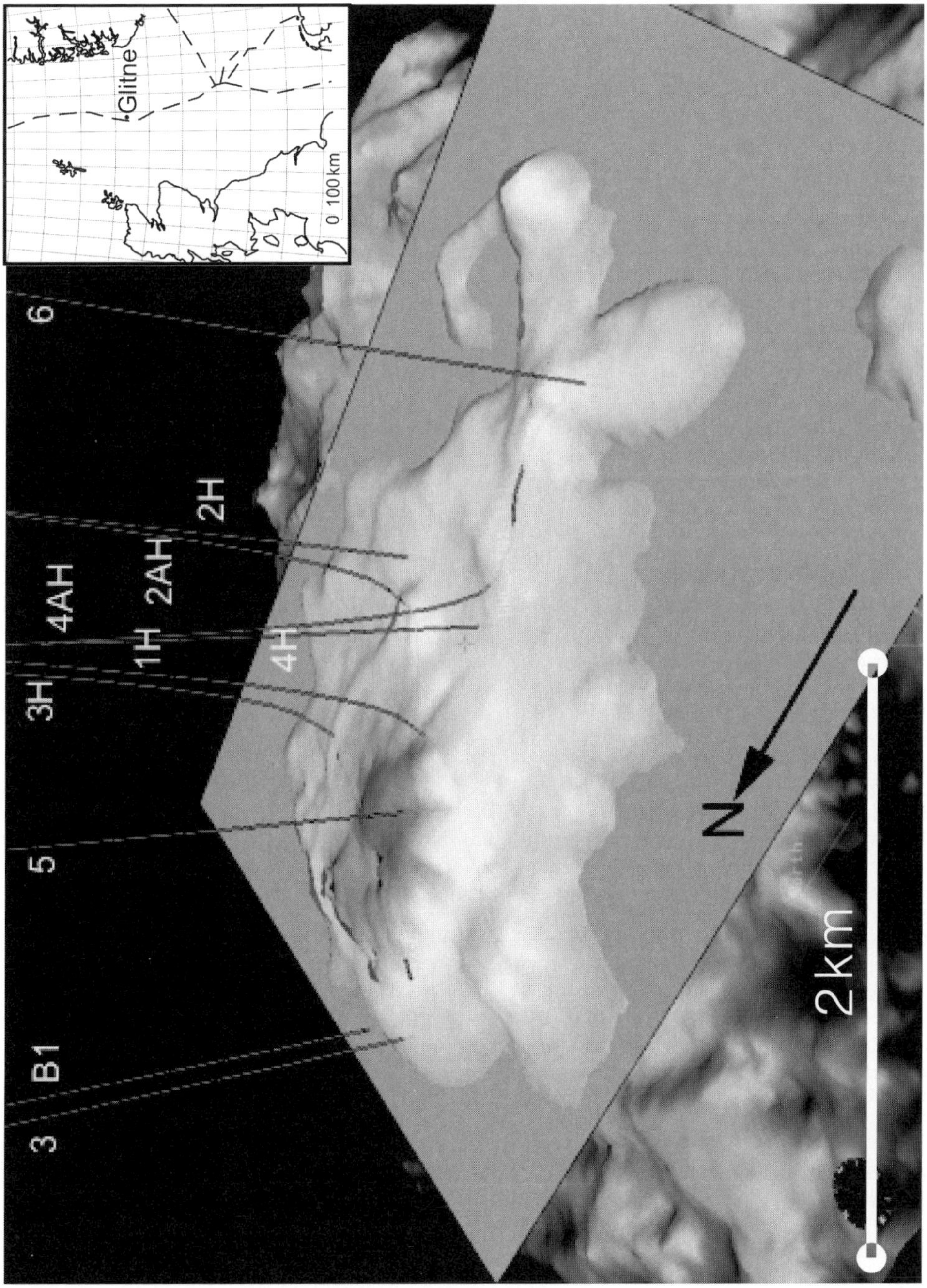

Fig. 10. Top structure map of the Glitne Field, which is located at the top of the Palaeocene Heimdal Formation in the South Viking Graben. The oil/water contact (horizontal plane) and some of the main injector and producer wells have been marked. The location of the Glitne Field is marked on the inset map.

(3) Successful integration of the extensive collection of outcrop log data with the results from the borehole cores and logs has been used effectively to show the compensational stacking and architectural variations within zones of each fan, and shows the way in which the existence of the fan itself modifies subsequent deposition via compensational stacking between zones and even between fan systems.
(4) Detailed mapping in 3-D has allowed features which may not necessarily be visible at the outcrop to be identified, because they are either too subtle (e.g. low angle confining topography as at the base of the fan) or at a much larger scale (e.g. overall topographic variations on the base of Fan 4 in the northern part of the model area). This information gives an improved understanding of how the fan systems develop on a variety of scales and illustrates the importance of pre-existing topography. In addition, the overall development of regionally important phases of fan progradation, aggradation and retrogradation has only been fully appreciated and accurately mapped through use of DGPS data-collecting techniques anchored to borehole control points.
(5) A reservoir model is a combination of many geological aspects (e.g. sedimentology, structural geology etc.). This data collection approach ensures the geology is treated in a holistic manner. The model area is regarded in its entirety rather than as a series of separate projects focusing on different disciplines (i.e. the structural geological model being collected and characterized as a completely independent project, divorced from the stratigraphy). This approach reduced the need for repeated work (e.g. re-examination of the same ground) at a later stage and has significantly increased the potential for using the data in a variety of different ways. Having collected *all* the available data and built a resultant high-resolution model that incorporates *all* of that data, it is now much easier to examine and ask questions of individual extracted subsets of this model.
(6) Generally, having the final data stored in a single, integrated dataset makes manipulation, analysis and navigation around the dataset much simpler and efficient. As subsurface reservoir-modelling software becomes more sophisticated, with larger and larger storage/memory capability combined with wide-ranging input/output abilities, it has become much easier to store large volumes of outcrop data with a wide variety of formats (e.g. ASCII files, Raster or JPEG image files, gridded surfaces etc.). Hence, within software such as Petrel it is now a simple procedure to extract spatial geostatistics derived from the outcrop models of particular geobodies and/or facies associations and export them and use them as input parameters in spatial population of particular facies objects during stochastic model building and multiple realization generation. This was the approach that was adopted by Statoil in the construction of their latest generation of the Glitne reservoir model (see below and Fig. 10).
(7) Digitally mapped datasets and the subsequent models provide an ideal test dataset for fluid-flow simulations and for synthetic seismic modelling because a large amount of prior information is known about the study area maintained within an accurate 3-D framework. This is really an inverse approach to that describe in point (6) above; in this example, the standard dynamic reservoir petrophysical parameters from an existing subsurface field are output as Eclipse reservoir simulator-format files. Having the geological outcrop model exist in an industry-standard software and database means that such parameters are easily imported and then superimposed onto the existing lithological model. As a result, extensive scenario testing and sensitivity analysis of particular features of interest (e.g. channel–overbank connectivity, shale barrier dimensions, reservoir heterogeneity etc.) can be conducted using a synthetic 'no-risk' approach to assess, for example, hydrocarbon migration pathways, depletion strategies, the effects of heavy oil etc. This work is currently underway as part of the NOMAD work effort.

Finally, the conceptual depositional model and quantitative data from the NOMAD models, particularly from the Los Kop area of the Tanqua Fan 3 system, have been implemented in building a 3-D geological model of the Glitne Field, South Viking Graben, North Sea (Fig. 11). The use of this data has been instrumental in providing a better understanding of how the different facies types should be distributed within the model and what dimensional constraints have to be placed on the

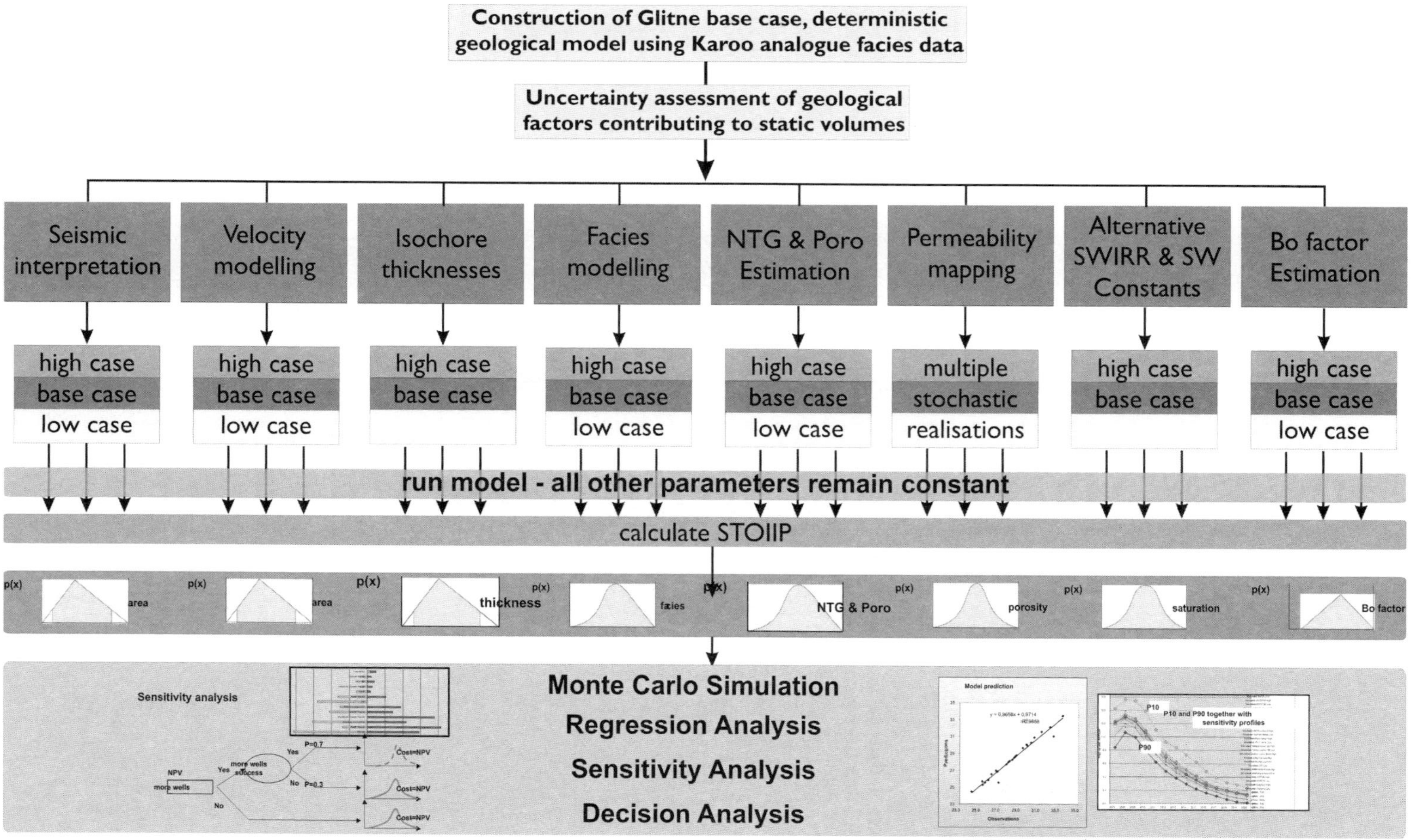

Fig. 11. Flowchart of the work process used in assessing the significance of individual geological factors that contribute to the uncertainty in static volumes on the Glitne Field, Norwegian North Sea.

individual facies. This approach has resulted in the construction of a single, deterministic facies model for the Glitne Field that is regarded as a base-case realization of the subsurface geology. However, inherent in the nature of deterministic models is the lack of expression of uncertainty associated with the facies-association distribution and facies-association dimensions that the model displays. Therefore, as part of an ongoing study into quantifying significant geological uncertainties and their impact on the variation in calculated STOIIP from the base-case model, the Glitne deterministic facies model has been used as a conditioning input to a stochastic modelling study.

So far this uncertainty study has indicated that, by using a well-understood analogue data set in quantitatively constraining the geological facies model of the Glitne Field, the uncertainty inherently associated with building the facies model has been estimated to lie in a range that is less significant than other geological uncertainties tested in the study.

Conclusions

Extensive data have been collected from turbidite outcrops of the Skoorsteenberg Formation in the Tanqua depocentre, southwestern Karoo Basin, using a combination of new and traditional field techniques. Integration of DGPS surveying, photogrammetry, outcrop and boreholes log descriptions has allowed 3-D models to be built of these outcrops. This DGPS mapping has allowed accurate multi-scale surveying over one to tens of kilometres, in this case within a 20×40 km field area at a high spatial resolution (typically 2.5 cm relative to the base station).

Total-station/laser rangefinder techniques have been used for smaller scale, more detailed studies in steeper areas where satellite coverage may be poor or rover units may prove to be too cumbersome. This has also been used in conjunction with digital photogrammetry, which has proven a useful tool in areas that are otherwise inaccessible, or in those areas of good exposure where high-resolution detailed data on lateral thickness/facies variations could be rapidly obtained. Over 350 outcrop logs have easily been converted to a digital form without loss of information by using the appropriate software tools; once in a digital form this facilitated many kinds of data analysis, including, e.g. palaeocurrent, bed thickness, power spectral density, architectural element spatial distribution. Furthermore, the GPS mapping provided insights into the model area's structural evolution not only through the identification and mapping of faults, with provision of information on displacement, but also on the degree and kind of folding in a scale of tens of metres to kilometres. These data have lead to a better understanding of the kinematic evolution of an area.

Integration of these data into high-resolution, 3-D geological models has provided a significantly improved understanding of the deep-water depositional systems that these models represent by presenting all the data in its true 3-D context. Only by collecting and analysing data in 3-D have the complexities of the geological regime under investigation become fully understood. The mapping of object geometries and internal facies distribution are far better understood from outcrop data than is possible from the subsurface; when this outcrop data is fully integrated and characterized it can significantly improve existing reservoir models of similar geometries in subsurface oilfields. The improved understanding of object and facies distribution down and along depositional dip within a deep-water fan system has also been applied to oilfield examples, helping to reduce the range in uncertainty of these parameters and subsequently aid increased recovery due to the improvement in existing models from the application of the NOMAD dataset.

This work was carried out as part of the consortium research project, NOMAD, which is sponsored by the European Union's 5th Framework Programme under the Energy, Environment and Sustainable Development thematic programme.

The NOMAD consortium is an international, joint industry/university venture comprising: Schlumberger Cambridge Research (UK); Statoil (Norway); Liverpool University's Stratigraphy Group (UK); Stellenbosch University (South Africa); and the Technical University of Delft (Netherlands).

The NOMAD objective is to help optimize the positioning, number and deliverability of production wells drilled in deep-water reservoirs, thus improving hydrocarbon recovery while reducing development costs. This will be achieved through improved reservoir characterization, using a 3-D geological and petrophysical model of the world's best outcrop analogue of deep-water sediments for guidance: the Permian Tanqua Basin turbidite fan complex in South Africa.

Many of the results presented in this paper are from the collected efforts of a large number of field geologists from the NOMAD partner institutions. The authors would like to thank them collectively. (See: www.nomadproject.org for further details.)

References

BANGERT, B., STOLLHOFEN, H., LORENZ, V. & ARMSTRONG, R. 1999. The geochronology and significance of ash-fall tuffs in the glaciogenic Carboniferous–Permian Dwyka Group of Namibia and South Africa. *Journal of African Earth Sciences*, **29**, 33–34.

BOUMA, A. H. & WICKENS, H. DE V. 1991. Permian passive margin submarine fan complex, Karoo Basin. South Africa: possible model to Gulf of Mexico. *Gulf Coast Association of Geological Societies, Transactions*, **41**, 30–42.

DROMGOOLE, P., BOWMAN, M., LEONARD, A., WEIMER, P. & SLATT, R. M. 2000. Developing and managing turbidite reservoirs – case histories and experiences: results of the 1998 EAGE/AAPG research conference. *Petroleum Geoscience*, **6**, 97–105.

JOHNSON, S. D., FLINT, S., HINDS, D. & WICKENS, H. DE V. 2001. Anatomy, geometry and sequence stratigraphy of basin floor to slope turbidite systems, Tanqua Karoo, South Africa. *Sedimentology*, **48**, 987–1023.

KOLLA, V., BOURGES, PH., URRUTY, J.-M. & SAFA, P. 2001. Evolution of deep-water Tertiary sinuous channels offshore Angola (west Africa) and implications for reservoir architecture. *Bulletin of the American Association of Petroleum Geologists*, **85**, 1373–1407.

OKA, N. 1998. Application of photogrammetry to the field observation of failed slopes. *Engineering Geology*, **50**, 85–100.

REID, T. R & HARRISON, J. P. 2000. A semi-automated methodology for discontinuity trace detection in digital images of rock mass exposures. *International Journal of Rock Mechanics & Mining Sciences*, **37**, 1073–1089.

ROZMAN, D. J. 2000. Characterization of a fine-grained outer submarine fan deposit, Tanqua-Karoo basin, South Africa. *In:* BOUMA, A. H. & STONE, C. G. (eds) *Fine-Grained Turbidite Systems*. American Association of Petroleum Geologists, Memoirs, **72** / SEPM (Society for Sedimentary Geology), Tuba, Special Publications, **68**, 291–298.

SLAMA, C. C. 1980. *Manual of Photogrammetry*. 4th edition. American Society of Photogrammetry, Falls Church, Virginia.

TANKARD, A. J., JACKSON, M. P. A., ERIKSSON, K. A., HOBDAY, D. K., HUNTER, D. R. & MINTER, W. E. L. 1982. *Crustal Evolution of South Africa: 3.8 Billion Years of Earth History*. Springer-Verlag, New York.

TURNER, B. R. 1999. Tectonostratigraphical development of the Upper Karoo foreland basin: Orogenic unloading versus thermally-induced Gondwana rifting. *Journal of African Earth Sciences*, **28**, 215–238.

VAN DER WERFF, W. & JOHNSON, S. (2003). High resolution stratigraphic analysis of a turbidite system, Tanqua Karoo Basin, South Africa. *Marine and Petroleum Geology*, **20**, 45–69.

VERGARA, L., WREGLESWORTH, I., TRAYFOOT, M. & RICHARDSEN, G. 2001. The distribution of Cretaceous and Paleocene deep-water reservoirs in the Norwegian Sea basins. *Petroleum Geoscience*, **7**, 395–408.

VISSER, J. N. J. 1991. Geography and climatology of the Late Carboniferous to Jurassic Karoo Basin in southwestern Gondwana. *Annual South African Musea*, **99**, 415–431.

WARNER, W. S. 1995. Mapping a three-dimensional soil surface with hand-held 35 mm photography. *Soil and Tillage Research*, **34**, 187–197.

WICKENS, H. DE V. 1994. *Basin floor fan building turbidites of the southwestern Karoo Basin, Permian Ecca Group, South Africa*. PhD thesis, University of Port Elizabeth.

WICKENS, H. DE V. & BOUMA, A. H. 2000. The Tanqua Fan Complex, Karoo Basin, South Africa: outcrop analog for fine-grained, deep-water deposits. *In:* BOUMA, A. H. & STONE, C. G. (eds) *Fine-Grained Turbidite Systems*. American Association of Petroleum Geologists, Memoirs, **72** / SEPM (Society for Sedimentary Geology), Tuba, Special Publications, **68**, 153–165.

Sensitive dependence, divergence and unpredictable behaviour in a stratigraphic forward model of a carbonate system

PETER M. BURGESS[1] & DAVID J. EMERY[2]

[1]*Shell International Exploration and Production B.V., Volmerlaan 8, Postbus 60, 2280 AB Rijswijk, The Netherlands (e-mail: Peter.Burgess@shell.com)*

[2]*School of Computing, University of Staffordshire, The Octagon, Beaconside, Stafford, ST18 0AD, UK*

Abstract: Although conceptual models of carbonate systems typically assume a dominance of external forcing and linear behaviour to generate metre-scale carbonate parasequences, there is no reason to preclude autocyclic and non-linear behaviour in such systems. Component parts of the carbonate system represented in this numerical forward model are entirely deterministic, but several parts are non-linear and exhibit complex interactions. Onshore sediment transport during relative sea-level rise generates autocyclic quasi-periodic shallowing upward parasequences but model behaviour is sufficiently complex that water depth evolution and parasequence thickness distributions are not predictable in any detail. The model shows sensitive dependence on initial conditions, resulting in divergence of two model cases, despite only a small difference in starting topography. Divergence in water-depth history at one point takes ~ 10 ka, and for the whole model grid takes ~ 100 ka. Fischer plots from the two cases show that divergence leads to entirely different parasequence thickness evolution in each case. Chaotic behaviour is a specific type of sensitive dependence, and calculation of trajectory divergence in a 3-D pseudo-phase space indicates that water depth evolution is not truly chaotic. If sensitive dependence, divergence and complex processes generating random products turn out to be common in real carbonate systems, predictions should be limited to elements of the system unaffected by these phenomena, or limited to cases where an element of periodic external forcing over-rides their affects. These results also suggest that increasingly complex and sophisticated stratigraphic forward models are not necessarily going to lead directly to more deterministic predictive power, although they may well be useful sources of statistical data on carbonate strata.

Introduction

Predictive conceptual models of carbonate systems and of sedimentary systems in general, emphasise simple linear behaviour and a dominance of control by external forcing (Haq *et al.* 1988; Goldhammer *et al.* 1990; Chen *et al.* 2001). Linear behaviour simply means that effect magnitude is directly proportional to cause. For example, it is widely assumed that small high-frequency changes in relative sea level create parasequences, and larger lower frequency changes create sequences. An assumption of external forcing means that features of the strata can be directly attributed to simple mechanisms such as sea-level change, external to the sedimentary system itself. These assumptions allow simple prediction of system behaviour and allow for inferences such as cycle correlation without direct physical evidence of age equivalence and/or rock equivalence (e.g. Read & Goldhammer 1988; Overstreet *et al.* 2003).

However, sedimentary systems may not behave in this manner; certainly they may not behave exclusively in this manner. An alternative possibility is autocyclic behaviour (Ginsburg 1971; Goldhammer *et al.* 1993; Demmico 1998) and non-linear behaviour, in which external forcing is not necessary to produce cycles, and in which effect may not be in proportion to cause. A key aspect of many non-linear systems is their sensitive dependence on initial conditions, whereby small and perhaps immeasurably small variations in some component of the system will, after some interval of time, have significant consequences for the products of the system. This is known as the 'butterfly effect' (Lorenz 1993) and also as divergence, because products of a system quickly become dissimilar for even very small differences in starting conditions, making detailed predictions difficult or perhaps

From: CURTIS, A. & WOOD, R. (eds) 2004. *Geological Prior Information: Informing Science and Engineering*. Geological Society, London, Special Publications, **239**, 77–94. 186239-171-8/04/$15.00

impossible. Systems in which divergence is particularly rapid are referred to as 'chaotic systems' (Williams 1997). The possibility of non-linear and deterministic chaotic behaviour in sedimentary systems has received little detailed attention, despite some general reviews (Slingerland 1989; Smith 1994; Bailey 1998), but in fact such behaviour seems quite likely since sedimentary systems are known to include many non-linear processes (e.g. Boscher & Schlager 1994; Gaffin & Maasch 1991; Paola 2000).

Unfortunately, identifying sensitive dependence and chaos in natural systems is difficult (Kantz & Schreiber 1997; Williams 1997) and will be particularly difficult with ancient sedimentary systems because of incompleteness and uncertainty in dating strata. Given this, determining the likelihood and consequences of such behaviour in a numerical forward model is a useful starting point that may guide the search for a similar signal in real systems in general and may establish theoretical limits of predictability in carbonate sedimentary systems in particular.

Model rationale and formulation

Model results presented here were generated with a numerical stratigraphic forward model modified from that described in Burgess *et al.* (2001) and Burgess & Wright (2003). The model is a representation of combined processes of carbonate sediment production and transport. It is clearly much simplified relative to real systems, and aspects of the formulation are speculative (e.g. the relationship between productivity and transport). The model results should therefore be interpreted accordingly; they represent a possible behaviour of carbonate systems, not a definitive statement that this is exactly how they behave.

Carbonate production is assumed to depend on water depth, sediment flux and the production mosaic such that:

$$p_{xy}(w_{xy} > 0.1) = M_{xy} \cdot P_{\max} \cdot \tanh(5 \cdot \exp(-0.1 \cdot (w_{xy} - w_{\text{prod max}}))) \cdot ((E_{\text{thresh}} - E)/E_{\text{thresh}}), \quad (1)$$

and

$$p_{xy}(w_{xy} \leqslant 0.1) = M_{xy} \cdot P_{\max} \cdot \tanh(5 \cdot \exp(-1.0 \cdot (w_{\text{prod max}} - w_{xy}))) \cdot ((E_{\text{thresh}} - E)/E_{\text{thresh}}), \quad (2)$$

where x and y are the spatial coordinates of model grid points, w is water depth in metres, M is the dimensionless production mosaic value (see below), $P_{\max}$ is the maximum carbonate production value in m/Ma, $w_{\text{prod max}}$ is the depth of maximum productivity in metres, E is the transported sediment thickness in m/Ma, and E_{thresh} is the threshold transported thickness, in m/Ma, which would produce a concentration of sediment in the water column at or beyond which production cannot occur. Variation of production with water depth (Fig. 1) follows the relationship outlined by Boscher & Schlager (1992). Production is modified by the production mosaic value to simulate biological controls (see below) and is also modified by the presence of any sediment in the water column that would block light penetration.

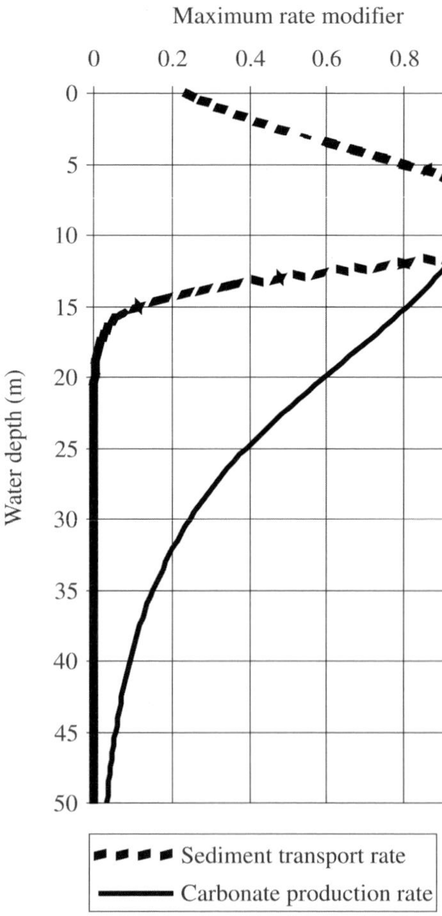

Fig. 1. Values of sediment transport and carbonate production rate modifiers plotted against water depth, as used in the model runs.

Sediment flux is calculated using a similar non-linear depth relationship in which erosion and sediment entrainment E is related to water depth such that

$$E_{xy}(w_{xy} > 0.1) = E_{\max} \cdot \tanh(5 \cdot \exp(-0.30 \cdot (w_{xy} - w_{t\max}))) \cdot F \quad (3)$$

and

$$E_{xy}(w_{xy} \leqslant 0.1) = E_{\max} \cdot \tanh(5 \cdot \exp(-0.75 \cdot (w_{t\max} - w_{xy}))) \cdot F \quad (4)$$

where F is a coefficient proportional to fetch distance in metres and $w_{t\max}$ is depth of maximum transport in metres. Erosion and entrainment increases with water depth to a maximum at 10 m (Fig. 1) in response to increasing wave energy with decreasing friction due to decreasing bottom contact, and then decreases with depth in response to decreasing wave orbital motion. Increasing fetch distance increases erosion and entrainment because wave size increases with fetch distance. If erosion at a particular point on the model grid is greater than zero, sediment is removed, entrained and transported along a transport path following an onshore direction modified locally due to wave refraction calculated using Snell's law. The same relationship with water depth is used to determine when deposition occurs; so if water depth decreases along the transport path, deposition will occur. However, increasing fetch distance tends to offset the effect of decreasing water depth and, in most cases, sediment is deposited in shallow water adjacent to a shoreline.

An important aspect of these representations of productivity and transport rate is that they are non-linear. The non-linearities arise both directly from the terms and functions in the equations, from discontinuities such as sea level where transport processes change instantly, and also from the presence of feedbacks between different variables (e.g. productivity depends on water depth and on transport rate, but transport rate also depends on water depth and therefore on productivity). Non-linear relationships are common in natural systems, and their presence creates the possibility of more complex dynamic behaviour than is typical in systems dominated by linear relationships.

In contrast to previous versions of the model described in Burgess *et al.* (2001) and Burgess & Wright (2003), spatial distribution of carbonate production in the model is controlled by a production mosaic calculated using deterministic cellular automata. A cellular automaton is a grid of values with simple rules specifying how the value at each point is calculated for the next time-step according to the value of adjacent points (e.g. Holland 1998). Cellular automata are useful modelling tools because they can be entirely deterministic and are very flexible (Wolfram 2002). They have been used previously by Tipper (1997) to model carbonate lag time and by Drummond & Dugan (1999) to simulate carbonate facies mosaics, with the important difference in both cases that a stochastic element was included to determine cell occupancy. In the model presented here, the deterministic cellular automata represents a simple ecology governed by rules that control the producing organism population, and therefore the potential carbonate productivity in each model cell (M in Equations 1 & 2).

The rules determining cell occupancy are specified in Table 1. Note that the increase in value of M in a particular cell is defined in the model as a rate (0.2 per 10 years), so that the time-step dependency of the automaton is similar to any other element in a model composed of discretized representations of continuous processes. At the beginning of the model, an initial condition exists where grid points on the edge of the model grid are populated in a pattern that will, in subsequent time steps, propagate into the interior of the grid, according to the rules in Table 1, and according to the distribution of subtidal areas on the model grid. Propagation of producing elements from the edge of the grid is intended to represent producing organisms migrating from areas external to the model into areas on the model grid devoid of producers. Because only subtidal cells can be colonized, migration of supratidal and intertidal islands influences the spatial evolution of colonization. However, distribution of intertidal and supratidal islands is determined by the distribution of water depth, which controls sediment transport, and water depth is in turn influenced by productivity and hence by the distribution of carbonate-producing elements. Thus there exists a feedback in the model between production and transport, and this adds to the non-linearity of model behaviour.

It is important to note, as with all similar numerical forward-modelling studies, that this cellular automaton production mosaic, coupled with depth-dependent sediment production and transport, is intended to replicate aspects of real carbonate-system behaviour in a broad, simplified sense. Any consideration of the model results should therefore be limited by the awareness of the degree of realism of the model

Table 1. *The rules employed in the cellular automata to determine the mosaic production factor per model grid cell*

Value of production mosaic element (M)	Number of occupied neighbours (n)	Action	Comment
$M=0$	$n<2$	None	
$M=0$	$n=2$	$M=0.2$	Optimal conditions, production commences
$M=0$	$n>2$	None	
$0<M<1$	$n<2$	$M=M-0.5$	Underpopulated, production decreases
$0<M<1$	$n=2$ or $n=3$	$M=M+0.2$	Population optimal, production increases
$0<M<1$	$n>3$	$M=M-0.5$	Overpopulated, population decreases
$M=1$	N/A	$M=0$	Population mature, production ceases

A neighbouring cell is considered occupied if its production mosaic value (M) is greater than zero, and only orthogonally adjacent cells are considered in the occupied neighbour (n) count.

formulation, so that the results are not over-interpreted or imbued with false predictive power. In this case, the purpose of the model is to investigate consequences of interactions between simulated biological and physical processes in a much simplified manner in order to determine whether sensitive dependence and complex and chaotic behaviour can result. In this context the model formulation seems reasonable.

In all other respects, the model is unchanged from that described in Burgess *et al.* (2001) and Burgess & Wright (2003); steady subsidence, carbonate production, and landward sediment transport create progradational shorelines and islands that, in turn, create shallowing-upward carbonate parasequences in a manner similar to that initially described by Ginsburg (1971). Although no tidal processes are calculated in the model, a tidal range of 1 m is assumed in order to apply supratidal, intertidal and subtidal classifications to deposited sediments and thus to simulate peritidal strata.

Results from model experiments

Results from two model cases are presented to demonstrate sensitive dependence, following a method outlined in Lorenz (1993) and used also by Meijer (2002). The two model runs had identical input parameters (Table 2), except for the initial water depth at one point at the centre (x = 50 km, y = 50 km) of the 100 by 100 point model grid (Fig. 2). In Case 1 the water depth at this point was 1 m; in Case 2 the water depth was 2 m. To ensure that model runs started from a typical topographical configuration, a topographical surface from a previous model run was used, with an existing system of intertidal and supratidal islands separated by subtidal areas. In both cases, supratidal islands are produced by deposition of sediment transported from subtidal producing areas. Complex distributions of islands occur due to spatially variable production caused by interactions between topography, sediment transport and sediment production. A series of maps illustrating the time evolution of the model topography and the production

Table 2. *Parameters used in model Cases 1 and 2*

Parameter	Parameter value	Source
Time step	10 a	
Chronological interval	2 ka	
Total subsidence rate	200 m/Ma	
Eustatic sea level	Constant	
Maximum carbonate point production rate (P_{max})*	3500	Enos (1991)
Depth of maximum carbonate production	0.1 m	Boscher & Schlager (1993)
Sediment transport threshold	1.0×10^5 m/Ma	
Maximum transport thickness (T_{max})*	3000 m/Ma	Enos (1991)
Depth of maximum transport	10 m	
Diffusion coefficient	1.0×10^8 m^2/Ma	Kenyon & Turcotte (1985)

*Note that marked parameters are maximums; actual values in model runs are often lower because of limiting influence of other factors.

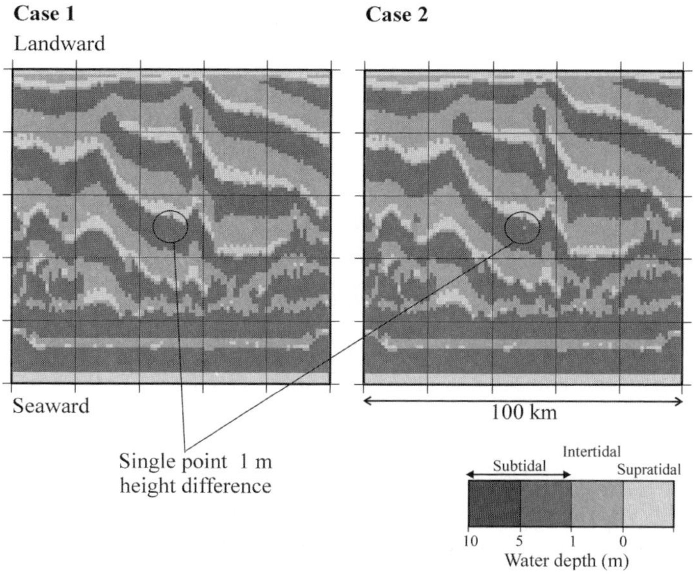

Fig. 2. Initial topographies used in model Case 1 and Case 2. The difference in elevation of 1 m in one model grid point is circled.

mosaic are shown in Figure 3. Prograding islands produce shallowing-upward parasequences (Burgess *et al.* 2001; Burgess & Wright 2003). Parasequence thickness is dependent on subsidence rate and sediment transport rate and, since transport rate is dependent on productivity rate, variations in the carbonate production mosaic can influence parasequence thickness (Burgess & Wright 2003). The model is realistic in the sense that it produces shallowing-upward parasequences, with thickness from one to several metres, of the type interpreted to be present in many ancient outcrop examples.

Intuitively one might not expect any significant dissimilarity in outcome of the two model cases since the difference in initial topography is small and limited to only one point. Note also that it would be virtually impossible to detect such a topographical difference from study of ancient strata. However, modelled sediment production and transport rates are sensitive to this difference, and the topography evolves differently in the two cases, in turn influencing development of the production mosaic. A chronostratigraphic diagram (Fig. 4) taken along a seaward to landward section at $x = 40$ km shows that island evolution initially creates apparently identical shallowing-upward parasequences, but differences are immediately obvious in the distribution of preserved subtidal strata, due to the evolution of the production mosaic. Because of the feedback between sediment production and sediment transport, differences in production influence transport and vice versa and, after $\sim$ 20 ka, significant divergence in parasequence development between the two cases is clear (Fig. 4). Differences are also visible in the cross-sections through the youngest part of the model strata (Fig. 5). Importantly however, despite this divergence, shallowing-upward parasequences of similar thicknesses continue to be generated and to prograde. Thus the model cases differ in detail, but not in broad general behaviour.

Differences in water depth between Case 1 and Case 2 are shown in time-series plots in Figure 6. Water depth at a single point ($x = 40$ km, $y = 70$ km) on the model grids follows a similar path in both cases for the first 8 ka (Fig. 6a), then diverges until 15.5 ka, follows the same path again for $\sim$ 2 ka, and finally diverges at 18 ka (Fig. 6a, b). Divergence of the two cases is also apparent when considering the whole model grid. Absolute difference in water depth at each point on the grid for the two cases was calculated, and the sum of the differences for all points on the grid plotted against elapsed model time. The resulting time series (Fig. 6c) shows relatively rapid divergence until $\sim$ 34 ka, continued but slower divergence until 95 ka, and then oscillations around a value of 4800 m, representing an average difference of 0.5 m per model cell. Divergence rate will probably change with different rates of production and transport

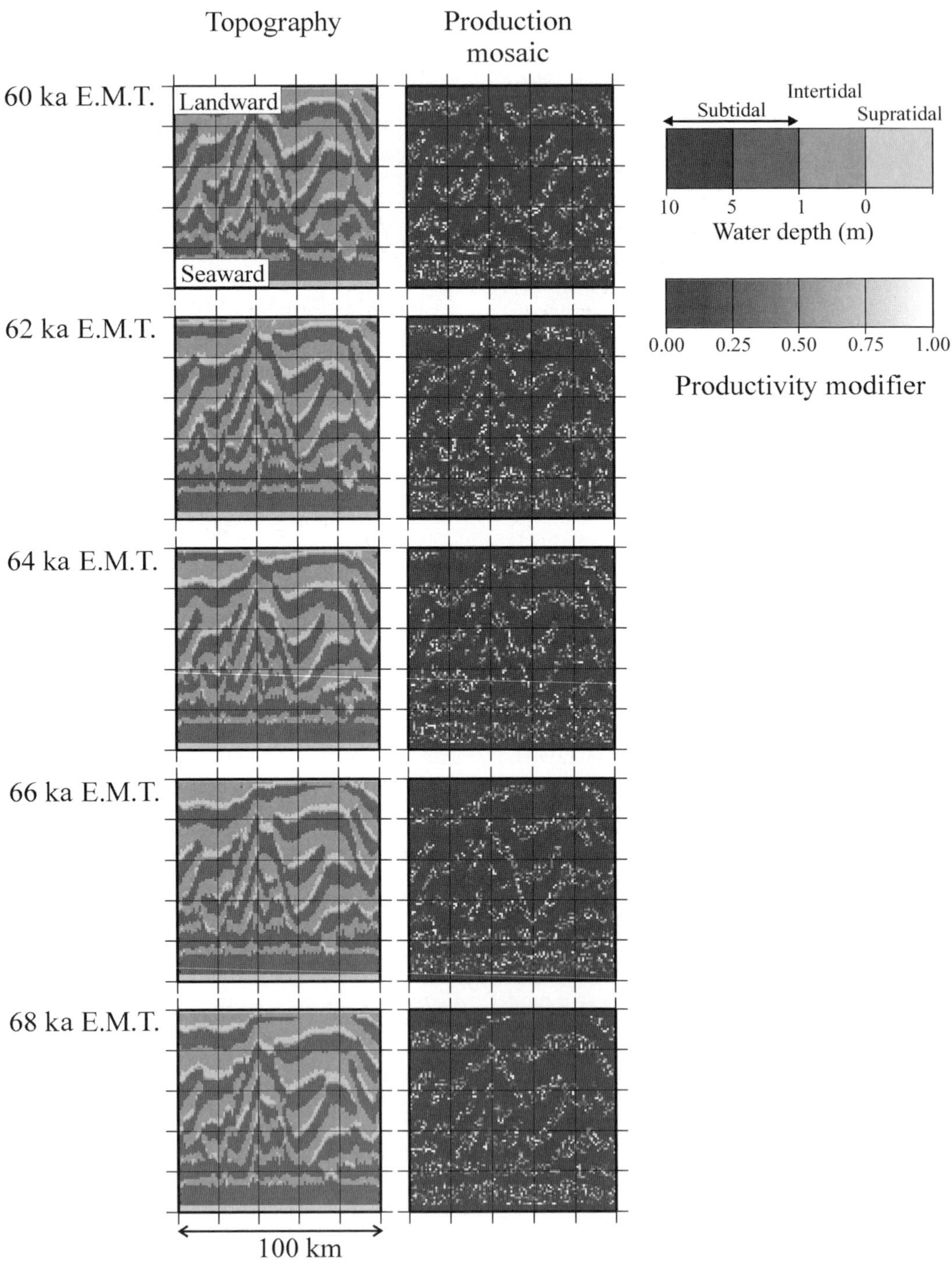

Fig. 3. Map views of a time series of model topographies and associated production mosaics for the time interval 60–68 ka in model Case 1. The maps illustrate how islands prograde and develop in a complex manner, and how the production mosaic is limited to subtidal areas and develops in a similarly complex manner.

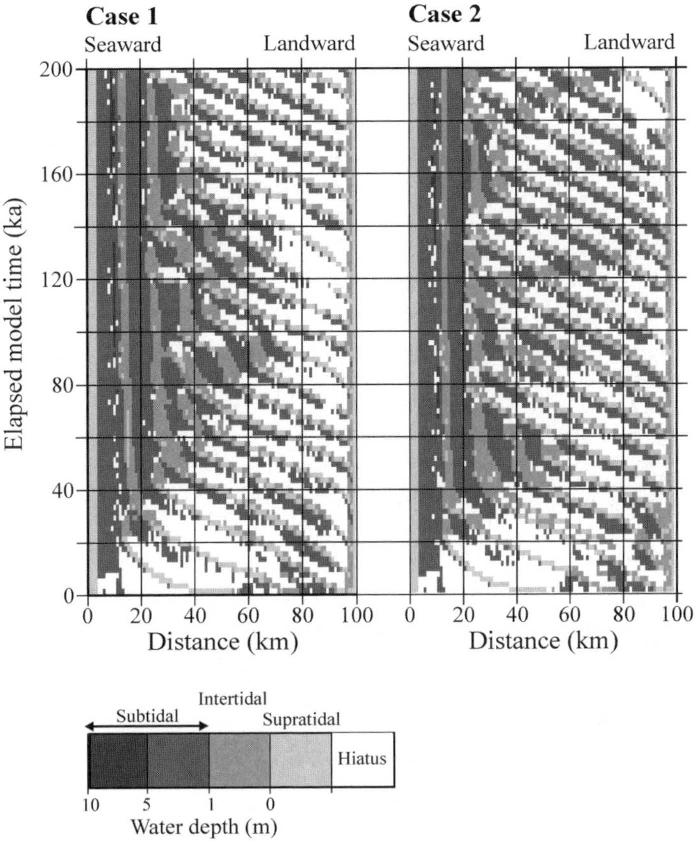

Fig. 4. Chronostratigraphic diagrams from model Cases 1 and 2. Diagrams are constructed at $x = 40$ km, along the y axis (dip sections) parallel to the direction of sediment transport. Preserved strata are coded according to water depth of deposition. Hiatuses, where strata were not deposited, or were eroded, are white. Diachronous parasequences are marked by shallowing-upward deposition and prograde seawards. Subtidal deposition is spatially and temporally discontinuous, due to the carbonate production mosaic. Divergence in the two model cases is illustrated by the different subtidal depositional patterns in the first 20 ka between 20 km and 60 km, and by the differences in the detail of parasequence evolution apparent from 20 ka onwards. Despite this, overall general patterns of parasequence progradation in the two cases are similar.

and different rules for the evolution of spatial distribution of productivity; this requires further investigation.

In stratigraphic forward models, all details of the history of the stratigraphic system can be recorded, but since real strata are known to be typically incomplete (Sadler 1981), it is important also to consider what portion of the strata will actually be preserved and how incomplete preservation influences the record of divergence. Vertical sections from the point $x = 40$ km and $y = 70$ km in Cases 1 and 2 have stratigraphic completeness of $\sim 19\%$ measured at a 20 a resolution. This incompleteness is a consequence of missing time on both inter- and intra-parasequence hiatuses (Burgess & Wright 2003) representing both non-depositional and erosional hiatuses. Figure 7a shows the preserved water depth history plotted against time from Cases 1 and 2. The time series records a series of shallowing-upward intervals separated by hiatuses with some intervals of water depth varying between 1.0 m and 1.5 m. Divergence in preserved water depth histories occurs after only 2.3 ka. Similarly rapid divergence is apparent in the same time series plotted against accumulated thickness (Fig. 7b). Both examples show that, with the additional uncertainty of incomplete preservation, divergence in the record of water depth history in the two model cases happens even more rapidly.

Although water depth evolution diverges in these two model cases, the two histories do share some common features. In both cases

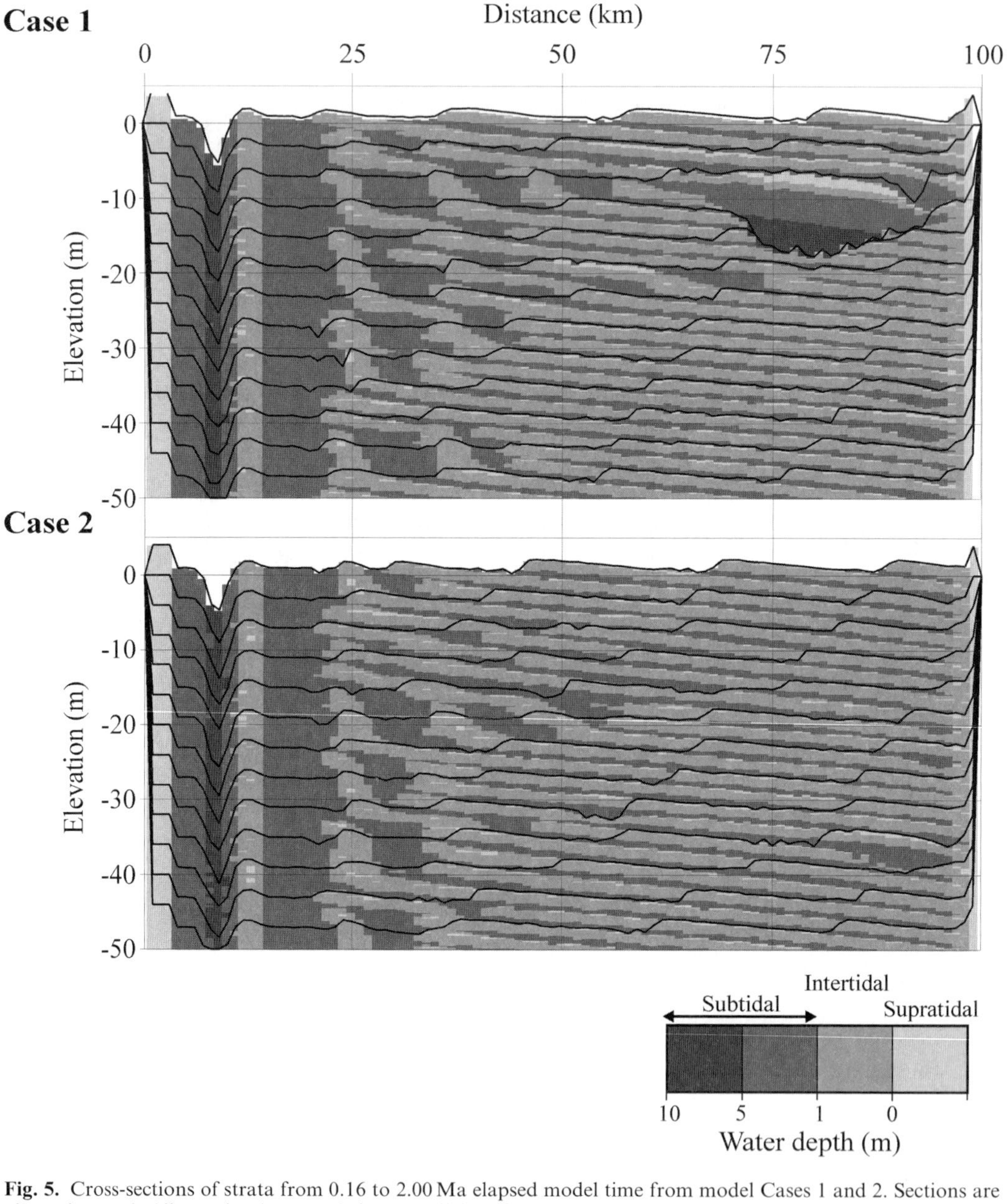

Fig. 5. Cross-sections of strata from 0.16 to 2.00 Ma elapsed model time from model Cases 1 and 2. Sections are along the y axis (dip sections) at $x = 40$ km, parallel to the direction of sediment transport. Strata are colour-coded according to water depth of deposition and time lines are drawn at intervals of 20 ka. Diachronous parasequences prograde seawards and are marked by shallowing-upward deposition. Small-scale progradational clinoforms are delineated by some of the time lines. Although overall patterns of development are similar, comparison of the two cases shows significant differences in vertical and horizontal distribution of water depths recorded in strata, and in parasequence thickness.

water depth values remain within the same approximate range of values, and in both cases there are clearly shallowing-upward cycles present. These cycles are autocycles resulting from shoreline and island progradation, and Markov analysis clearly shows unambiguous order present in similar models (Burgess & Wright 2003). Thus divergence is limited, and

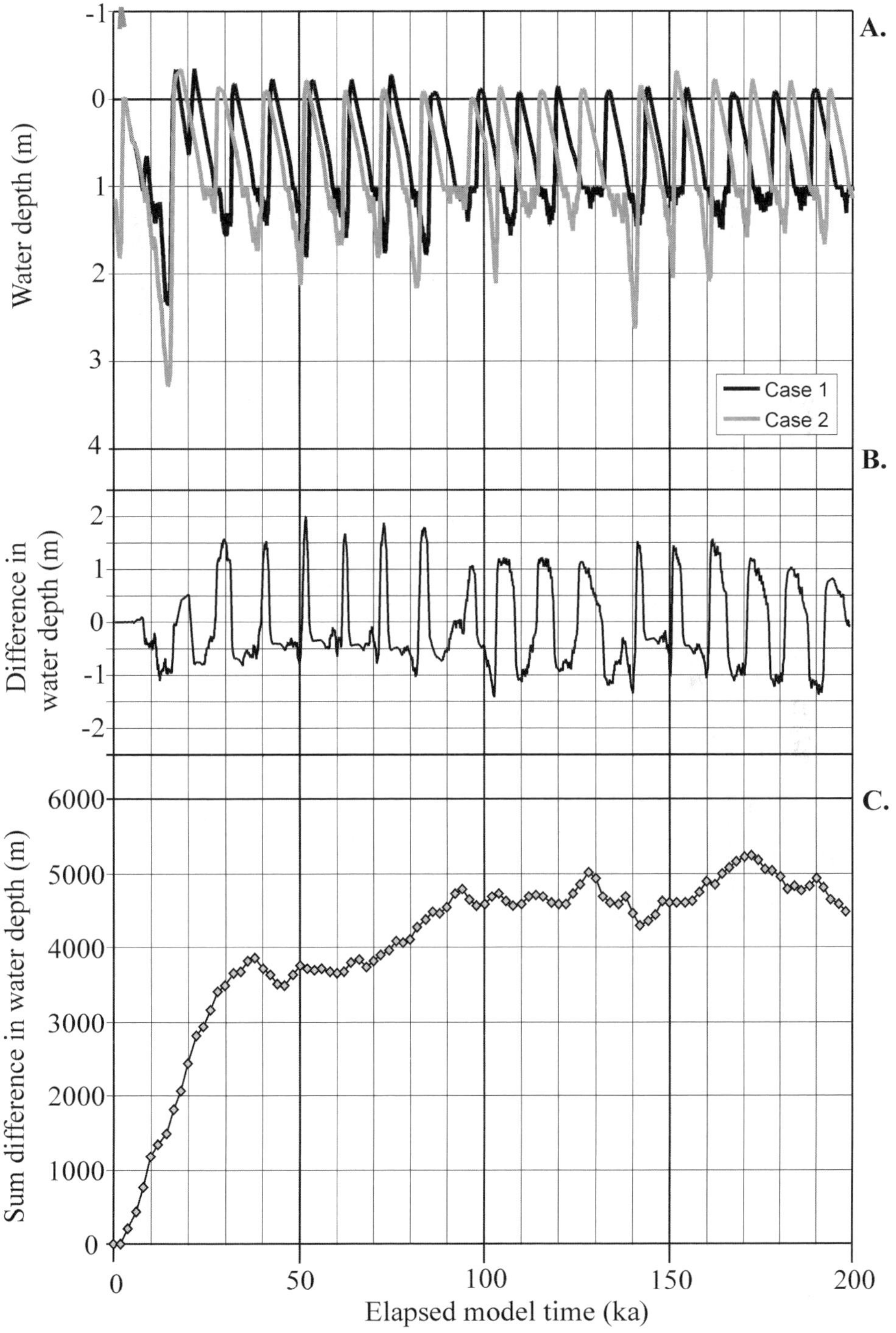

Fig. 6. (**a**) and (**b**) Water depth history from point $x = 40$ km, $y = 70$ km in model Cases 1 and 2, showing initially similar evolution, followed by divergence. Despite the divergence, water depth remains within the same overall limits in both cases. (**c**) The sum of the differences in water depth, for all points on the grid, for the two cases, plotted against elapsed model time. Divergence due to sensitive dependence is marked by the difference increasing relatively rapidly until ~ 34 ka and then increasing at a slower rate until 95 ka.

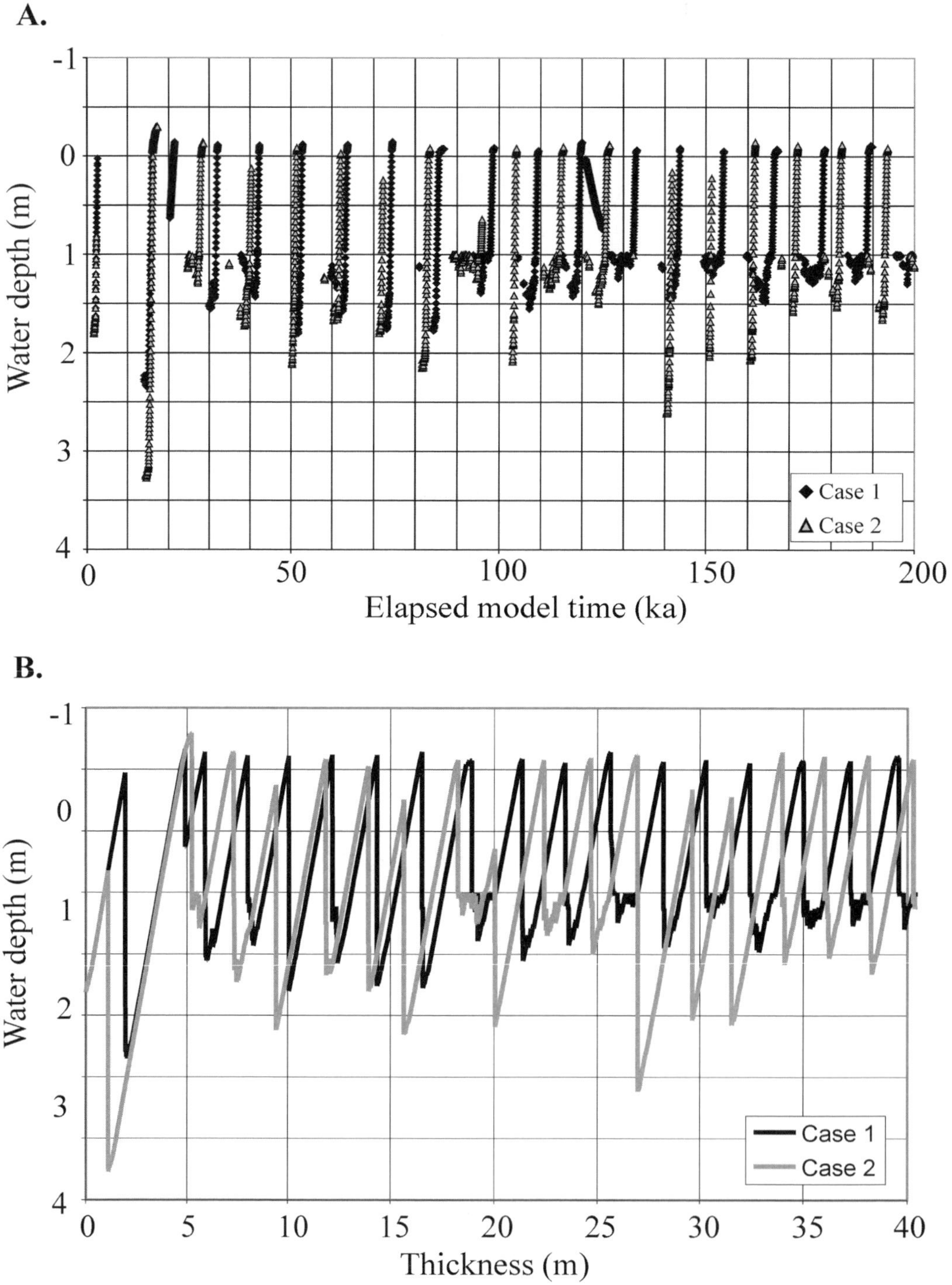

Fig. 7. (**a**) Water depth history recorded by preserved strata for Cases 1 and 2 plotted against age of strata. Divergence occurs after only 23 ka of record as a consequence of lack of preservation of Case 2 strata. (**b**) Water depth history of preserved strata for Cases 1 and 2 plotted against accumulated thickness. Divergence occurs after only 1.1 m of accumulated thickness.

therefore the model behaviour is not totally unpredictable; general predictions regarding the range of water depths likely to occur and the consistent presence of shallowing-upward trends will be reliable. However, there are other aspects of the model behaviour that limit prediction.

Firstly, although modelled water depth history is cyclical, it is not periodic. Fourier analysis of the water depth history of Case 1 demonstrates that there are no dominant frequencies in the power spectrum (Fig. 8); the Fourier coefficients are indistinguishable from a set drawn at random from a normal distribution with mean 0 and standard deviation 0.1. This shows that the time series does not conform to any simple periodic pattern of development, at least over the whole 1 Ma duration of the model run, and can therefore be considered to be random in the sense that prediction of future states from any particular state is not possible. This randomness occurs despite the fact that all the individual components of the model are deterministic.

Secondly, in both model cases parasequence thickness does not appear to follow any pattern of development. This is shown in Figure 9 where the thickness of one parasequence is plotted against the thickness of the next. Any patterns or order present in thickness distributions should be apparent as trends on the plot showing systematic transition from one parasequence thickness to the next, but no such trends are present. To test this more rigorously, systematic trends, also referred to as directionality, can be identified using a serial autocorrelation method such as the Durban-Watson test. Calculating the Durban-Watson test statistic on parasequence thicknesses for Cases 1 and 2 gives the values 1.75 and 1.94 respectively. Both values fall in the field of randomness, so the null hypothesis of randomness cannot be rejected. Thus it is clear that a deterministic model can create apparently random development of parasequence thickness simply as a consequence of complex interactions between model processes; no external forcing or stochastic element in the processes creating parasequences is necessary.

Parasequence thickness evolution and preservation in Case 1 and Case 2 also show divergence and this is visible in the Fischer plots in Figure 10. All the variations in parasequence thickness shown in these plots are a consequence of varying distributions of productivity and varying sediment transport rates, and they demonstrate sensitive dependence on initial conditions. As already demonstrated, parasequence thickness variations are undistinguishable from random. Note that the long negative runs visible on Fischer plots from both cases are not indications of ordered variation, but simply a consequence of a positively skewed thickness distribution. There is no external forcing, and the model is entirely deterministic, i.e. without any stochastic components. Thus apparently

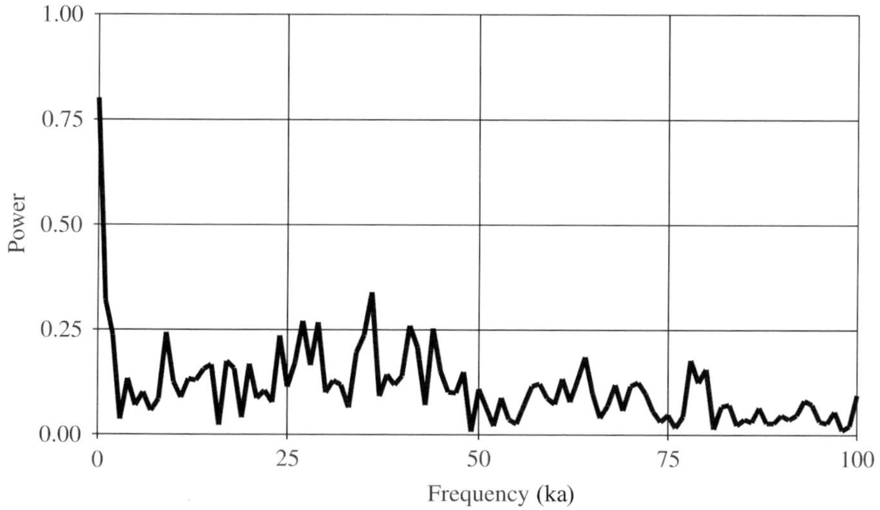

Fig. 8. A power spectrum calculated by Fourier analysis performed on the complete time series from point $x = 40$ km, $y = 70$ km from Case 1. There are no clearly dominant frequencies apparent in the plot, suggesting that the time series does not conform to any simple periodic pattern of development.

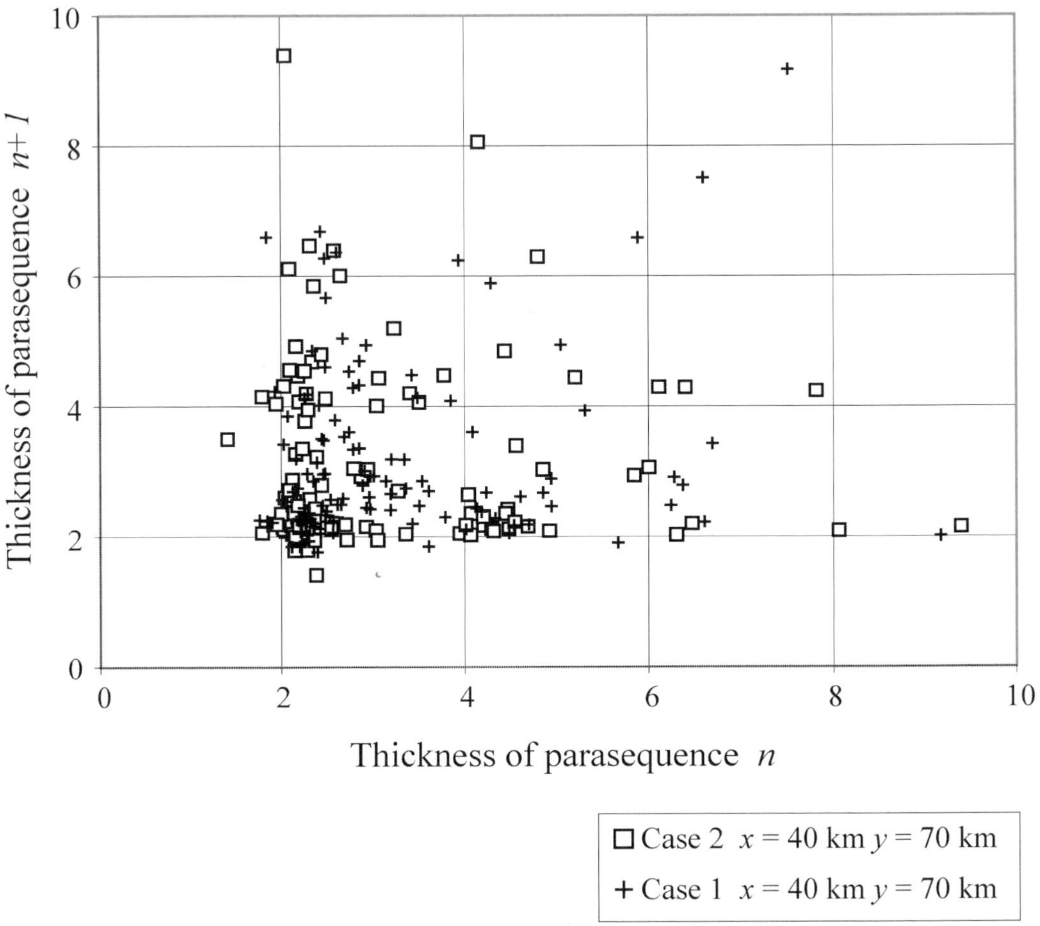

Fig. 9. Thickness of one parasequence (parasequence n) plotted against the thickness of the next parasequence in the vertical succession (parasequence $n + 1$) for both model cases. Obvious scatter on the plot suggests that there is no simple relationship between one parasequence and the next. (See text for discussion.)

complex signals, often interpreted as due to external forcing (e.g. Walkden & Matos 2000; Overstreet *et al.* 2003), or attributed to stochastic effects (Wilkinson *et al.* 1999), can be produced simply by deterministic internal dynamics of a modelled carbonate system.

If real carbonate systems behave in this manner, true chronostratigraphic correlation of ancient strata on the basis of parasequence thickness trends (e.g. Read & Goldhammer 1988; Overstreet *et al.* 2003) is not possible unless an external signal strongly overprints such randomness and sensitive dependence. Note also that the criteria to identify shallowing-upward cycles are based on water depth classifications only; greater complexity present in the strata (i.e. more than one facies per water depth group) is ignored in this definition of the parasequences.

But is it chaos?

As discussed above, comparison of the water depth time series for the two model cases demonstrates divergence, Fourier analysis demonstrates that there are no dominant frequencies in the power spectrum (Fig. 8), and Durban-Watson analysis demonstrates that no order can be identified in the parasequence thickness history. Although this divergence and lack of ordered periodic behaviour is significant, it still does not establish the presence or absence of chaotic behaviour. A system can only be described as chaotic if it shows sensitive

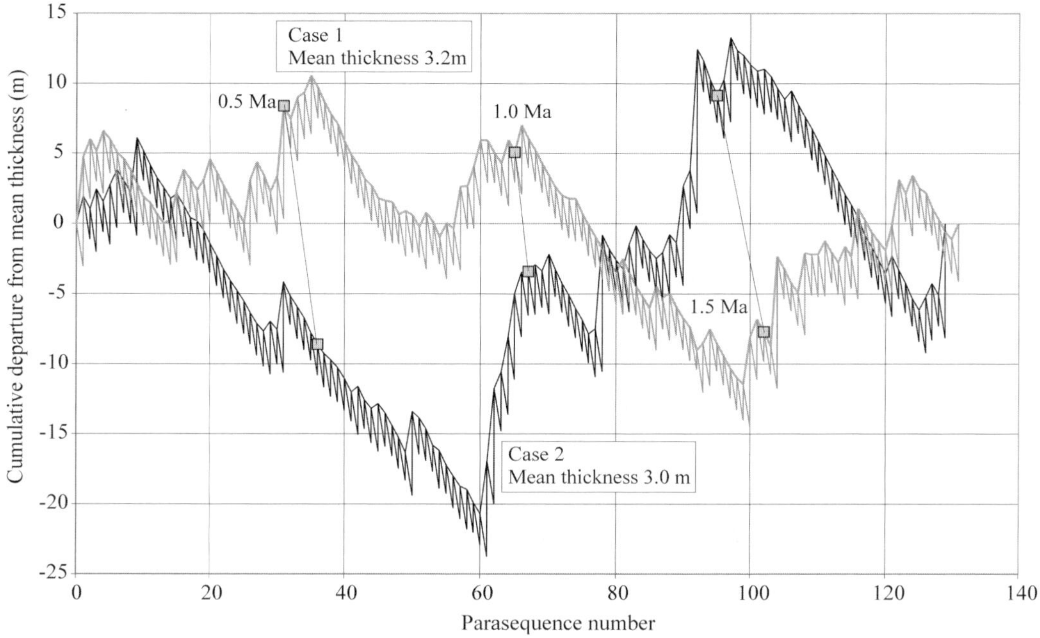

Fig. 10. Fischer plots for Cases 1 and 2 from point $x = 40$ km, $y = 70$ km showing divergent parasequence evolution and patterns of parasequence thickness typical of ancient carbonate strata. Three points on each train of parasequences are labelled with their age. Similar thickness variations observed in ancient strata are usually interpreted to be due to changes in accommodation driven by relative sea-level oscillations. In this case they are due to unforced, autocyclic and entirely deterministic variations in carbonate productivity and sediment transport. Differences in Fischer plots for the two cases are due to divergence driven by sensitive dependence. Note that long negative runs present on both plots arise as a consequence of the positively skewed thickness distribution giving a mean thickness value higher than the mode.

dependence that manifests itself as exponential divergence of trajectories in a pseudo-phase space (Kantz & Schreiber 1997; Williams 1997). It is meaningless to describe a system as chaotic without an analysis to establish such exponential trajectory divergence.

To determine the presence or absence of such exponential divergence we analysed the water depth time series from point $x = 40$ km, $y = 70$ km for Case 1 using the concepts and methods described in detail in Williams (1997) and Kantz & Schreiber (1997). The first step was construction of a pseudo-phase space, which is equivalent to the state space of the underlying dynamical system. This process, known as a pseudo-phase space reconstruction, was accomplished by a time-delay embedding of the data. As a result of examining the autocorrelation of the time series to determine the offset at which the autocorrelation coefficient approached zero, an appropriate magnitude of time delay was determined to be 250 time steps, or 2.5 ka. Output from an algorithm computing the number of false nearest neighbours (Kantz & Schreiber 1997) indicated a low embedding dimension of 3, 4 or possibly 5. Graphical representation of the time series plotted in phase space suggested that an embedding dimension of three is sufficient to completely unfold the trajectories of the system so that self-intersections do not occur and the structure of the system attractor is apparent (Fig. 11).

Typically in a chaotic system, trajectories on the attractor in a pseudo-phase space initially separated by a small distance will diverge from one another exponentially fast until separated by distances comparable to the size of the systems attractor. This effect can be measured in a pseudo-phase space by taking a reference point $\mathbf{s}_{n0}$ embedded time series and calculating

$$S(\Delta n) = \frac{1}{N}\sum_{n_0=1}^{N} \ln\left(\frac{1}{|u(\mathbf{s}_{n_0})|}\sum_{\mathbf{s}_n \varepsilon u(\mathbf{s}_{n_0})} |s_{n_0+\Delta n} - s_{n+\Delta n}|\right), \quad (5)$$

where $u(\mathbf{s}_{n_0})$ is the neighbourhood of $\mathbf{s}_{n_0}$ with diameter ε, N is the number of sample points from the time series and Δn is a time span

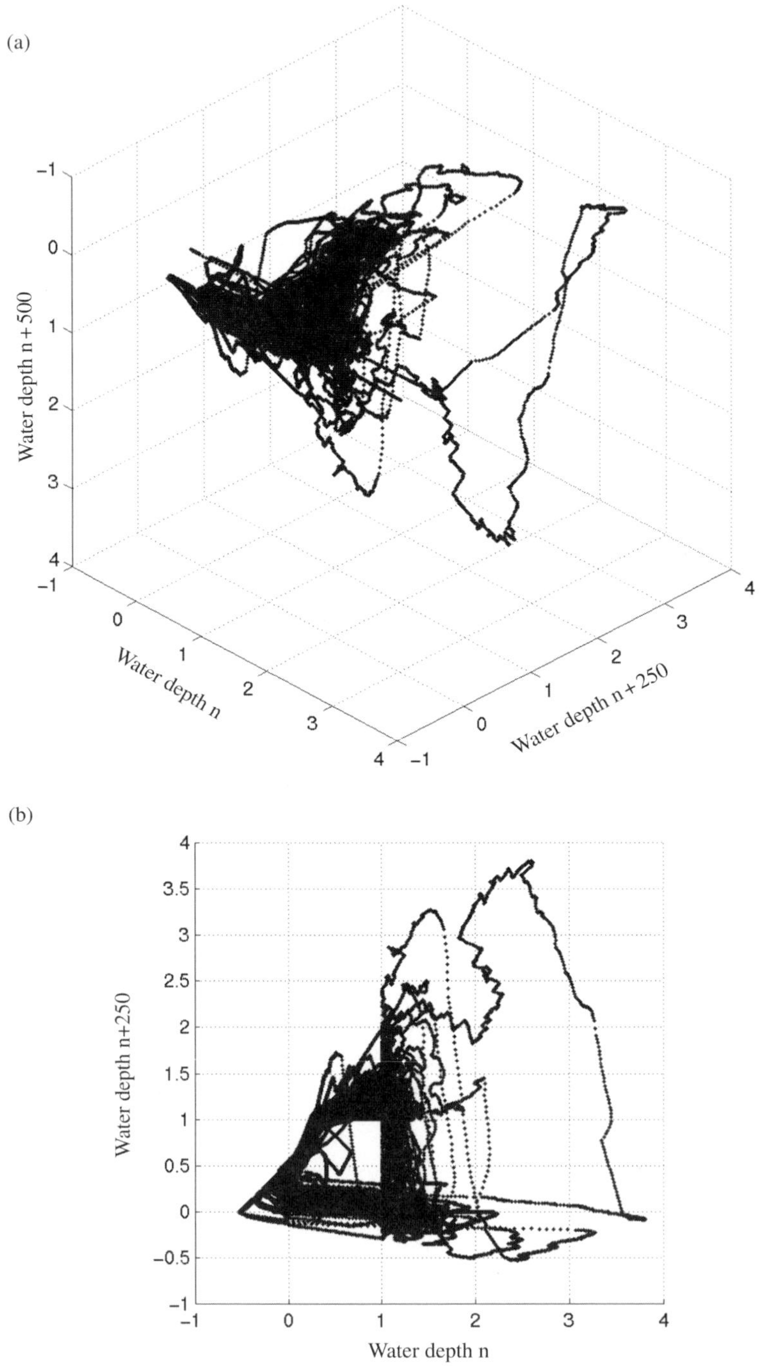

Fig. 11. The water-depth time series for Case 1 from point $x = 40$ km, $y = 70$ km plotted as an embedding in a 3-D pseudo phase space. (**a**) The structure of the attractor is visible in the 3-D plot, with a large cluster of points and trajectories occupying a small area of the volume and a series of excursions from this cluster marking periods of deeper water. Similar excursions into deeper water are visible in Figure 5 and occur when island geometry blocks the colonization of the given area so that production ceases and subsidence and erosion create relatively deep water. (**b**) A 2-D plot of the attractor showing the cluster and the excursions. Note that there is a sharp delineation at the 1 m water depth value, caused by inability of carbonates to grow *in situ* into the intertidal zone. Trajectories showing decreasing water depths are produced by transport processes accumulating strata in the intertidal and supratidal zones.

(Kantz & Schreiber 1997). Plotting $S(\Delta n)$ against Δn for different values of embedding dimension m (Fig. 12) allows calculation of the rate of trajectory divergence. If the initial portion of the plot forms a straight line, the gradient of the straight line gives a value referred to as the maximal Lyapunov exponent, which is a measure of exponential trajectory divergence (Goodings 1991; Kantz & Schreiber 1997). A positive value of the maximal Lyapunov exponent derived in this way provides strong evidence of chaos. Figure 12a shows this plot for the complete time series from point $x = 40$ km, $y = 70$ km in Case 1. The initially positive gradients of the lines demonstrate divergence but the absence of an unambiguous straight-line section in the plot suggests that the time series do not exhibit exponential divergence, and the same appears to be the case for the plot derived from the preserved portion of the time series (Fig. 12b). Consequently, in both cases it seems that true chaotic behaviour is not present. Further work is required to determine whether this is also the case for other parameter values in the model because systems can show sudden transition into chaotic behaviour as parameter values change (Williams 1997). For example, higher rates of sediment transport might trigger onset of chaotic behaviour.

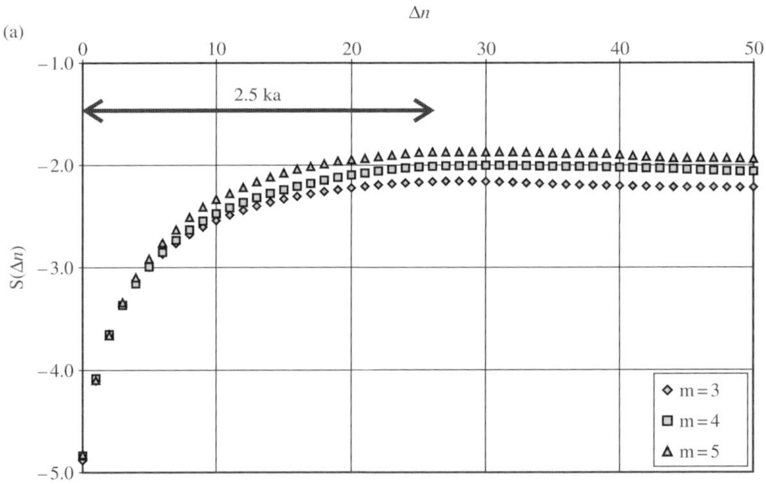

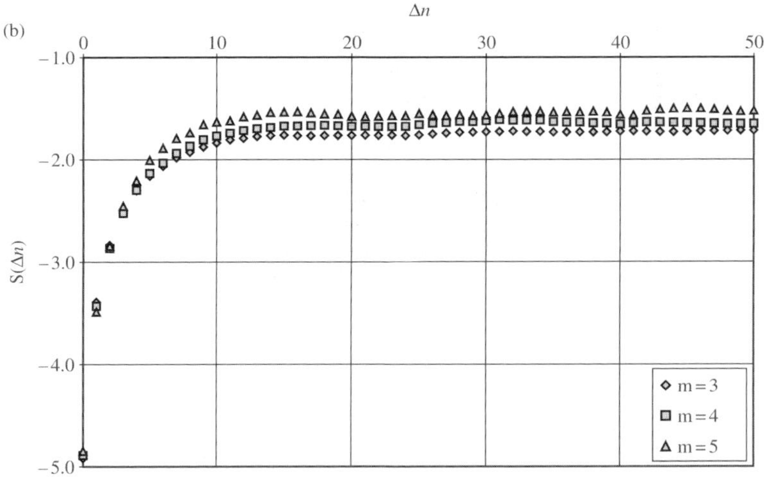

Fig. 12. Plots of $S(\Delta n)$ against Δn showing trajectory divergence for: (**a**) the complete time series, and (**b**) the preserved portion of the time series. (m is the embedding dimension) The data lack an unambiguous initial linear section that would be indicative of exponential divergence and truly chaotic behaviour. (See text for discussion.)

Consequence for prediction of carbonate stratal patterns

Although it is clearly important to remember that the forward model is merely a simplified representation of real carbonate systems, and to remember therefore that model results should not be over-interpreted, such simplified models are still useful ways of improving our understanding of ancient sedimentary systems (Paola 2000), and these results have some important implications for understanding carbonate systems and carbonate strata. In this particular case, sensitive dependence, divergence and effectively random behaviour have potentially profound implications for predictive ability in ancient systems. If real carbonate systems include non-linear relationships between water depth, sediment production (e.g. Boscher & Schlager 1992) and sediment transport, plus some spatially variable ecological control on sediment production, they are likely to behave as this model behaves and exhibit similar sensitive dependence and apparently random behaviour. Note that, in terms of its component parts, this is an entirely deterministic model. However, the non-linear components and complex interactions generate a level of complex behaviour and divergence that makes its evolution at any one point in space unpredictable in detail and therefore random (cf. Wolfram 2002). Therefore, even though the model exhibits a clearly cyclical, quasiperiodic component in the form of autocyclic shallowing-upward parasequences, development of variables such as water depth and individual parasequence thickness through time are effectively random and perhaps impossible to predict.

Given this, how can numerical forward models achieve a meaningful best fit between model output and subsurface and outcrop data for this type of system? As models become more complex, and perhaps more realistic, they are increasingly likely to exhibit this kind of behaviour. Furthermore, automated inversion methods that attempt to condition models in order to achieve a best fit based on an objective function that measures error between model output and observed strata may be entirely impractical if the minima in parameter space are highly localized due to sensitive dependence (cf. Tetzlaff 2004). Consequently, in systems where sensitive dependence operates, conceptual and numerical models should be limited to general predictions known to be unaffected by divergence, perhaps relying on statistical properties of the strata. More work is required to determine what processes and products of modelled and real carbonate systems are likely to be affected by divergence, but results from this modelling exercise suggests that water depth history and parasequence-thickness distributions are likely candidates.

Although these results illustrate how sensitive dependence can significantly reduce predictive ability, it is quite possible that external forcing may limit divergence, increase order and allow more predictions to be made. In other words, sensitive dependence and random behaviour inherent to the sedimentary systems may be overcome by a periodic external signal such as sea level or climatic oscillations (Barnett *et al.* 2002; Teztlaff 2004) if such ordered, periodic oscillations have indeed occurred. In this case, some sequence stratigraphic models predicting parasequence thickness patterns may remain valid, but further work is needed to investigate this; it should not be assumed *a priori*.

Model behaviour in these examples is complex, but it could be more complex. It is important to note that model strata are coded simply according to water depth at time of deposition, and it is this that leads to the obvious element of cyclicity. However, as demonstrated by Rankey (2004), there may be no simple link between water depth and facies in many carbonate systems. Consequently the level of variation in lithofacies distributions may be significantly higher than the distribution of water depth classifications in this model. If realistic facies distributions were included in the model, the apparent disorder and divergence may increase significantly.

Conclusions

The carbonate system represented in this numerical forward model shows autocyclic but not periodic behaviour, sensitive dependence on initial conditions manifested as divergence of water depth, and divergence of parasequence thickness development in two model cases that have only a very small difference in their initial starting topographies. Furthermore there is no detectable order in development of parasequence thicknesses through time. These results have some important implications for understanding carbonate systems:

- Divergence in water depth history at one point takes $\sim$ 10 ka, and for the whole model grid takes $\sim$ 100 ka. Incomplete preservation of strata tends to reduce apparent divergence time. More work is required to investigate how divergence time varies with different model parameter values (e.g. transport rate).

- A general pattern of autocyclic quasi-periodic shallowing-upward parasequence development is present in both model cases, despite divergence, but Fischer plots from the two cases show that divergence leads to entirely different parasequence thickness histories.
- Although the model is entirely deterministic, non-linear components and complex interactions between processes make the model behaviour sufficiently complex for water depth and parasequence thickness evolution to be effectively random and therefore not predictable, despite a clearly present element of autocyclic behaviour responsible for the shallowing-upward parasequences. This illustrates that models can generate complex unpredictable behaviour without any explicitly stochastic model elements, and that aspects of order and disorder can coexist.
- Chaotic behaviour is a specific type of sensitive dependence, and calculation of trajectory divergence in a 3-D pseudo-phase space indicates that model behaviour in this case is not truly chaotic.
- If sensitive dependence and divergence turn out to be common in real carbonate systems, these modelling results suggest that predictions should be limited to general statistical properties of the strata, to elements of the system unaffected by these phenomena, or to cases where an element of periodic external forcing over-rides their affects.
- These results also suggest that increasingly complex, sophisticated and perhaps realistic stratigraphic forward models are not necessarily going to lead directly to more simple, deterministic predictive power.

PMB wishes to thank P. Wright for an inspirational introduction to carbonate systems and M. Eller and P. Hirst for first discussions of chaos. A. Barnett provided comments on an earlier version of the paper and some valuable insight into modern carbonate systems. We both thank C. Emery for her initial introduction. Reviews by B. Wilkinson, D. Waltham and A. Curtis significantly improved the clarity and rigour of the manuscript.

References

BAILEY, R. J. 1998. Review: stratigraphy, meta-stratigraphy and chaos. *Terra Nova*, **10**, 222–230.

BARNETT, A. J., BURGESS, P. M. & WRIGHT, V. P. 2002. Icehouse world sea-level behaviour and resulting stratal patterns in Late Visean (Mississippian) carbonate platforms: integration of numerical forward modelling and outcrop studies. *Basin Research*, **14**, 417–438.

BOSCHER, H. & SCHLAGER, W. 1992. Computer simulation of reef growth. *Sedimentology*, **39**, 503–512.

BURGESS, P. M. & WRIGHT, V. P. 2003. Numerical forward modelling of carbonate platform dynamics: an evaluation of complexity and completeness in carbonate strata. *Journal of Sedimentary Research*, **73**, 637–652.

BURGESS, P. M., WRIGHT, V. P. & EMERY, D. 2001. Numerical forward modeling of peritidal carbonate parasequence development: implications for outcrop interpretation. *Basin Research*, **13**, 1–16.

CHEN, D., TUCKER, M. E., JIANG, M. & ZHU, J. 2001. Long-distance correlation between tectonically-controlled, isolated carbonate platforms by cyclostratigraphy and sequence stratigraphy in the Devonian of South China. *Sedimentology*, **48**, 57–78.

DEMMICO, R. V. 1998. Cycopath 2D: a two dimensional, forward model of cyclic sedimentation on carbonate platforms. *Computers & Geosciences*, **24**, 405–424.

DEUTSCH, C. V. 2002. *Geostatistical Reservoir Modelling*. Oxford University Press, Oxford.

DRUMMOND, C. N. & DUGAN, P. J. 1999. Self-organizing models of shallow water carbonate accumulation. *Journal of Sedimentary Research*, **69**, 939–946.

ENOS, P. 1991 Sedimentary parameters for computer modeling. *In:* FRANSEEN, E. K., WATNEY, W. L. & ROSS, W. (eds) *Sedimentary Modelling: Computer Simulations and Methods for Improved Parameter Definition.* Geological Survey of Kansas, Bulletin, **233**, 63–99.

GAFFIN, S. R. & MAASCH, K. A. 1991. Anomalous cyclicity in climate and stratigraphy and modelling non-linear oscillations. *Journal of Geophysical Research*, **96**, 6701–6711.

GINSBURG, R. N. 1971. Landward movement of carbonate mud: new model for regressive cycles in carbonates. *American Association of Petroleum Geologists Annual Meeting Abstract Programs*, **55**, 340. [Abstract]

GOLDHAMMER, R. K., DUNN, P. A. & HARDIE, L. A. 1990. Depositional cycles, composite sea-level changes, cycle stacking patterns, and the hierarchy of stratigraphic forcing: examples from Alpine Triassic platform carbonates. Bulletin of the *Geological Society of America*, **102**, 535–562.

GOLDHAMMER, R. K., LEHMANN, P. J. & DUNN, P. A. 1993. The origin of high-frequency platform carbonate cycles and third-order sequences (Lower Ordovician El Paso Group, West Texas): constraints from outcrop data and stratigraphic modelling. *Journal of Sedimentary Petrology*, **63**, 318–359.

GOODINGS, D. 1991. Nonlinear differential equations and attractors. *In:* MIDDLETON, G. V. (ed.) *Nonlinear Dynamics, Chaos and Fractals, With Applications to Geological Systems.* Geological Association of Canada Short Course Notes, **9**, 23–33.

HAQ, B. U., HARDENBOL, J. & VAIL, P. R. 1988. Mesozoic and Cenozoic chronostratigraphy and cycles of sealevel change. *In:* WILGUS, C. K., HASTINGS, B. S., POSAMENTIER, H., VAN WAGONER, J., ROSS, C. A. & KENDALL, C. G. St. C (eds) *Sea-Level Changes: An Integrated Approach.* SEPM (Society for Sedimentary Geology), Special Publications, **42**, 71–108.

HOLLAND, J. H. 1998. *Emergence, from Chaos to Order.* Oxford University Press, Oxford.

KANTZ, H. & SCHREIBER, T. 1997. *Nonlinear Time Series Analysis.* Cambridge Nonlinear Science Series, Cambridge University Press, **7**.

KENYON, P. M. & TURCOTTE, D. L. 1985. Morphology of a prograding delta by bulk sediment transport. *Bulletin of the Geological Society of America*, **96**, 1457–1465.

LORENZ, E. 1993. *The Essence of Chaos.* UCL Press, London.

MEIJER, X. 2002. Modelling the drainage evolution of a river-shelf system forced by Quaternary glacio-eustasy. *Basin Research*, **14**, 361–378.

OVERSTREET, R. B., OBOH-IKUENOBE, F. E. & GREGG, J. M. 2003. Sequence stratigraphy and depositional facies of Lower Ordovician cyclic carbonate rocks, Southern Missouri, U.S.A. *Journal of Sedimentary Research*, **73**, 421–433.

PAOLA, C. 2000. Quantitative models of sedimentary basin filling. *Sedimentology*, **47**, 121–178.

RANKEY, E. (2004) On the interpretation of shallow shelf carbonate facies and habitats: how much does water depth matter? *Journal of Sedimentary Research*, **74**.

READ, J. F. & GOLDHAMMER, R. K. 1988. Use of Fischer plots to define third-order sea-level curves in Ordovician peritidal carbonates, Appalachians. *Geology*, **16**, 895–899.

ROSENSTEIN, M. T., COLLINS, J. J. & DE LUCA, C. J. 1993. A practical method for calculating largest Lyapunov exponents from small data sets. *Physica D*, **65**, 117.

SADLER, P. M. 1981. Sediment accumulation rates and the completeness of stratigraphic sections. *Journal of Geology*, **89**, 569–584.

SLINGERLAND, R. 1989. Predictability and chaos in quantitative dynamic stratigraphy. *In:* CROSS, T. A. (ed.) *Quantitative Dynamic Stratigraphy.* Prentice Hall, Englewood Cliffs, New Jersey, 45–53.

SMITH, D. G. 1994. Cyclicity or chaos? Orbital forcing versus non-linear dynamics. *In:* DE BOER, P. L. & SMITH, D. G. (eds) *Orbital Forcing and Cyclic Sequences.* International Association of Sedimentologists Special Publications, **19**, 531–544.

TETZLAFF, D. M. 1989. Limits to the predictive ability of dynamic models that simulate clastic sedimentation. *In:* CROSS, T. A. (ed.) *Quantitative Dynamic Stratigraphy.* Prentice Hall, Englewood Cliffs, New Jersey, 55–65.

TETZLAFF, D. M. 2004. Input uncertainty and conditioning in siliciclastic process modelling. *In:* CURTIS, A. & WOOD, R. (eds) *Geological Prior Information: Informing Science and Engineering.* Geological Society, London, Special Publications, **239**, 95–109.

TIPPER, J. C. 1997. Modeling carbonate platform sedimentation-lag comes naturally *Geology*, **25**, 495–498.

WALKDEN, G. M. & MATOS, J. D. 2000. 'Tuning' high-frequency cyclic carbonate platform successions using omission surfaces: Lower Jurassic of the U.A.E. and Oman. *In:* ALSHARHAM, A. S. & SCOTT, R. W. (eds) *Middle East Models of Jurassic/Cretaceous Carbonate Systems.* SEPM (Society of Sedimentary Geology), Tulsa, Special Publications, **69**, 37–52.

WILKINSON, B. H., DRUMMOND, C. N., DIEDRICH, N. W. & ROTHMAN, E. D. 1999. Poisson processes of carbonate accumulation on Paleozoic and Holocene platforms. *Journal of Sedimentary Research*, **69**, 338–350.

WILLIAMS, G. P. 1997. *Chaos Theory Tamed.* Taylor and Francis, London.

WOLFRAM, S. 2002. *A New Kind of Science.* Wolfram Media Incorporated, Champaign, Illinois, pp. 1197.

Input uncertainty and conditioning in siliciclastic process modelling

DANIEL M. TETZLAFF

Schlumberger-Doll Research, 320 Bent Street, Cambridge MA 02141, USA

Abstract: Deterministic forward sedimentary process models enhance our quantitative understanding of sedimentary systems. They are also being used increasingly to assist in the reconstruction of the geological past and the inference of the present configuration of sedimentary deposits. Such usage presents the challenge of having to establish the initial and boundary conditions that will cause the model's output to match present-day observations. This constitutes an inversion problem that can sometimes be handled by traditional optimization methods. Clastic sedimentation models, however, often incorporate complex non-linear relationships that preclude the use of these techniques. The problem must then be handled statistically by relaxing the requirement of honouring exact observations and matching only the spatial variability of the observed deposits. Recent advances in control of non-linear dynamic systems may also shed light on possible solutions to the inversion problem in siliciclastic models. This paper reviews known approaches to problems related to input uncertainty and conditioning, and presents original preliminary results on control of sedimentation models.

Introduction

A model is a simplified representation of a real system. To be useful, it must permit easier manipulation and visualization than the system it represents, while still emulating the system's main behaviour and components. The terms 'geological modelling' and 'reservoir modelling' refer to a representation of a rock volume and its properties in the form of a computer dataset. The goal of geological modelling is to understand the real rock volume through visualization (slicing, enhancing, zooming, highlighting data relationships etc.) and analysis (geostatistics, volumetrics etc.), and to use this understanding for the estimation and prediction of rock and fluid properties in the subsurface. This form of modelling is essentially static in that its goal is to represent the current status of a geological system for the purposes of, for example, predicting petroleum reservoir volume, reservoir performance and planning the drilling of wells.

Geological *process* (or 'forward') modelling, on the other hand, incorporates the representation of past geological processes, such as sedimentation, heat flow, compaction and hydrocarbon generation and migration. It is dynamic in that representation along the time dimension is a crucial part of the model. The basic tenet of geological sciences is to explain the present as a result of past processes. Therefore, it is only natural that, as geological process models have become more realistic, they have also been increasingly sought as a tool in the petroleum industry for determining the present-day properties of basins and reservoirs, thereby complementing static models. In this new role, however, process models encounter limitations imposed by lack of knowledge of past conditions. These limitations do not preclude their practical applications, but impose a need for additional techniques to make them effective. Some of these techniques are described below.

Siliciclastic sedimentary process modelling

Geological process modelling can be understood to incorporate laboratory as well as theoretical representations of natural systems, conceptual or rigorously mathematical descriptions and probabilistic or deterministic assumptions. The scope of this discussion is limited to deterministic numerical models that operate on a digital computer (Fig. 1). Geological process models may represent one or more of a whole spectrum of geological phenomena. 'Basin models' that represent heat flow, compaction, diagenesis, hydrocarbon generation, maturation and migration have been in use for several decades. Structural models have evolved from simple 2-D back-stripping techniques to full 3-D folding and complex faulting models. Sedimentary processes on the other hand, have only recently been modelled at geological time scales using physical principles of flow and sedimentation (Tetzlaff & Priddy 2001).

From: CURTIS, A. & WOOD, R. (eds) 2004. *Geological Prior Information: Informing Science and Engineering*. Geological Society, London, Special Publications, **239**, 95–109. 186239-171-8/04/$15.00

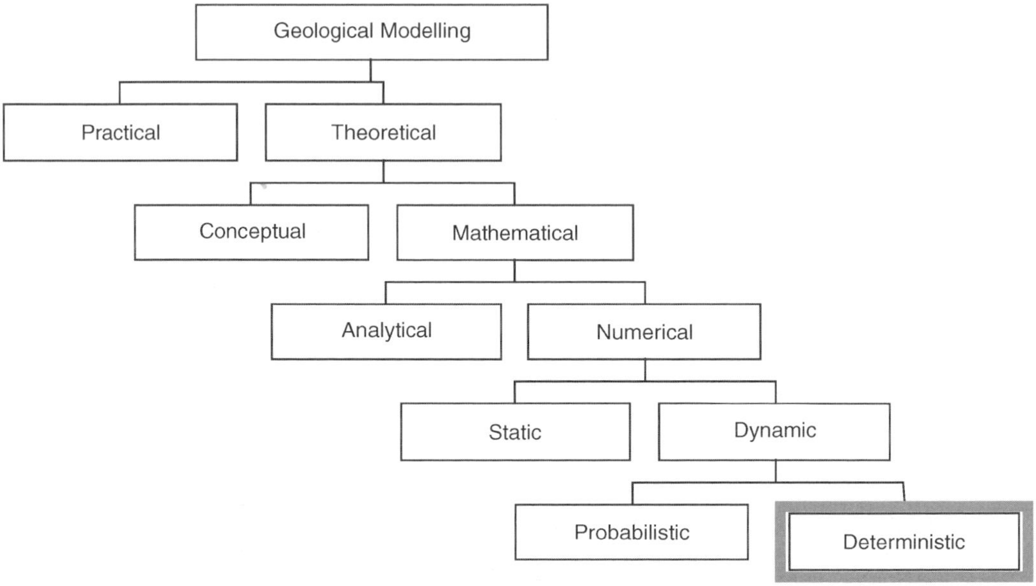

Fig. 1. Classification of geological process models according to methodology.

The reason for the relatively late, and still incomplete, entry of quantitative models of sedimentary processes into the realm of practical applications is usually attributed to their complexity and apparently different behaviour at many scales. At the small end of the spatial scale, for example, the laws that govern the movement of a single solid particle through water are relatively well known. On the large scale, atmospheric and oceanic circulation has been quantitatively modelled. Mid-scale models of sedimentary systems, which cover scales of a few hundred to a few thousand metres overlap the two extremes. Bars and meanders in rivers and submarine channels, shoreline deposits, deltas and other features of sedimentological interest develop in shapes that ultimately are controlled by forces acting on individual grains and flowing fluids, as well as the continental-scale forces that determine their overall location and evolution. Additionally, they exhibit complex behaviour that appears to arise from internal relationships between their components, without a necessary external forcing. Thus, mid-scale sedimentary processes are particularly challenging to model deterministically.

There is interest in the modelling of sedimentary processes because the depositional geometry and the spatial distribution of primary properties of sediments (grain composition, initial porosity, initial permeability) are the main control on subsequent processes such as compaction, diagenesis and heat flow. A model that represents flow in porous media, for example, be it a hydrogeological model, an oil migration model or a hydrocarbon reservoir simulator, requires a detailed picture of the distribution of petrophysical properties (especially porosity and permeability) that stem directly from primary depositional properties. Such a detailed picture is rarely available from observations only, because of the sparse sampling density, but can sometimes be improved by forward sedimentary modelling (Koltermann & Gorelick 1992).

Furthermore, modelling a sedimentary sequence via the physical processes that shaped it can give insights and constrain interpretations in ways not otherwise possible, and can complement geostatistical techniques, which generally incorporate little knowledge of geological processes. For example, certain petroleum reservoir or aquifer architectures, formation thicknesses or other properties may be sedimentologically implausible for palaeogeographical conditions known to have existed at the time of deposition. Traditional interpolation and even geostatistical techniques (if based on two-point variograms and stationarity assumptions) may have no constraint to prevent implausible scenarios. Quantitatively unlikely scenarios (e.g. an interpretation of an ancient delta containing more sand than the amount its river could have possibly transported in its lifetime through the observed channel remnants) may escape even the trained eye.

Only a handful of siliciclastic sedimentary process models have been successfully used as predicting tools (Cross 1990; Franseen *et al.* 1991; Slingerland *et al.* 1994; Harbaugh *et al.* 1999; Merriam & Davis 2001). Some of the results discussed in this paper were produced with an experimental model, which is in turn based on similar principles as SEDSIM, a sedimentation package developed at Stanford in the 1980s and later further developed and presently used by the Australian consortium CSIRO (Commonwealth Scientific and Industrial Research Organization) (Griffiths 2001; Tetzlaff & Priddy 2001). These packages incorporate a quantitative representation of open-channel flow over arbitrary topographical surfaces, coupled with erosion, transport and deposition of sediment. These simulations offer a virtual laboratory to explore 'what if' scenarios in natural systems and are free from the scaling problems associated with laboratory experiments. Numerical representations may be scaled at will. However, the user must be aware of gridding effects, numerical instabilities, edge effects and other pitfalls associated with numerical simulation on a computer. Also, calibration using laboratory data and field measurements is still a necessity.

Before searching for solutions to the particular problem of conditioning a model to match observations, it is important to consider the broader issue of what knowledge can be used as input to a model in practical applications, and what knowledge the model is expected to generate. A model is not only a simplification of a real system. It is also an isolation of a portion of reality which interacts with the external world through specified boundary and initial conditions, assumed to be known by the modeller. A model's ability to produce output that resembles observations is by no means a measure of its predictive ability or its usefulness. Some models may require input that basically constitutes a description of the desired output (e.g. some early cross-sectional models developed in the 1980s), whereas others function more like a laboratory in which the user can only provide an initial set-up, while the model emulates physical principles. The latter will be harder to condition, but will probably provide unexpected outcomes and better insights.

Input uncertainty

Every model has limitations in its ability to emulate a natural system (Watney *et al.* 1999). A partial list of reasons (in no particular order) for these limitations might include the following:

- lack of understanding of the processes represented (use of empirical approximations to represent complex sedimentary processes, or incomplete understanding of multiscale phenomena);
- discretization of time and space that occurs when the model is reduced to numerical representation on a computer;
- lack of computer power to represent all phenomena relevant to the system;
- lack of knowledge of initial and boundary conditions.

The last of these is critical in siliciclastic process modelling over time spans of geological significance. While sedimentation models used in engineering (e.g. for predicting erosion and sedimentation in harbours, beaches, riverbanks and water reservoirs) generally have abundant data about initial conditions, geological models involve the use of a (generally not simply reversible) forward model to work backwards from observations.

Spatial scale

Input and boundary conditions that control a model must be considered in the context of the spatial and time scales represented. What a model should predict at one scale may be best considered input at a coarser or longer scale. For example, when trying to establish the depositional geometry of a delta, the sediment and flow regime, as well as slope and sea-level oscillations, whether known or inferred, would be input to the model. The model would predict, for example, the distribution of channels. When trying to understand the geometry of a single channel, however, the channel entry point, the slope and the flow in the channel may be provided as input, while sea-level oscillations would be irrelevant due to the short time span involved.

The relationship between spatial scale and model input can be better illustrated by the example shown in Figure 2. The figure shows a set of resistivity borehole images successively enlarged from left to right. Figure 2a shows 50 m of a clastic sequence consisting of sands (light colour) and shaly sands (dark colour). The sands have been interpreted to represent alluvial fans,

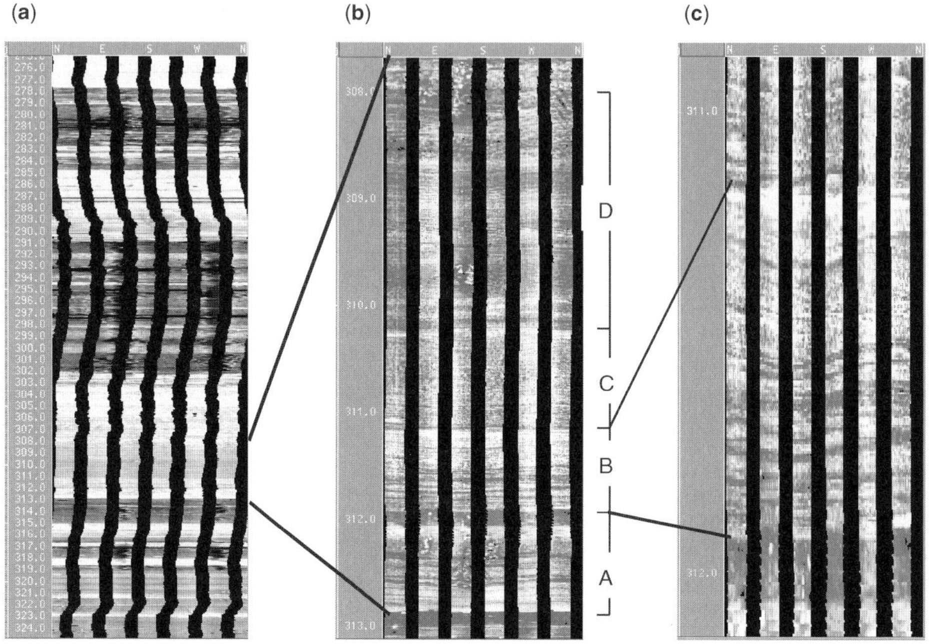

Fig. 2. Electric resistivity borehole images showing a sequence of alluvial fans, successively magnified from left to right. Images cover: (**a**) 50 m; (**b**) 5 m; (**c**) 1.25 m.

whereas the shaly sands probably represent fan deltas and lacustrine deposition. An enlargement of one of the sandy intervals interpreted as fans (Fig. 2b) reveals several channel deposits (A, B, C and D, and others not shown), separated by erosional surfaces.

Suppose we want to determine the 3-D architecture of these deposits. If the goal is the geometry of the entire sequence (Fig. 2a), the climatic variations that caused the changes between lacustrine and alluvial environments must be considered and determined by adjusting the model to match observations, or be obtained from other sources. In combination with palaeogeographical information, the model should then predict a plausible distribution of fans and lacustrine deposits. If the goal is to predict or reproduce the geometry of one fan (Fig. 2c), establishing and using long-term climatic variations may be irrelevant, as they probably do not control the internal architecture of the fan. Rather one should use a model that takes as input the location of the channel feeding the fan, the slope, flood event statistics and sediment input, while the model would hopefully predict channel-shifting and depositional events that control the resulting geometry.

The desired detail of the output of a sedimentary process model should, of course, determine the resolution of the discretization of spatial variables (gridding resolution). Siliciclastic sedimentary models often represent systems that exhibit variability at all scales. Therefore, grid resolution must be sufficient not only to represent the output at the desired detail, but also to represent the physical phenomena that produce the output with adequate detail. In the alluvial-fan example mentioned in the preceding paragraph, if alluvial fans a few hundred metres across are to be modelled, it may be assumed that a grid with cells of 100 m a side may be sufficient. If the grid is regular (all cells of the same size and shape), however, this model would not be capable of modelling channels any narrower than about 200 m, which is probably much larger than the actual channels involved, rendering the model's predictions for the fan itself suspect, despite the fact that the grid matches the target resolution. Local grid refinement may offer a better approach by providing detail where it is needed without requiring undue computer power. In the case of alluvial fans, however, channels may shift rapidly across the fan's surface, and the refinement would have to be adaptive (i.e. changing through time) in order to be useful.

Time scale

Like spatial extent and resolution, the time span and time resolution of a siliciclastic model must be carefully considered for the model to yield useful answers. A model that successfully represents a certain time span may need major adjustments to extend it to represent, for example, ten times longer. Many geological systems are dominated by events that are 'rare' (i.e. infrequent and short in duration) but fundamental to the model in the long run. For example, the 'thousand-year' flood may be more important in shaping the course of a river or preserved fluvial deposits than all the minor floods that occur between these major events. Therefore, a fluvial simulation that represents 0.5 ka may require a completely different input than a simulation that runs for 5 ka. For even larger time spans, systematic climate variations would have to be incorporated.

If we only want to model a few typical large floods that are responsible for most of the deposition in a given sedimentary interval, then an appropriate value of river stage to use would be the expected (median) largest flood stage for the number of years modelled. A statistical analysis of river-stage variation must be use to answer questions like:

> 'How should the river-flow model input be changed as the interval represented increases?'

An example of an answer to this question can be found by making the assumption that the statistical density distribution of river stage follows an exponential distribution, namely:

$$f(x) = \lambda e^{-\lambda x} (for\ x \geqslant 0), \qquad (1)$$

where x is the river water level or stage relative to an arbitrary reference, $f(x)$ is the statistical density function describing the river stage values, and λ is a constant.

We are interested in finding the maximum stage that is likely to occur during a period of N years. We can denote this value as $S(N)$. It is a random variable whose distribution depends on N, and on $f(x)$ (which in this case depends on λ). Because the maximum stage is a random variable, a reasonable value to use as maximum stage for N years may be the median of $S(N)$. (The probability that the actual N-year maximum is larger than the median of $S(N)$ is the same as the probability of it being smaller.) This value can be shown to be:

$$\text{Median of } S(N) = \ln(2N)/\lambda \qquad (2)$$

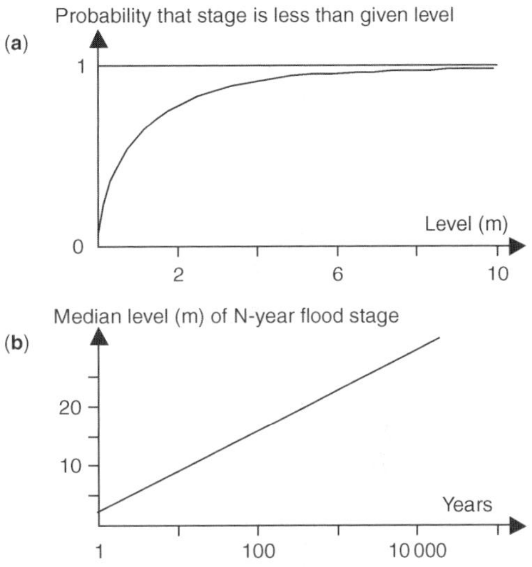

Fig. 3. (**a**) Cumulative statistical distribution of river stage; (**b**) median of maximum flood stage for a given period as a function of period duration in years.

Data for present-day analogs may yield, for example, a value of λ of approximately 0.35. The graphs for the cumulative exponential distribution and for the expected maximum flood stage for this particular example are given in Figure 3. In one year, the expected maximum flood would be 2 m. If we simulate a period of 10 years, the expected maximum flood would be 8.56 m (in the sense that, for any 10-year period, there is an equal chance that the 10-year maximum may be higher or lower than this value). If we simulate 1000 years it would be 21.7 m. For much longer periods of time (greater than 10 000 years) the value gets larger, but climatic variations would have to be taken into account.

Time scale and spatial scale are closely related. Random variables that are spatially correlated, such as rainfall, require statistical tools for determining the proper input to a model. Daily rainfall over an entire continent, for example, shows relatively little percentual variation day after day. Daily rainfall at any single point, however, typically shows large daily variations and a completely different statistical distribution (Linsley *et al.* 1975). Changing the areal extent of a model, therefore, may require changes to the time distribution of its input. The 'correct' input for deterministic models must often be based on statistical assumptions and is highly dependent on the time and space scales represented.

Extreme sensitivity due to non-linear dynamics

Another need for statistical approaches in deterministic modelling arises from the complex behaviour of some deterministic models. A model may be deterministic in that a set of input variables and the numerical methods built into the program would lead to a single possible outcome after a given simulated time. Repetitions of the experiment always yield the same result. However, highly non-linear systems often exhibit pseudo-cyclic quasi-random behaviour called 'chaos' (Schuster 1975; Slingerland 1990; Crutchfield *et al.* 1994). This property may cause modelling runs that differ infinitesimally from one another in input conditions to differ significantly in output (a classic example is the idea that changing one grain of sand in a meandering river causes a significantly different channel path after many years have gone by). This behaviour is not complete disorganization, but rather unpredictability of future states of the system, because of the amplification of small perturbations.

The cross-sections of the 3-D model in Figure 4a shows a turbidite fan created by 100 successive flows (each represented by a single layer in cross-section) and yielding a complex sand–shale interfingering, even though sea level, flow and sediment supply were the same for every flow. The dynamic of the system has two significant effects: (1) the model produces alternating deposits, in a pattern that is not cyclic (it never repeats itself exactly), but is not totally disorganized either; (2) a small change in the input changes the output relatively dramatically, as explained below.

Figure 4b shows another run of this model that differs from the first one very slightly in input conditions (by 0.1% in the volume of each of the 100 turbidite flows represented). The behaviour of the model is initially similar in both cases (the first two or three flows), but later the state of the system starts to differ significantly between the two runs. However, the final overall 'pattern' of behaviour (e.g., average meander wavelength, sinuosity etc.) is still similar in both cases. Another illustration of this behaviour is given by the output of the simple fluvial-modelling program explained below and shown in Figure 9. Though neither of these models have been rigorously proven to possess the property of chaos, they do show both a quasi-cyclic, not externally forced evolution, as well as extreme sensitivity to initial conditions that makes strict prediction difficult, despite their deterministic nature.

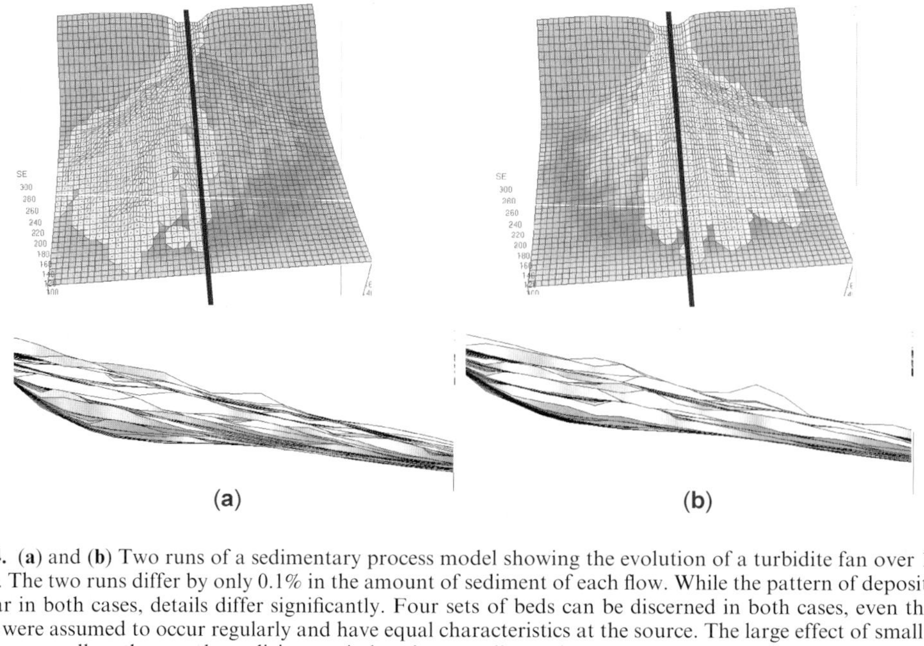

Fig. 4. (**a**) and (**b**) Two runs of a sedimentary process model showing the evolution of a turbidite fan over 10 000 years. The two runs differ by only 0.1% in the amount of sediment of each flow. While the pattern of deposition is similar in both cases, details differ significantly. Four sets of beds can be discerned in both cases, even though flows were assumed to occur regularly and have equal characteristics at the source. The large effect of small input changes as well as the pseudo-cyclicity not induced externally are both characteristic of complex non-linear dynamic systems.

Such behaviour may be unwelcome by modellers eager to carefully adjust input conditions to match a given data set (e.g. well data). It makes it impossible to use inversion techniques to adjust a model. However, it is an unavoidable property of many natural systems, and of numerical systems that closely imitate them. It is also the root cause of much of the variability that we see in sedimentary deposits, particularly at the sub-basin scale. Although it may be a hindrance to useful practical applications, it is appears to be an intrinsic property of most clastic sedimentary systems and their deposits.

Conditioning

When used for characterizing the geometry and properties of present-day sedimentary sequences from seismic, well log and perhaps sparse outcrop information, sedimentary process models have the advantage over other interpolation and prediction techniques in that they are driven quantitatively by known physical and geological principles. Their disadvantage lies in the fact that it is not easy to condition the output of these models to match observations in detail. While the conditioning problem has not been solved in general, there are various techniques that have been used or are being researched which can help geological process models generate useful practical applications. Some of these techniques are described below.

'Manual' adjustment

The simplest way to adjust a model to observations is to perform repetitive runs adjusting the model's parameters so that the output matches observations. The choice of variables to adjust, as well as the direction and amount by which they are adjusted, is simply dictated by personal experience. While this method is decidedly primitive and intuitive, it is widely used in practice, not only in siliciclastic sedimentation modelling, but also in basin modelling, structural modelling, hydrocarbon reservoir simulation and hydrogeological modelling. Even when a good match is found, there is no certainty that the model is unique. Still, the modelled output represents one plausible scenario, and it may give a good indication of the extent, geometry, internal architecture and other properties of the actual deposit. Also, the method is not as time consuming as it may appear, because a user familiar with the sedimentary environment being modelled and with the computer program itself can make intelligent input adjustment decisions more efficiently than most automatic methods.

Minimization

The process of trial and error can be formalized by defining an error or mismatch between observations and model results. This error is then treated as a function (the 'objective' function) of the model's input parameters. The goal is to find the combination of parameters that minimizes the function. One possible error function could be, for example, the sum of the squares of the differences between values of an observed variable (e.g. porosity of a formation at several locations) and the values of the variable as predicted by the model. Finding the minimum of this function yields the best fit to observations. Fast determination of the derivatives of the objective function at any point would facilitate the process. Unfortunately, this is almost always impossible in complex siliciclastic process models. The rate of change in the objective function must be evaluated by finding additional function points, with each point requiring a complete run of the model. Although a large variety of mathematical methods exists to treat this type of problem (Brent 1973), the procedure is computationally expensive. It has been applied successfully in simple models of up to two dimensions. Models that are based on diffusion (Hanks *et al.* 1984), even if three-dimensional, are good candidates for minimization methods, because the output is a relatively 'smooth' function of the input. However, the number of variables and computational expense of more complex models precludes its widespread application.

One indirect advantage of minimization is that it provides information on the characteristics of the objective function. If the model properly represents reality, then there should be a set of input variables that reduces the error function to zero. Figure 5a represents this case for a single variable. Often the minimum is not zero. The error cannot be reduced further by changing the model's input (Fig. 5b). The residual mismatch probably indicates inadequacy of the model to represent the processes involved in producing the observed data. Also, the minimum could be relatively 'flat' (Fig. 5c). A change in the input variables neither reduces nor increases the error, indicating that the model is relatively insensitive to that combination of variables.

An even more common occurrence is that the objective function has many local minima (Fig. 5d). While mathematical tools to attack this case exist, they only add to the computational expense of the minimization process. An extreme case of the 'multiple local minima' occurrence

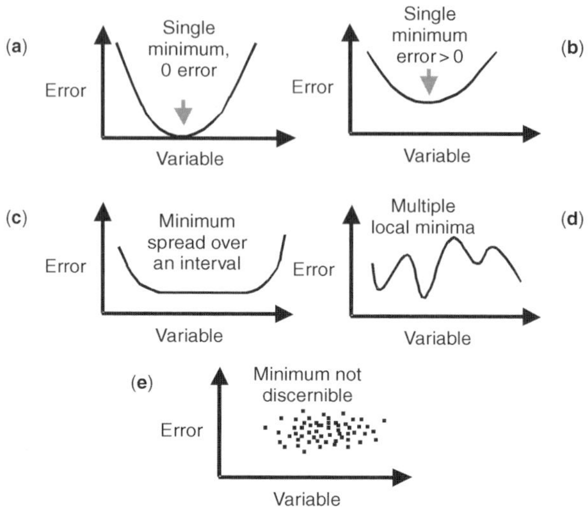

Fig. 5. (**a**) – (**e**) Schematic illustration of possible error functions showing a single model-input variable on the horizontal axis and the corresponding mismatch or error relative to hypothetical field observations on the vertical axis.

often takes place in siliciclastic models (Fig. 5e) when the model embodies highly non-linear interaction between variables. Mathematical and physical systems that are truly chaotic have been shown to produce large departures in model state when varying the input by even a small amount anywhere in the input range (Lorenz 1963*a*, *b*), thereby yielding an error function of the type shown in Figure 5e. This error function, for practical purposes, is similar to that of a random model. Poincaré (1908) first described this property of certain complex dynamic systems as follows:

> A very small cause that escapes our notice determines a considerable effect that we cannot fail to see, and then we say that the effect is due to chance . . . It may happen that small differences in initial conditions produce very great ones in the final phenomena. A small error in the former will produce an enormous error in the latter.

In siliciclastic sedimentary models, this situation may occur when trying to condition a model that generates a meandering river so that the river passes through one point at a given time. Assume that only the regional slope is used to try to minimize the error. Even the smallest change in slope might cause the river to follow a completely different path after a long simulation time. Even though the error expressed as a function of slope is mathematically a continuous function, in practice (numerically and computationally) it behaves as a function that is discontinuous at every point. No conventional search mechanism will indicate in which direction or by how much the input should be changed to produce a better match. One is better off trying random values of the input.

Exhaustive search

Given a complex dynamic model, such as that of a meandering river, one could follow exactly the strategy mentioned above, namely search randomly in the vicinity of a plausible combination of variable values for inputs that produce good matches with observations. This becomes *de facto* a Monte Carlo technique. Even though the model may be completely deterministic, its input is varied randomly, producing a large set of outputs that can be treated statistically.

One drawback to the method is that it is computationally expensive: many runs are necessary, and conditioning essentially consists of picking the right matches. The more data that the model has to honour, the harder it is to find matches. The search, however, does not always require a run for every outcome. Because there is usually uncertainty as to the spatial location of the conditioning point, or points, in relation to the boundaries of the model, a single model run could be searched spatially to find areas that appear to match the data. Specifically, one can translate or rotate a particular outcome of a model to match the observations, because the

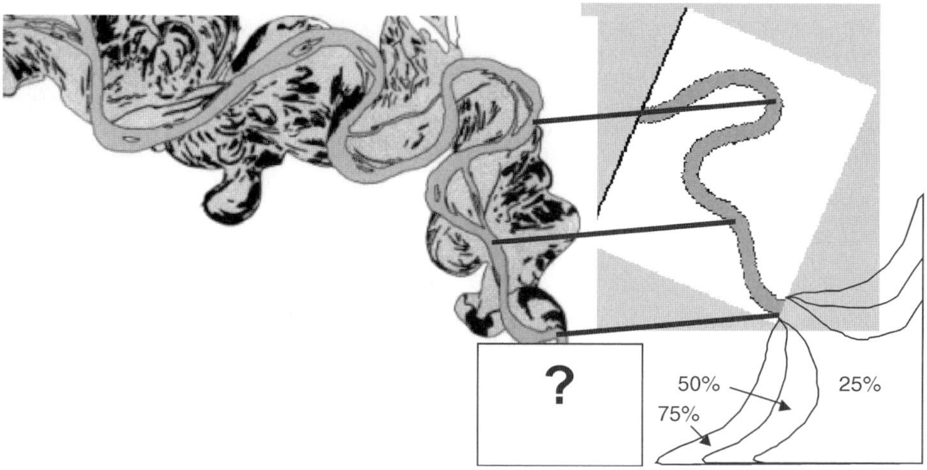

Fig. 6. Method of estimating probabilities of channel position based on exhaustive search of partially matching simulated channels.

absolute position and orientation of the model relative to the wells is not known.

To illustrate this point, assume we have data showing a meandering channel (Fig. 6). We want to use a deterministic forward model to try to find the channel to the southeast (question mark in Fig. 6) of the data. The model can first be adjusted to match the general characteristics of the meandering channel, and then many runs could be made in Monte Carlo fashion. In each run, the entire length of the river could be searched for portions that match the data. Every matching portion (within a given tolerance) is considered an outcome that matches data. The set of successful outcomes is then used to draw probabilistic conclusions. In Figure 6, this has been done with a simple meandering river model that produced a set of 20 matching outcomes for one portion of the known channel. Outputs were considered to be 'matching' if they differed from the conditioning channel portion by less than twice the channel's width throughout the portion's length. For a point away from the data, the probability of finding the channel is simply derived by treating the matching outcomes as a sample set. Figure 6 shows curves of equal probability of finding the channel in the area in question. Although common sense may yield an equally good result in this simplified case, the method may be used in more complex cases, and it permits rigorous treatment of probabilities that are often needed for risk analysis in industry and all too often left to wild guesses.

This technique helps the conditioning to a small number of data points (two or three points, or a small stretch of channel), but it may still be computationally too expensive or downright unfeasible for matching a large number of data points.

Statistics and geostatistics

Another approach to the problem of rough objective error functions is to relax the definition of mismatch to include only differences in the statistical characteristics of observations and model predictions. This may produce a 'smoother' objective function whose minimum (or minima) may be found by standard techniques. For example, in the case of modelling turbidite fans, the error could be defined as the squared difference between modelled and observed *average* flow deposit thickness plus the squared difference between modelled and observed shaliness (each appropriately weighted). Such an error function is typically very 'smooth', allowing rapid manual or automatic optimization. However, the result will only match the statistical characteristics included in the error function, and not the data at specific points, which reduces the usefulness of this method for many practical problems.

Another statistical approach to the problem of exact data matching with deterministic models of unpredictable behaviour is to exploit the spatial variability of the predicted model outcome. Existing geostatistical conditional simulation techniques can produce outcomes that match data exactly, taking spatial variability (variogram) information from the output of a process model. This use is an ideal symbiosis between geostatistics and deterministic modelling: the process model provides the geostatistical model with variability information (often hard to obtain, especially horizontally between wells) and geological constraints, while the geostatistical method provides a means to honor data (Tetzlaff 1991). A use of this technique is illustrated in Figure 7. The upper section (Fig. 7a) shows the outcome of a model that generates a sand/shale fluvial sequence (the model is similar to the fluvial model described below (see 'Controlled Non-linear Dynamics'). The lower part (Fig. 7b) shows a simulated section that was generated by a geostatistical method using statistical spatial variability information obtained from the upper section and conditioned to the two wells at the ends of the section. The intermediate wells have been placed for comparison.

One drawback of this technique is that geostatistical methods may not capture the complex variability produced by sedimentary process models, thereby 'wasting' some of the information produced by the forward model. Although this is not the case in the example of Figure 7, the problem is shown by Figure 8. The upper part (Fig. 8a) shows a section through modelled meandering river deposits. The lower section (Fig. 8b) shows an unconditioned geostatistical simulation in which the variogram was obtained from the upper section. Both sections have exactly the same variogram. Clearly, the lower section does not reproduce the U-shaped channel pattern observed in the upper section. Furthermore, the hydraulic characteristics of the deposits are different in each case, even though they have the same spatial variability as expressed by a variogram.

A promising improvement to this situation may lie in the use of geostatistical methods that capture more complex variability than the conventional variogram (Doyen *et al.* 1989; Caers & Zhang 2002; Arpat *et al.* 2002; Strebelle *et al.* 2002). These methods require a large number of data, often not available from the field, particularly where only a few wells are available. Their combination with forward modelling would let each method provide what the other lacks (the deterministic model would provide the outcomes based on physical processes, and the statistical model would provide the ability to match data). Research in this area is still vastly open.

Controlled non-linear dynamics

The extreme sensitivity to initial conditions exhibited by complex non-linear dynamic models hinders their ability to match observations. At the same time, this characteristic implies that extremely small adjustments to the input conditions suffice to 'steer' the model in

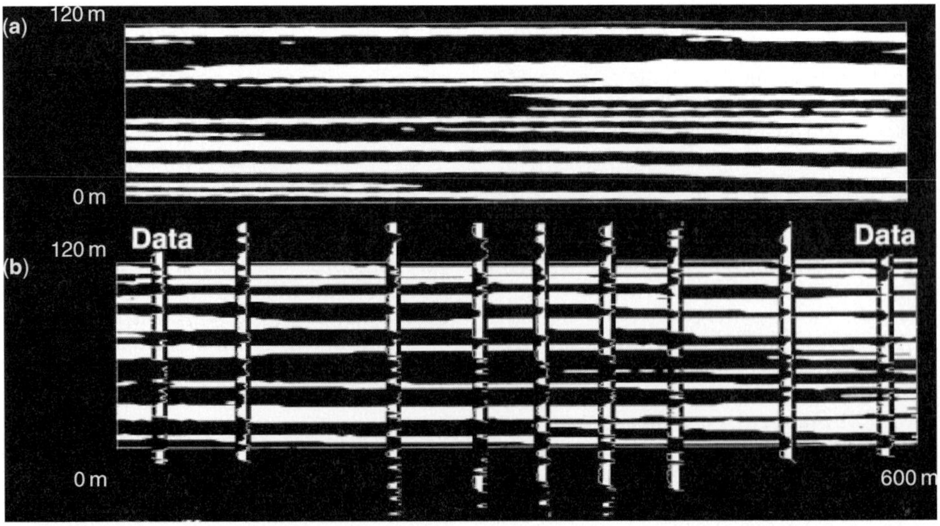

Fig. 7. (**a**) Cross-section showing sands (white) and shales (black) obtained by modelling a braided-river sequence with a fluvial forward model. (**b**) Geostatistical conditional simulation output conditioned to the two wells labelled 'Data'. Logs shown are resistivity well logs from an actual densely drilled shallow oilfield. All wells between the two extremes were not used in the simulation, and were placed to show the plausibility of the correlations found by the method. Extent of sections: vertical, 120 m; lateral, 600 m.

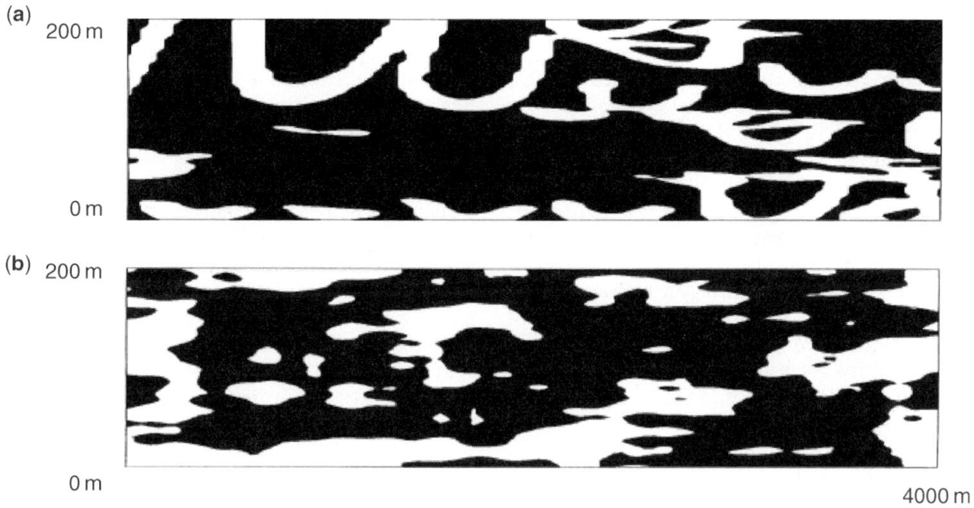

Fig. 8. (**a**) Outcome of a fluvial forward model representing a meandering-river sequence (U-shaped structures are not individual channels but meandering-channel sets migrating across the plane of the section). (**b**) Geostatistical conditional simulation generated using a variogram extracted from the upper section. White, sands; black, shales. Although both sections have exactly the same variogram, they differ significantly in overall appearance and in hydraulic properties, thus evidencing the limitations of variogram-based methods as a complement to sedimentation forward models. Extent of sections: vertical, 200 m; lateral, 4000 m.

the desired direction so as to cause it to match specific data. Furthermore, the adjustment can be small enough not to significantly affect the adherence to the known paleogeographical conditions. The problem is to find out exactly what variables must be adjusted and by what amount. In other words, although we may know that changing one grain of sand in a river could significantly alter its course thousands of years later, we do not know exactly what change has to be applied to a sand grain in order to achieve the desired effect. Although some successes in the area of conditioning siliciclastic systems have been achieved for random walk models (Karssenberg *et al.* 2001), purely deterministic non-linear models of clastic sedimentary systems that exhibit pseudo-random behaviour remain difficult to condition.

Hope may lie in advances in the study of controlled non-linear dynamics. It has been known for some time that some mechanical and mathematical chaotic systems will become periodic when perturbed by even a minute periodic input (Ott *et al.* 1990; Langreth 1991; Shinbrot *et al.* 1993). Also, methods have been developed to synchronize two similar dynamic systems initially evolving along different paths (Lai & Grebogi 1994). Can we use these methods to condition models of natural systems?

To illustrate and study control in a siliciclastic system, we have used an extremely simple planoform, geometric model of a meandering river. This is based on similar principles of several previous meandering-river models (Ikeda *et al.* 1981; Ferguson 1984; Howard 1984; Howard & Knutson 1984; Sun *et al.* 1996). Our model assumes that a river can be represented by a line that moves sideways at a rate proportional to its curvature immediately upstream. This is justified by the fact that channel curvature promotes helical flow and excess shear stress on the outside of the bend, both of which cause the channel to shift. Simultaneously, the general valley slope tends to straighten the channel in the downslope direction. Mathematically, the river is represented by a parametric curve $X(P, t)$, where X represents a point on the plane, P is a positional parameter which increases uniformly along the river's path, and t is time. At a point $X(P, t)$, the river is assumed to shift laterally according to the following equation:

$$r = \int_0^P [ck_c \exp(-q_c(P-p)) + (1-\cos(\alpha)) - sign(\sin(\alpha))k_s \exp(-q_s(P-p))]dp \tag{3}$$

where:

r = rate of lateral channel movement (positive to the right of channel direction)

c = signed curvature (inverse of curvature radius, positive when turning left)
k_c = constant, rate of decay of curvature effect
q_c = constant, amplitude of curvature effect
α = angle between general valley slope and channel heading
k_s = constant, rate of decay of slope effect
q_s = constant, intensity of slope effect (dependent on valley slope angle).

Sharp kinks in the river path can be handled by the equation because, although the curvature may be infinite at a point, it occurs over zero distance, and the formula works with an integral of curvature over distance. In addition to this equation, meander avulsion or 'capture' is modelled by assuming that, when the river shifts in such a way that it intersects itself, cut-off occurs. The channel portion between the two intersection points is abandoned and the channel is retraced, skipping the abandoned portion. The channel is assumed to have a finite width, and 'intersection' occurs as soon as the margins touch. It is also possible to assume an aggradation rate and make the model fully three-dimensional (albeit very simple).

The initial condition consists of any non-intersecting river path and values for each one of the four constants. A few additional details are needed for a complete mathematical definition, but this simple description captures the main aspects of the model and allows its implementation as a computer program. This model exhibits characteristics typical of more complex siliciclastic models and of natural systems themselves, including pseudo-random behaviour and extreme sensitivity to initial conditions throughout the range of possible states.

In this model, an initially straight channel will remain straight forever, but even the slightest departure from a straight line causes the river to start meandering. One way to show the behaviour of the model is to display the evolution of a pair of variables through time: position across regional slope of one point in the channel, and channel direction at that point. These two variables constitute a small subset of the 'state space' of the system (the set of all variables that define the model's state at a given time), but suffices to illustrate the model's behaviour in a qualitative way. Figure 9 shows the evolution of a river (shown in plan in Fig. 9a) accompanied

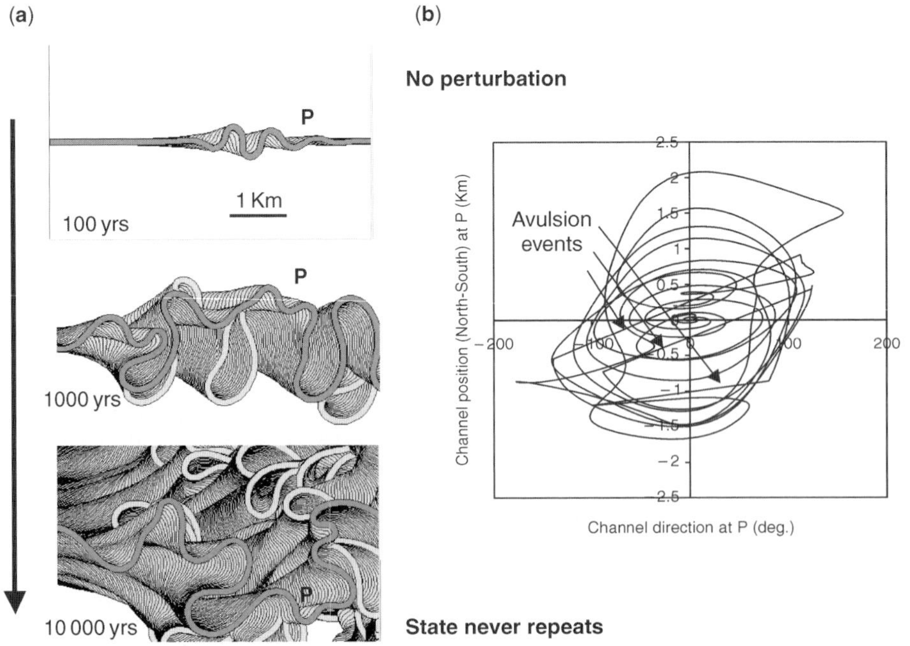

Fig. 9. (**a**) Successive plan views of a meandering channel modelled by the planoform river program (flow is from left to right). Dark grey, active channel; light grey, previous abandoned channels. (**b**) State diagram showing the evolution of the system by means of two variables: channel heading and channel position in the north–south direction (across the regional slope direction). Straight segments in the state diagram correspond to jumps in the system's state caused by meander avulsion.

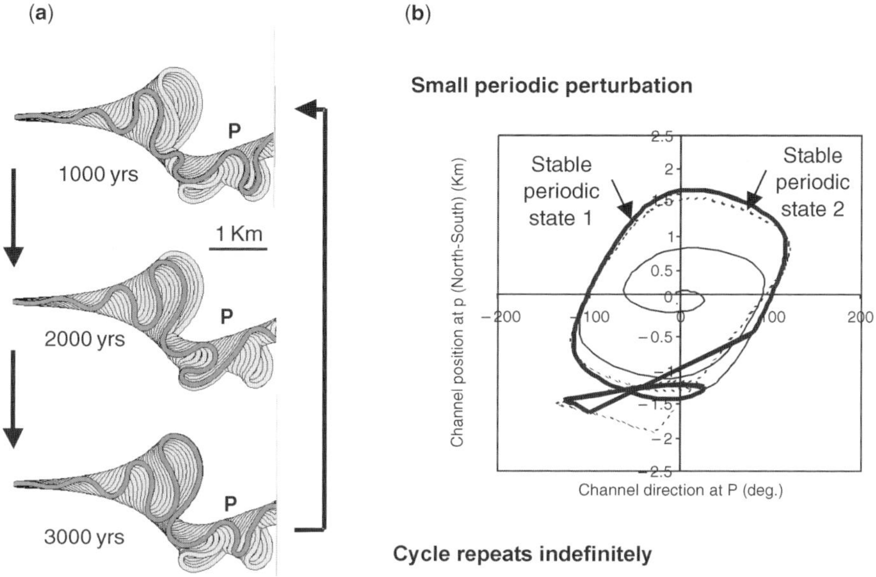

Fig. 10. (**a**) Plan views and (**b**) state diagram similar to those in Figure 9, but obtained after introducing a small periodic perturbation in the channel direction at its source. The perturbation consists of an oscillation of 1 degree in channel direction over a channel stretch of 50 m, with a period of 150 years. Although such perturbation is barely physically significant, it is sufficient to lock the system into a periodically repeating state (thick line, state 1). A different perturbation (with a period of 140 years), locks the system into a slightly different periodic state (dotted line, state 2).

by a plot showing the system's evolution (Fig. 9b). Along the x axis, the plot shows the heading (compass direction) of the river channel at a point that in the channel. Along the y axis, the plot shows the position of the point in the north–south direction, or across the regional slope, which is assumed to be towards the east. Time increases along the path of the curve (although time is not labelled on the plot).

The system evolves from the centre of the plot (straight channel), first through oscillations of increasing amplitude and then through a path that, though never repeating, remains within a confined area of the plot. The plot contains occasional discrete discontinuities or jumps in the path because of the built-in meander avulsion mechanism.

Two runs that differ by extremely small amounts in any of the input conditions will be similar at first, but quickly diverge into separate trajectories. The introduction of any number of small perturbations changes the evolution of the system, but causes no change in the overall appearance and pattern of behaviour. Furthermore, the introduction of a white-noise perturbation of small amplitude in the position or direction of the channel at one point does not affect the overall picture. In other words, although the model is very sensitive to changes in initial conditions, the statistical characteristics of the outcome (average meander amplitude and wavelength, cut-off rate) appear to be rather insensitive to the introduction of random perturbations.

The scene changes when a periodic perturbation (even of very small amplitude) is introduced. When varying the position of one point sinusoidally, for example, at first there is no obvious departure from chaotic behaviour, but later the model locks into a periodic pattern. Figure 10 shows such a pattern when the channel direction is shifted at the source by a sinusoidal perturbation of amplitude 1 degree and period 150 years. The initial trajectory of the system (before it locks into a periodic pattern) is shown in Fig. 10b by a thin continuous line. The system quickly becomes periodic (thick continuous line labelled 'Stable periodic state 1'). The period of the systems state is a multiple of that of the perturbation.

Achieving periodic behaviour is not an accomplishment in itself, as natural rivers do not repeat their pattern exactly after a fixed number of years. The significance lies in that the response of the system to a small change in the periodic perturbation (e.g. a slight change in period or amplitude of the perturbation) now causes a relatively small change in the periodic trajectory. In this example, when the period of the perturbation is decreased from 150 to 140 years, the closed trajectory of the system is slightly but significantly changed (Fig. 10b, dotted line labelled 'Stable periodic state 2'). This should allow the use of traditional minimization techniques that rely on continuity of the objective function and its first derivatives to solve the conditioning problem. Thus, if we wish to match the channel position and direction with the model, provided that the desired position and direction represent a point that lies within the possible sets of the system, they can presumably be matched by finding the right periodic perturbation through a conventional minimization technique. The perturbation can be applied for the relatively short period of time that the system takes to lock into the trajectory that causes it to pass through the desired point.

Nonetheless, this result is so far of theoretical interest only, and many unanswered questions remain in this ongoing line of research. For example:

> How can we find all possible perturbations that cause the trajectory to pass through the desired point in order to obtain a statistically significant sample of outcomes?
>
> What is the effect of superimposing periodic perturbations of different amplitudes and periods?
>
> Will this be equally simple if the number of variables represented is extended (e.g. if many points need to be matched?

Further research on these and other issues could lead to a practical and rigorous method.

Even if a satisfactory and theoretically justified procedure to use this approach is found for this particular model, the problem may still be formidable for more complex systems. From this research, however, it is clear that, at least in this simple model, this method greatly reduces the number of model runs necessary to obtain a match with observations by converting the method from a blind search to a forward minimization procedure. While rigorous prediction may be impossible in complex non-linear models of physical systems (as envisioned by Poincaré nearly a century ago), conditioning through control may be feasible.

Presently, the choice of technique for conditioning a model should ultimately be determined by the nature of the problem to be solved. One should not model for the sake of modelling, but to achieve practical goals, ranging from an understanding of a natural system to effective exploitation of natural resources.

Conclusions

Deterministic forward models of siliciclastic processes must incorporate phenomena that occur at a wide range of spatial and temporal scales. The scale and resolution of the model must be adequate to represent not just the desired sedimentary structures, but also the phenomena that originate them.

Quasi-periodic behaviour may arise in a deterministic siliciclastic model in the absence of external forcing, due to complex non-linear interaction between internal variables.

Complex non-linear models of siliciclastic systems may also exhibit extreme sensitivity to initial conditions, which makes prediction and inversion difficult.

Although complex siliciclastic forward models may be difficult to condition to hard data, several proven techniques help approach this goal to the extent of making siliciclastic models useful for solving practical problems.

Variogram-based geostatistics do not always capture the complex variability provided by sedimentation models. Therefore, advanced geostatistical techniques (such as 'multiple-point' geostatistics) may be necessary in order to combine a process-based deterministic model with a data-honouring geostatistical method.

In the long term, research on control in non-linear dynamic systems may yield a rigorous way to condition deterministic models without resorting to geostatistical methods.

References

ARPAT, B., CAERS, J. & STREBELLE, S. 2002. Feature-based geostatistics: an application to a submarine channel reservoir. *Proceedings of the Society of Petroleum Engineers 68th Annual Conference and Exhibition, San Antonio, Texas, Paper* **77426**.

BRENT, R. P. 1973. *Algorithms for Minimization Without Derivatives*. Englewood Cliffs, New Jersey, Prentice Hall Series on Automatic Computing.

CAERS, J. & ZHANG, T. 2002. Multiple-point geostatistics: a quantitative vehicle for integrating geologic analogs into multiple reservoir models. In: GRAMMER, G. M., HARRIS, P. M. & EBERLI, G. P. (eds) *Integration of Outcrop and Modern Analog Data in Reservoir Models*. American Association of Petroleum Geologists, Memoirs, **80**, 373–394.

CROSS, T. A. (ed.) 1990. *Quantitative Dynamic Stratigraphy*. Prentice Hall, Englewood Cliffs, New Jersey.

CRUTCHFIELD, J. P., FARMER, J. D., PACKARD, N. H. & SHAW, R. S. 1986. Chaos. *Scientific American*, **255** (6), 46–57.

DENNIS, J. E. & SCHNABEL, R. B. 1983. *Numerical Methods for Unconstrained Optimization and Nonlinear Equations*. Prentice Hall, Englewood Cliffs, New Jersey.

DOYEN, P. M., GUIDISH, T. M. & DE BUYL, M. H. 1989. Seismic determination of lithology in sand/shale reservoirs, a Bayesian approach. *Proceedings of the 1989 Convention of the Society of Exploration Geophysicists*, Dallas, 719–722.

FERGUSON, R. I. 1984. Kinematic model of meander migration. *In:* Elliot, C. M. (ed.) *River Meandering*. Proceedings of the 1983 Conference on Rivers, 1983, New Orleans, American Society of Civil Engineers, 942–951.

FRANSEEN, E. K., WATNEY, W. L., KENDALL, C. G. ST. C. & ROSS, W. (eds) 1991. *Sedimentary Modelling, Computer Simulations and Methods for Improved Parameter Definition*. Kansas Geological Survey Bulletin, **233**.

GRIFFITHS, C. M. 2001. SEDSIM in hydrocarbon exploration. *In:* MERRIAM, D. F. & DAVIS, J. C. (eds) *Geologic Modelling and Simulation, Sedimentary Systems*. Kluwer Academic/Plenum, New York.

HANKS, T. C., BUCKNAM, R. C., LAJOIE, K. R. & WALLACE, R. E. 1984. Modification of wave-cut and faulting-controlled landforms. *Journal of Geophysical Research*, **89**, 5771–5790.

HARBAUGH, J. W., WATNEY, W. L., RANKEY, E. C., SLINGERLAND, R., GOLDSTEIN, R. H. & FRANSEEN, E. K. (eds) 1999. *Numerical Experiments in Stratigraphy: Recent Advances in Stratigraphic and Sedimentologic Computer Simulations*. SEPM (Society for Sedimentary Geology), Tulsa, Special Publications, **62**.

HOWARD, A. D. 1984. Simulation model of meandering. In: ELLIOT, C. M. (ed.) *River Meandering*. Proceedings of the 1983 Conference on Rivers, 1983, New Orleans, American Society of Civil Engineers, 952–953.

HOWARD, A. D. & KNUTSON, T. R. 1984. Sufficient conditions for river meandering: a simulation approach. *Water Resources Research*, **20**, 1659–1667.

IKEDA, S., PARKER, G. & SAWAI, K. 1981. Bend theory of river meanders. Part 1: Linear development. *Journal of Fluid Mechanics*, **112**, 363–377.

KARSSENBERG, D., TÖRNQVIST, T. E. & BRIDGE, J. S. 2001. Conditioning a processed-based model of sedimentary architecture to well data. *Journal of Sedimentary Research*, **71**, 868–879.

KOLTERMANN, C. E. & GORELICK S. M., 1992. Paleoclimatic signature in terrestrial flood deposits. *Science*, **256**, 1775–1782.

LAI, Y. C. & GREBOGI, C. 1994. Synchronization of spatiotemporal chaotic systems by feedback control. *Physical Review*, **E50**, 1894.

LANGRETH, R. 1991. Engineering dogma gives way to chaos. *Science*, **252**, 776–778.

LINSLEY, R. K., KOHLER, M. A. & PAULHUS, J. M. H. 1975. *Hydrology for Engineers*. Water Resources and Environmental Engineering, Series, McGraw-Hill, New York.

LORENZ, E. N. 1963*a*. Deterministic nonperiodic flow. *Journal of Atmospheric Sciences*, **20**, 130–148.

LORENZ, E. N. 1963*b*. The mechanics of vacillation. *Journal of Atmospheric Sciences*, **20**, 448–464.

MERRIAM, D. F. & DAVIS, J. C. 2001. *Geologic Modelling and Simulation: Sedimentary Systems*. Kluwer Academic/Plenum, New York.

OTT, E., GREBOGI, C. & YORKE, J. A. 1990. Controlling chaos. *Physical Review Letters*, **64**, 1196.

POINCARÉ, J. H. 1908. *Science et Méthode*. Flammarion, Paris. *Translated by HALSTEAD, G. B. In: The Foundations of Science*. University Press of America, Washington DC 1982. First edition 1946.

SCHUSTER, R. E. 1975. *Deterministic Chaos: An Introduction*. VCH Publishers, New York.

SHINBROT, T., GREBOGI, C., OTT, E. & YORKE, J. A. 1993. Using small perturbations to control chaos. *Nature*, **363**, 411.

SLINGERLAND, R., 1990. Predictability and chaos in quantitative dynamic stratigraphy. *In:* CROSS, T. A. (ed.) *Quantitative Dynamic Stratigraphy*. Prentice-Hall, Englewood Cliffs, New Jersey.

SLINGERLAND, R., HARBAUGH, J. W. & FURLONG, K. 1994. *Simulating Clastic Sedimentary Basins: Physical Fundamentals and Computer Programs for Creating Dynamic Systems*. PTR Prentice Hall, Englewood Cliffs, New Jersey.

STREBELLE, S., PAYRAZIAN, K. & CAERS, J. 2002. Modelling of a deepwater turbidite reservoir conditional to seismic data using multiple-point geostatistics. *Proceedings of the Society of Petroleum Engineers 68th Annual Conference and Exhibition, San Antonio, Texas Paper* **77425**.

SUN, T., MEAKIN, P. & JOSSANG, T. 1996. A simulation model of meandering rivers. *Water Resources Research*, **32**, 2937–2954.

TETZLAFF, D. M. 1991. The combined use of sedimentary process modelling and statistical simulation in reservoir characterization. *Proceedings of the Society of Petroleum Engineers 66th Annual Conference and Exhibition*, Paper **22759**, 937–945.

TETZLAFF, D. M. & PRIDDY, G. 2001. Sedimentary process modelling, from academia to industry. *In:* MERRIAM, D. F. & DAVIS, J. C. (eds) *Geologic Modelling and Simulation: Sedimentary Systems*. Kluwer Academic/Plenum, New York, 45–69.

WATNEY, W. L., RANKEY, E. C. & HARBAUGH, J. W. 1999. Perspectives on stratigraphic simulation models: current approaches and future opportunities. *In:* HARBAUGH, J. W., WATNEY, W. L., RANKEY, E. C., SLINGERLAND, R., GOLDSTEIN, R. H. & FRANSEEN, E. K. (eds) *Numerical Experiments in Stratigraphy: Recent Advances in Stratigraphic and Sedimentologic Computer Simulations*. SEPM (Society for Sedimentary Geology), Tulsa, Special Publications, **62**, 3–21.

Classical logic and the problem of uncertainty

CYRIL A. PSHENICHNY

Levinson-Lessing Earthcrust Institute (NIIZK), St Petersburg State University, 7/9 Universitetskaya Naberezhnaya, St Petersburg 199034, Russian Federation

Present address: Environmental Engineering Department, St Petersburg State Technological Institute, Technical University, 26 Moskovsky Prospect, St Petersburg 198013, Russian Federation (e-mail: pshenich@kp1306.spb.edu; pshenich@pochtamt.ru)

Abstract: The uncertainty of knowledge, in contrast to that of data, can be assessed by its probability in the logical sense. The logical concept of probability has been developed since the 1930s but, to date, no complete and accepted framework has been found. This paper approaches this problem from the point of view of logical entailment and natural sequential calculus of Classical logic. It is shown herein that probability can be comprehended in terms of a set of formal theories built in similar language. This measure is compliant with general understanding of probability, can be both conditional and unconditional, accounts for learning new evidence and complements Bayes' rule. The approach suggested is practically infeasible at present and requires further theoretical research in the domain of geoscience. Nevertheless, even within the framework of existing methods of expert judgement processing, there is a way of implementing logic that will improve the quality of judgements. Also, to reach the state of formalization necessary to use logical probability, techniques of knowledge engineering are required; this paper explains how logical probabilistic methods relate to such techniques, and shows that the perfect formalization of a domain of knowledge requires them. Hence, the lines for future research should be: (1) the development of a strategy of co-application of existing expert judgement-processing techniques, knowledge engineering and classical logic; and (2) further research into logic enabling the development of formal languages and theories in geoscience.

Among the sources of uncertainty reported in literature, several relate to scientific reasoning and language. Probabilistic methods used to handle them require conditions that are difficult to meet, for example, an absolutely objective and bias-free supervisor of expert judgements, a statistically representative number of experts, or a kit of test datasets and questions that are guaranteed to be previously unknown to the tested experts and to be absolutely perfect themselves (Aspinall & Woo 1994; Aspinall & Cooke 1998).

Another option, in the author's view, is the study of reasoning itself, which is known to be governed by formal rules that are similar for humans and computers. This paper explores the applicability of Classical logic in assessing and reducing various kinds of uncertainty. To approach this goal, it is necessary to:

(1) give an overview of known sources and measures of uncertainty;
(2) investigate how logic can cope with uncertainty;
(3) discuss the role and place of logic among other uncertainty-reducing methods.

Each of these tasks is addressed in this paper.

Sources and measures of uncertainty

Science has developed a wide vocabulary of hesitation (e.g. uncertainty, probability, possibility, inaccuracy, imprecision, fuzziness, error, disagreement) and there is little consensus in the understanding of these terms (e.g. Woo 1999; Virrantaus 2003; Baddeley *et al.* 2004). Nonetheless, to the author's knowledge, most researchers are inclined to use 'uncertainty' as a more-or-less umbrella term for situations in which confidence in a scientific result is lacking. In this paper, it will be used as the general term for all kinds of measurable (at least principally or hypothetically) lack of confidence.

The principle of the complex formal approach to treating information (Pshenichny 2003) seems

From: CURTIS, A. & WOOD, R. (eds) 2004. *Geological Prior Information: Informing Science and Engineering*. Geological Society, London, Special Publications, **239**, 111–126. 186239-171-8/04/$15.00

to serve as an appropriate basis for correct and concise classification of sources and measures of uncertainty. According to this approach, any information about the object of study can be regarded as data or knowledge. *Data* are anything expressed as a singular statement (in which the predicate is related to a singular subject), for example, 'The sample 72039 contains 65 wt% silica'. It is commonly accepted in science that data are the result of observation or measurement. *Knowledge* is anything expressed as a general statement (in which the predicate is related to a general subject), for example, 'The rock [meaning '*all* or *some* of the studied samples'] contains (contain) 65 wt% silica'. As shown in Pshenichny (2003), this distinction is actually context-dependent and what is considered knowledge at a local scale (e.g. the scale of an outcrop) can become data at greater scales (e.g. the scale of a region), and vice versa. Nevertheless, in every particular case, knowledge and data can (and should) be clearly separated. Data and knowlege are then treated by different formal approaches.

Sources of uncertainty that refer to data and means of their processing include: sampling, observational and measurement errors, errors of mathematical evaluation of data and propagation of errors (Bardossy & Fodor 2001); estimation errors, scale issues, ignorance and human error (Bowden 2004); temporal, structural, metric-related and translational uncertainties (Rowe 1988); and possibly some others. A spectrum of probabilistic, possibilistic (fuzzy) and joint probabilistic-possibilistic methods (Bardossy & Fodor 2001), including computer applications (e.g. UncertaintyAnalyzer, see http://www.isgmax.com) can be applied to analyse such sources. Here the probability is meant in a statistical or frequentist sense (Baddeley *et al.* 2004; and Woo 1999, respectively).

Consideration of data-related uncertainty and its measures in more detail is beyond the scope of the present paper. The sources of knowledge-related uncertainty are reasoning and language. Woo (1999) recognized two different kinds of such uncertainty: *conceptual* (*epistemic*) and *aleatory*. Epistemic uncertainty is rooted in the knowledge itself, and aleatory uncertainty is rooted in the belief in knowledge. This can be illustrated by the following examples from Bowden (2004):

> How can we express the likelihood of a new volcano forming in non-volcanic regions or of a new active fault forming in regions far from current active faults?
>
> How can we determine whether or not a volunteer site located in a region outside the 'obvious' exclusion zones for the site selection criteria should be included or excluded from the next investigation stage?

In the former case, for instance, it is implicitly supposed that there is a model or models (called 'probability models' in the literature) saying whether a volcano could form in a given non-volcanic region. Every model has its own intrinsic approximations and associated uncertainty: the epistemic uncertainty. In addition, every volcanologist may trust each model more or less: this is the aleatory uncertainty of a model attributed by individual scientists.

The degree of trust is determined by the researcher's *bias* (e.g. Halpern 1996; Baddeley *et al.* 2004; and references therein). It is accepted that personal or group bias erodes the quality of expert judgements and should be reduced where possible, e.g. by elicitation procedures (see reviews in Baddeley *et al.* 2004; Curtis & Wood 2004). However, bias is a virtue of geology and other largely hermeneutic (interpretative) sciences (e.g. history, psychology, medicine, economics).

> In effect, hermeneutics rejects the claim that facts can ever be completely independent of theory ... we always come to our object of study with a set of prejudgements: an idea of what the problem is, what type of information we are looking for, and what will count as an answer (Frodeman 1995).

Among these prejudgements, Frodeman lists as 'crucial ... and often discounted ... the social and political structures of science'. He acknowledges that exactly these structures and personal preferences, and not a pursuit of absolute truth, route the science.

Once bias is not only unavoidable but even determines the direction of research, then the very idea of decreasing bias becomes questionable, even though there is the obvious involvement of quite diverse circumstances (e.g. from rock composition to standard of living) into any single geoscientific rationale. At best we can substitute personal bias with the most appropriate collective one. Apparently, the most appropriate bias is not the one that is most logically correct or statistically proven, but the one shared by the largest and/or strongest group, for example, the bias that the entire world should adopt democratic values or that humankind should survive; the latter has given ground to the whole study of risk assessment. In general, science seems to have only two options: (1) to ignore bias and fight for objective knowledge, as prescribed by analytical philosophy, or (2) to follow the Continental philosophy apprehension and accept bias, reveal, assess and manage it, but not try to decrease it. This point will be dealt with below.

Discussing present measures of uncertainty, Woo (1999) considers three notions of probability: (1) *frequentist*, which relates to data in the classification above; also termed 'objective' (Baddeley *et al.* 2004) or statistical (Carnap 1950); (2) *subjective* (Baddeley *et al.* 2004); and (3) *logical*. Bardossy & Fodor (2001) add fuzzy measures.

The measure of both epistemic and aleatory uncertainty in most cases is subjective or epistemic probability (Woo 1999; Fitelson 2003), i.e. probability formulated in probabilistic judgements (Baddeley *et al.* 2004). Fitelson (2003) states:

> On epistemic interpretations of probability, $Pr_M(H)$ is (roughly) the degree of belief an epistemically rational agent assigns to H, according to a probability model M of the agent's epistemic state. A rational agent's background knowledge K is assumed ... to be 'included' in any epistemic probability model M, and therefore K is assumed to have an unconditional probability of 1 in M. $Pr_M(H \mid E)$ is the degree of belief an epistemically rational agent assigns to H upon learning that E is true (or upon the supposition that E is true ...), according to a probability model M of the agent's epistemic state ... So, roughly speaking, (the probabilistic structure of) a rational agent's epistemic state evolves (in time t) through a series of probability models $\{M_t\}$, where evidence learnt at time t has probability 1 in all subsequent models $\{M_{t'}\}$, $t'>t$.

A probabilistic framework for this evolving state of knowledge in the geosciences is described in Wood & Curtis (2004).

However, neither the rationality of the agent (i.e. the expert) nor the structure of K, M and E are explained by Fitelson (2003), and involvement of time (t) brings a healthy spirit of physiology into mathematical consideration, because evolution of epistemic state through time would depend, *inter alia*, on the expert's metabolism which determines the rate of digestion of new evidence.

The main criteria of quality of probabilistic statements (and, hence, the probability of subjective probability values) are qualification and objectivity of experts, as, for example, in the approach of Cooke (1990). In this approach, applied by Aspinall & Woo (1994) to assess the expert's qualification and objectivity, each expert is asked some test questions. It is supposed that correct answers for these questions are somehow known, and any other answer can be calibrated against these. Discussing the possible ways to obtain such 'bullet-proofed' answers, Aspinall & Cooke (1998) desire 'an experienced technical facilitator' of expert judgement elicitation procedure. The question remains unanswered as to who evaluates his or her answers.

One of the weakest points of subjective probability, to the author's mind, is that it is rather artificially and voluntarily normalized to unity. Probability, be it frequentist or any other form, requires the condition of additivity (e.g. Bardossy & Fodor 2001) that implies mutually exclusive cases or objects (populations). This readily entails normalization as the ratio n/N, where n is the number of cases or objects for which something is true, and N is the total number of said entities, for example, number of tails v. total number of tossings of a coin. However, if one expert says that he feels A has the probability of 0.7 while another expert gives 0.6, or if an expert is asked to give a probability distribution based on his intuition, it is unlikely that this can be considered 'true' normalization.

Nevertheless, this fits fairly well with the concept of a membership function of fuzzy logic. This has made Bardossy & Fodor (2001) express a preference for the use of possibilistic rather than probabilistic approaches to processing expert judgements. One problem, however, is that these judgements are commonly used, firstly, in Bayesian approaches for prior probability values where no distinction is made between them and frequentist values, and secondly, in probability trees where they are processed according to the theorems of probability theory. Hence, there is a need for probabilistic, not just fuzzy, estimation of knowledge-related uncertainty. As was pointed by Aspinall & Woo (1994):

> It has been shown theoretically ... that there is no essential distinction between probability assignments based on numerical frequencies and those based on (general – *CP*) judgments: both can be incorporated into a computation of hazard (or any other item of interest – *CP*) if the correct procedures are adopted.

Further pursuit of these procedures should be considered an important task for future research.

A possible alternative might be one which has been rejected with enthusiasm by the proponents of probabilistic and then fuzzy thinking in geoscience: the deterministic model. Once knowledge is not data, probability assessments based on general judgements must be of a different nature than those based on frequencies. On first sight, a 'deterministic probability model' is nonsense. Nevertheless, Woo shows that even the most strict and objective knowledge has, or may have, its own genuine uncertainty. The simplest example is division by zero in arithmetic. Woo also quotes: the Fourier expansion in a Hamiltonian system describing the planet orbits, which, in some cases, also reduces to a necessary division by zero; Heisenberg's principle of uncertainty in quantum mechanics; deterministic chaos in meteorology; and other examples where the uncertainty 'emerges from the midst of determinism' (Woo 1999, p. 72). Perhaps, one of the sources of uncertainty in geoscience cited by Bowden (2004), that of modelling constraints, also falls in this field.

The measure of this 'deterministic sort' of epistemic uncertainty, by Woo, is the third and last kind of probability that he lists: logical probability. Likewise, Virrantaus (2003, fig. 1) mentions among the sources of uncertainty the 'validity of the model', which includes, *inter alia*, 'the logic of algorithm'. However, in contrast to frequentist and subjective notions of probability considered in much detail, neither of these authors explain *how* logical probability can work. Woo only mentions the principle of indifference (or equiprobability) in absence of additional information.

The logical concept of probability has been developed since the 1930s by Johnson (see references in Fitelson 2003) Carnap 1950 and later works, and others. To evaluate this concept, Classical logic must be introduced. This is discussed in the next section.

To summarize this section, using the complex formal approach of Pshenichny (2003), uncertainty can be classified as that of knowledge and that of data. The uncertainty of data can be considered: to satisfy additivity and be measured by probability in the frequentist sense; to be 'fuzzy' and be measured by a variety of fuzzy parameters; or both. Reasoning and language contribute to the uncertainty of knowledge: the epistemic uncertainty and its refinement, the aleatory uncertainty. As for the measure of epistemic uncertainty, if one approaches knowledge intuitively, the measure may only be fuzzy; if one treats knowledge rationally the measure, if it exists, should be probabilistic in a logical, but not in a subjective, sense.

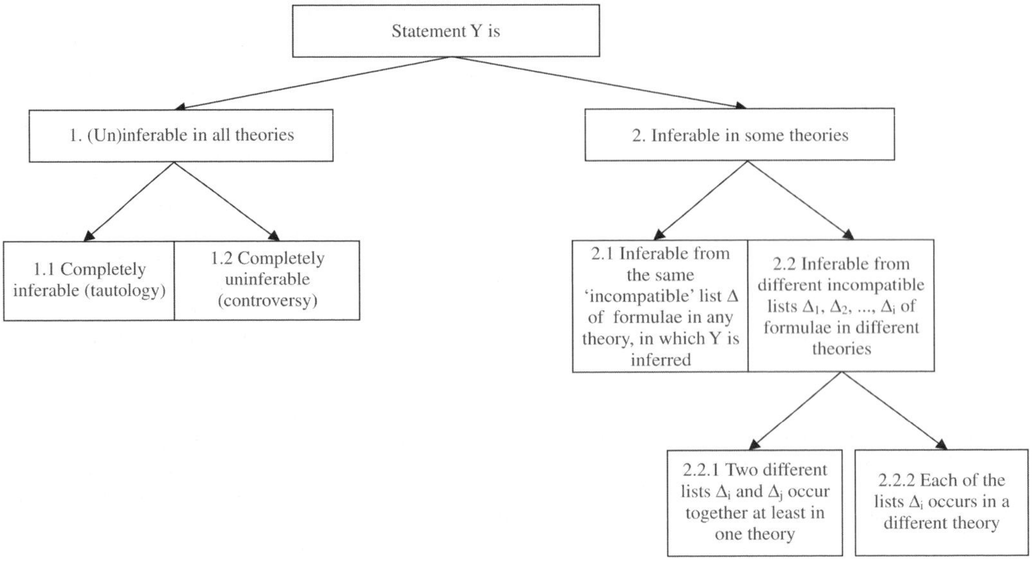

Fig. 1. Inferability of a statement Y in a set of theories built in a similar language (see comments in the text).

The study of logic and its relation to uncertainty

Classical logic, first described by Aristotle, was elaborated in its modern form in the second half of the nineteenth to the first half of the twentieth century by Frege (1896), Whitehead & Russell (1910), Kleene (1952), Hilbert & Bernyce (1956) and some others. A brief account will be given here, based dominantly on Kleene (1952). More details can be found in Pshenichny *et al.* (2003).

Essentially, the logic consists of two parts: propositional logic and predicate logic. They differ in the mode of record of simple statements (or narrative sentences) of natural language (English, Russian, Chinese etc.). By 'simple statements' we mean those grammatically consisting of one subject and one predicate, for example, 'Some magmas ascend to the Earth surface' or 'All volcanoes erupt'.

Propositional logic takes them as indivisible elements (*propositional variables*, or *propositions*), that necessarily have one, and only one, of two *logical* (or *truth*) *values*: *TRUE* and *FALSE*. From propositions *compound statements* are formed by the *logical connectives* ('not', 'and', 'or', 'either, or', 'if, then', 'is equivalent to' and possibly some others – see denotations and definitions in Table 1). For example:

Some magmas ascend to the Earth surface – p

All volcanoes erupt – q

Some magmas do not ascend to the Earth surface – $\neg p$

Some magmas ascend to the Earth surface and all volcanoes erupt – $p \,\&\, q$

If some magmas ascend to the Earth surface, then all volcanoes erupt – $p \supset q$.

Predicate logic, on the contrary, treats such statements as constructions consisting of an *individual variable* ('magma' – x; 'volcano' – y), the *predicate* ('ascend to the Earth surface' – P; 'erupt' – Q) and *quantifier* (*existential* one meaning 'some' – $\exists$, and *universal* one meaning 'all' – $\forall$). Then the same simple statements will be recorded as follows:

Some magmas ascend to the Earth surface – $\exists x P(x)$

All volcanoes erupt – $\forall y Q(y)$.

Literally, these statements mean: 'For some magmas it is true that they ascend to the Earth surface' and 'For all volcanoes it is true that they erupt', respectively. Obviously, predicate logic offers a better opportunity to study the anatomy of a statement. However, any concept can be expressed as an array of statements lined up by *logical inference* (see below), and basic rules of inference are the same for both types of logic.

Let us define both types of record of simple statements – $p, q, \ldots$, and $\exists x P(x), \forall y Q(y), \ldots$ – as *logical formulae* (henceforth generally denoted: A, B, C, ...) and postulate that, if A and B are logical formulae, then ¬A, A&B, A ∨ B, A⊃B, A≡B (see Table 1) are formulae too.

The truth value of a formula is that of its main connective. There are formulae that can have only one truth value (i.e. are always true or always false) with any values of variables; for example, the formula $(p \vee \neg p)$ is always true and $(p \& \neg p)$ is always false (see Table 2).

If a formula is always true, it is called a *logical law*, or *tautology*. If it is always false, it is a *controversy*. (This is not to be confused with controversy as the relationship between any two statements, one of which is the negation of the other; the conjunction of such statements gives the controversy in the sense meant here. To avoid ambiguity, this relation is also called *contradiction*.) All the remaining formulae may take both truth values and are called *neutral*, or *satisfiable*.

This gives us, according to Kleene (1952), a strict and definite answer to the question 'What means "to follow" in relation to thinking?' If A, B, C, ..., X, Y are formulae (i.e. statements or thoughts), then A, B, C, ..., X is a *finite list* of formulae. *Y follows from A, B, C, ..., X, if and only if* the following condition is met: *provided A, B, C, ..., X are true, Y is true*. The same is

Table 1. *General truth table for basic logical connectives*

A	B	¬A	A&B	A ∨ B	A ⊃ B	A ≡ B
True	True	False	True	True	True	False
False	True	True	False	True	True	True
True	False		False	True	False	True
False	False		False	False	True	False

Table 2. *Truth table for most simple tautology* $(p \vee \neg p)$ *and controversy* $(p \& \neg p)$

p	$\neg p$	$p \vee \neg p$	$p \& \neg p$
True	False	True	False
False	True	True	False

meant by the expression 'Y is inferred from A, B, C, . . . , X' and denoted by the arrow: A, B, C, . . . , Y → X.

A number of types of inference have been suggested in the logical literature. Here we will briefly describe the *natural sequential calculus* elaborated by Gentzen (1934).

(1) *Sequence* is expression A_1, A_2, . . . , $A_m \to B$, where A_1, A_2, . . . , A_m, B are formulae. A_1, A_2, . . . , A_m are front members of the sequence, and B is a back member. There may be no front members at all, but the back one must always be present: → B.

(2) *Inference in natural-sequential calculus* consists of a number of sequences. Each of these is either a *main sequence* or is derived from a previous one by a structural transformation or a rule of inference (see below). The last sequence of inference has no front members and its back member is a finite formula.

(3) There are two types of *main sequences*, called *logical* and *mathematical*. A *logical main sequence* is a sequence of general form C → C, where C is a formula (this sequence arises if the inference is based on an assumption expressed by C). A *mathematical main sequence* is a sequence of general form → D, where D is an axiom of mathematics.

(4) Allowed structural transformations (a horizontal line means that, if the above sequence is present, the below one is correct) are:

(4.1) Transposition of two front members:

$$\frac{C, D, \Gamma \to \Delta}{D, C, \Gamma \to \Delta}$$

(4.2) Withdrawal of a front member, which is the same as another front member:

$$\frac{C, C, \Gamma \to \Delta}{C, \Gamma \to \Delta}$$

(4.3) Addition of any formula to front members:

$$\frac{\Gamma \to \Delta}{C, \Gamma \to \Delta}$$

(5) Rules of inference. Let A, B and C denote any propositional formulae and Γ, Δ and Θ – any (possibly empty) lists of formulae divided by commas. The formulae of these lists are front members of some sequences. The following rules of inference of natural-sequential calculus are applicable both to propositional and predicate logic.

- Introduction of conjunction (henceforth Rule IC):

$$\frac{\begin{array}{c}\Gamma \to A \\ \Delta \to B\end{array}}{\Gamma, \Delta \to (A \& B)}$$

- Elimination of conjunction (henceforth Rule EC):

$$\frac{\Gamma \to A \& B}{\Gamma \to A}$$

$$\frac{\Gamma \to A \& B}{\Gamma \to B}$$

- Introduction of disjunction (henceforth Rule ID):

$$\frac{\Gamma \to A}{\Gamma \to A \vee B}$$

$$\frac{\Gamma \to B}{\Gamma \to A \vee B}$$

- Elimination of disjunction (henceforth Rule ED):

$$\frac{\begin{array}{c}\Gamma \to A \vee B \\ A, \Delta \to C \\ B, \Theta \to C\end{array}}{\Gamma, \Delta, \Theta \to C}$$

- Introduction of implication (henceforth Rule II):

$$\frac{A, \Gamma \to B}{\Gamma \to A \supset B}$$

- Elimination of implication (henceforth Rule EI):

$$\frac{\begin{array}{c}\Gamma \to A \\ \Delta \to A \supset B\end{array}}{\Gamma, \Delta \to B}$$

- Introduction of negation (henceforth Rule IN):

$$\frac{A, \Gamma \to B \quad A, \Delta \to \neg B}{\Gamma, \Delta \to \neg A}$$

- Elimination of double negation (henceforth Rule EN):

$$\frac{\Gamma \to \neg\neg A}{\Gamma \to A}$$

The following rules apply to predicate logic only:

- Introduction of universal quantifier ∀:

$$\frac{F \to G(x)}{F \to \forall x G(x)}.$$

- Introduction of existential quantifier ∃:

$$\frac{G(x) \to F}{\exists x G(x) \to F}$$

Here x is an individual variable; F is any correctly built formula independent of x, G(x) is predicate.

In the above expressions, symbols put to the left of the arrow can be regarded as the 'memory' of the inference, and those to the right are its 'working part'.

Hence, the conclusion is not a matter of scientist's intuition, but a certain operation verifiable by an objective tool of logical inference. This can be illustrated by the following example relevant to a situation of volcanic crisis like that in Montserrat.

A lava dome is growing; the direction of growth may change with time. Hot avalanches originate on the growing lava dome. Old domes occur nearby. X is a settlement located in the vicinity of the lava domes, very accessible to avalanches. Query, will the avalanches reach X or not? Volcanologists say, if the new dome 'chooses' a direction toward one of the old domes and reaches it, hot avalanches may stop, but only if the growing dome does not overwhelm the older one. Otherwise avalanches may resume and reach X. Logic allows us to verify whether this rationale is correct. If we define:

p – fresh portion of magma extrudes

q – block of old lava dome occurs on the fresh portion's way

r – hot avalanches reach X

s – fresh portion of magma overwhelmes the block of old dome,

then the whole rationale takes the form

$$(p \supset r) \vee ((p \& q) \supset (s \supset r)).$$

There are at least two ways to verify this. The first is using a truth table (Table 3). As is seen from the table, the statement is satisfiable and true except in two cases, when p, q, and s are

Table 3. *Calculation of truth values of the satisfiable formula* (p ⊃ r) ∨ ((p&q) ⊃ (s ⊃ r)) *and tautology* r ⊃ ((p ⊃ r) ∨ ((p&q) ⊃ (s ⊃ r))) *(see comments in the text)*

p	q	r	s	p ⊃ r	p&q	s ⊃ r	(p&q) ⊃ (s ⊃ r)	(p ⊃ r) ∨ ((p&q) ⊃ (s ⊃ r))	r ⊃ ((p ⊃ r) ∨ ((p&q) ⊃ (s ⊃ r)))
T	T	T	T	T	T	T	T	T	T
F	T	T	T	T	F	T	T	T	T
T	F	T	T	T	F	T	T	T	T
F	F	T	T	T	F	T	T	T	T
T	T	F	T	F	T	F	F	F	T
F	T	F	T	T	F	F	F	T	T
T	F	F	T	F	F	F	F	F	T
F	F	F	T	T	F	F	F	T	T
T	T	T	F	T	T	T	T	T	T
F	T	T	F	T	F	T	T	T	T
T	F	T	F	T	F	T	T	T	T
F	F	T	F	T	F	T	T	T	T
T	T	F	F	F	T	T	T	T	T
F	T	F	F	F	F	T	T	T	T
T	F	F	F	F	F	T	T	T	T
F	F	F	F	F	F	T	T	T	T

true and r is false, and when p and s are true and q and r are false.

Another method of verification is to use logical calculus. This is necessary when we are interested in inferability of a formula from a given set of premises. Premises can be any correctly built formulae. For instance, let us assume that:

(1) $p \rightarrow p$
(2) $q \rightarrow q$
(3) $r \rightarrow r$
(4) $s \rightarrow s$.

Then the inference of the considered statement (continued from line 5, as lines 1–4 are assumptions) is:

(5) $p, r \rightarrow r$ Addition of formula assumed in line 1 to front members for line 3
(6) $r, p \rightarrow r$ Transposition of two front members for line 5
(7) $r \rightarrow p \supset r$ Rule II for line 6
(8) $s, r \rightarrow r$ Addition of formula assumed in line 4 to front members for line 3
(9) $r, s \rightarrow r$ Transposition of two front members for line 8
(10) $r \rightarrow s \supset r$ Rule II for line 9
(11) $p, q \rightarrow p\&q$ Rule IC for lines 1, 2
(12) $p\&q,\ r \rightarrow s \supset r$ Addition of formula inferred in line 11 to front members for line 10
(13) $r \rightarrow (p\&q) \supset (s \supset r)$ Rule II for line 12
(14) $r, r \rightarrow (p \supset r) \vee ((p\&q) \supset (s \supset r))$ Rule ID for lines 7, 13
(15) $r \rightarrow (p \supset r) \vee ((p\&q) \supset (s \supset r))$ Withdrawal of a front member for line 14
(16) $\rightarrow r \supset ((p \supset r) \vee ((p\&q) \supset (s \supset r)))$ Rule II for line 15. Statement proved.

The statement $r \supset ((p \supset r) \vee ((p\&q) \supset (s \supset r)))$ is a tautology, i.e. true with any values of the variables (see Table 3). Similarly, one can add the next assumption (p, q or s) left of the arrow and move it to the right adding the implication, for example:

$\rightarrow r \supset ((p \supset r) \vee ((p\&q) \supset (s \supset r)))$
$s \rightarrow r \supset ((p \supset r) \vee ((p\&q) \supset (s \supset r)))$
$\rightarrow s \supset (r \supset ((p \supset r) \vee ((p\&q) \supset (s \supset r))))$.

and so forth, up to the sequence $\rightarrow p \supset (q \supset (s \supset (r \supset ((p \supset r) \vee ((p\&q) \supset (s \supset r))))))$ including all the premises (assumptions). Starting with $r \supset ((p \supset r) \vee ((p\&q) \supset (s \supset r)))$, only tautologies will be yielded at every new step of this inference. This inference indicates that the considered rationale is correct with the given premises.

However, as one might notice, nothing changes in the above inference if one were to substitute location X, say, with Edinburgh. This is because propositional logic ensures only formal correctness of thinking regardless of the contents of statements (their meaning in the real world). To deal with the contents, it should be 'enhanced' with the armoury of predicate logic. All rules of propositional logic are maintained in predicate logic but, in addition, contents-determined predicates are defined, special axioms can be formulated in terms of these predicates and individual variables and quantifiers can be involved (see last two inference rules above).

All this opens an opportunity to formulate an *ad hoc strict language* for a domain of knowledge and to construct in this language a *strict theory*, in which every statement is either an axiom or a theorem inferred from axioms by the rules of inference, either general (see above) or specific. To become a theory, a set of statements should comply with the following conditions: (1) the language for the theory is defined, (2) a correctly built formula is defined for this theory, (3) axioms are identified among the formulae of the theory, (4) rules of inference are defined in the theory. The set of axioms must ensure the theory be self-consistent (which implies the impossibility to infer A and not-A from any set of axioms and/or theorems in the theory), complete (meaning that any logical tautology is inferrable in the theory) and deducible (i.e. that there is an algorithm to identify whether A is a tautology in this theory, or controversy or neutral for any statement A of the theory). In addition, the set of axioms must be independent (Takeuti 1975).

Hilbert & Bernyce (1956) presented a strict theory of arithmetics. Tarski (1959) formalized Euclidean geometry. Another important example is Kolmogorov's axiomatization of probability theory (e.g. Woo 1999). An example of strict theory, prone to formalization by predicate logic, is Mendeleev's periodic system of elements and its implications. Also, attempts have been made to apply this method in physics (e.g. Dirac 1964). However, these examples refer to exact and experimental sciences, which, by Frodeman (1995), can satisfy the condition of bias-free observation and objective knowledge posed by analytic philosophy.

However, there has been a temptation in logic to conquer vaster terrains of mental activity than the artificial world of abstraction or experiment. Bacon (1620) in *The Novum Organon* first claimed that logic must be a tool for obtaining new knowledge and suggested a scheme for inductive entailment. Later this idea was developed in the eighteenth to nineteenth centuries by Hume, Mill and many others (historical background outlined in Fitelson 2003) and evolved into the project of inductive logic, aiming to build an inference from particular to general, relative to which the deductive inference would be an extreme case (Carnap 1950; Fitelson 2003). So far, such a formalism has not been constructed, and inductive logic as-is can generalize data but not introduce logical connectives as do logical calculi.

Another idea was to account for natural variability of the world by prescribing more than two truth values to statements. Thus three-, four-, n-, infinite-valued and, finally, fuzzy logics appeared from the work of Vasiliev, Heiting, Post, Levin, Zadeh and others (see, e.g., system and bibliography at http://www.earlham.edu/~peters/courses/logsys/nonstbib.htm). Despite their wide application in technique, medicine, geoscience and other fields, none of them has suggested its own inference allowing the introduction of connectives. Hence, these, so-called non-Classical logical systems are not logic *sensu stricto*. The same can be said, in the author's opinion, about the modal logic (see bibliography mentioned above). In the author's understanding, inductive, multi-valued and even modal logics all serve to process data and proceed, in an essentially inductive manner, from data to knowledge but not from knowledge to strict theory (see Pshenichny 2003). Knowledge remains as crisp and black-white, two-fold and true-false as it was in Aristotelean times, and data are as incomplete and prone to non-unique interpretation as they used to be.

The only actual attack on this concept, to the author's knowledge, was Brouwer's approach in mathematics, professing that thinking is essentially intuitive and can be best understood not through logical formalisms but through particular cases, one of which is logic itself. Consideration of this idea, so sensational in mathematics in the 1950s, and its relationship to (discordance with?) the approach of Pshenichny (2003) followed here, is outside of the scope of the present paper.

Another challenge to deal with the 'real' world was a logical approach to probability developed by Carnap (1950 and later works) and yet earlier by Johnson (see references in Fitelson 2003). Carnap formulated unconditional and conditional probabilities in the language of classical (i.e. the first-order predicate) logic to make them fit Bayes' formula. However, problems arose with formulation of conditional probability to account for acquisition of new evidence that made Carnap elaborate more and more complicated theories, reviewed by Fitelson (2003). In general, this problem was not solved. The methodological reason for the lack of success perhaps could be under-estimation of the difference between data and knowledge and inapplicability, or quite limited applicability, of formalisms developed to process ideas (i.e. predicate logic) to handle data – an incorrectness symmetrical to the application of probabilistic methods to study reasoning (see above).

The approach of Gentzen (1934) was free from this desideratum. Developing his logical calculi, he simply tried to simulate thinking in the natural science by virtue of one of them: the natural sequential calculus (see above). As has been demonstrated in this paper, (1) this calculus does work for its purpose, but (2) on its own it has little sense outside of a strict theory (e.g. in the real world).

The author deems that it is exactly the logical concept of strict theory that may give us the clue to understanding logical probability. Herewith, we should keep in mind the distinction between knowledge and data and resist the temptation to speak about the *actual* process of learning that refers to data acquisition and processing, not to logic *sensu stricto*.

Learning in a logical sense can be paralleled with the concept of inference (see above). If A, B, C, ..., X is a finite list of formulae (statements), then Y may be *learnt* from this list if, and only if, Y follows from it. The latter is unequivocally identified by a logical calculus. In the simplest case, axioms follow from themselves, $A \rightarrow A$.

If a number of theories are built in one language, then, as shown in Figure 1, a similar statement Y may be inferable in: all of them (if Y is tautology – case 1.1); in some of them (if Y is satisfiable – case 2); or in none of them (if Y is a controversy – case 1.2). The second case is of interest to us. In this case Y follows from some lists of formulae. Let us consider only the lists where: (1) no operation of addition of formulae to front members has been made, (2) no similar formulae are present (that is, there is nothing to reject), and (3) transposition of formulae is ignored, which means that lists A, B and B, A are considered to be the same.

These three conditions allow to us to consider (and define) such lists as *incompatible* with each other (even if there happens to be only one such list over the whole set of theories; Fig. 1, case 2.1). Incompatible lists can be quite different. For instance, if $Y \equiv (Y_1 \& Y_2)$, it may be deduced in one theory from a formula $(Y_1 \& Y_2) \vee Z$ by the rule of elimination of disjunction, and in another theory, say, from independent premises $Y_1 \rightarrow Y_1$ and $Y_2 \rightarrow Y_2$ by introduction of conjunction, and in yet another theory be taken for the axiom itself, $Y_1\&Y_2 \rightarrow Y_1\&Y_2$.

Given a language (roughly, the set of predicates and individual variables), a number (N) of mutually exclusive strict theories can be constructed. For instance, in addition to Euclidean geometry, there is Lobachevsky's geometry, as well as a few other geometries. All of them operate with similar objects (points and lines, from which angles, figures and other entities are derived) and predicates (betweenness, equidistance, parallel lines, etc.; see, e.g., Nam 1995) but are incompatible with each other. Yet simpler examples of an infinite set of theories from a similar language are Smullyan's puzzles about knights, knaves and werewolves (Smullyan 1978).

Let us denote one incompatible list A, B, C, ..., X, from which Y is inferred, by Δ_1, another list by Δ_2, yet another one by Δ_3 and so forth, up to Δ_k, k being a natural number. Each of Δ_i occurs in n_i theories in the given language. Then three cases are possible, 2.1, 2.2.1 or 2.2.2 (see Fig. 1). Case 2.2.1 is rather complicated for further consideration; to begin with, let us assume for simplicity that if Δ_i and Δ_j occur in one theory, then either of them is arbitrarily excluded, so the theory is considered solely Δ_i-containing or Δ_j-containing to make the alternatives incompatible. For cases 2.1 and 2.2.2, if the total number of theories in which Y is inferred from any incompatible list of formulae Δ_i is n, $n = n_1 + n_2 + n_3 + \cdots + n_k$, the *total logical probability* of *Y over a set of theories in a given language* can be defined exactly in the same way as a frequentist probability, i.e.

$$\begin{aligned} P_{LT}(Y) &= n/N = P_L(\Delta_1 \rightarrow Y) \\ &\quad + P_L(\Delta_2 \rightarrow Y) + \cdots + P_L(\Delta_k \rightarrow Y) \\ &= n_1/N + n_2/N + \cdots + n_k/N \\ &= (n_1 + n_2 + \cdots + n_k)/N. \end{aligned}$$

The additivity condition is satisfied by incompatibility of theories and lists of formulae. A more accurate extension of this consideration to case 2.2.1 (Fig. 1) will be reported elsewhere.

The same can be said not only about Y but about every formula in every list Δ_i as well. Each of them would have a probability over a set of theories in a given language. The probability of a list, apparently, should be considered as the sum of probabilities of all its formulae B_1, B_2, ..., B_h, i.e.

$$\begin{aligned} P_L(\Delta_i) &= P_{LT}(B_1) + P_{LT}(B_2) + \cdots + P_{LT}(B_h) \\ &= n_{B1}/N + n_{B2}/N + \cdots + n_{Bh}/N \\ &= (n_{LT\text{-}B1} + n_{LT\text{-}B2} + \cdots + n_{LT\text{-}Bh})/N, \end{aligned}$$

the probabilities $P_{LT}(B_1)$, $P_{LT}(B_2)$, ..., $P_{LT}(B_h)$ defined in the way shown in previous formula. *A priori logical conditional probability* of Y given any of Δ_i, $P_{LC}(Y \mid \Delta_i)$, is nothing else but the probability of logical inference $\Delta_i \rightarrow Y$, which is obviously equal to 1. Hence, *a posteriori logical conditional probability* of any (new) Δ_i given Y, $P_{LC}(\Delta_i \mid Y)$, in accordance with Bayes' theorem, would take the following form:

$$\begin{aligned} P_{LC}(\Delta_i \mid Y) &= P_L(\Delta_i)P_{LC}(Y \mid \Delta_i)/P_{LT}(Y) \\ &= P_L(\Delta_i)/P_{LT}(Y) \\ &= (P_{LT}(B_1) + P_{LT}(B_2) \\ &\quad + \cdots + P_{LT}(B_h))/P_{LT}(Y) \\ &= (n_{B1} + n_{B2} + \cdots n_{B_h})N/ \\ &\quad N(n_1 + n_2 + \cdots + n_k) \\ &= (n_{B1} + n_{B2} + \cdots + n_{Bh})/ \\ &\quad (n_1 + n_2 + \cdots + n_k), \end{aligned}$$

with the relative likelihood ratio expressed by the simple term $1/P_{LT}(Y)$. Thus, the posterior logical probability of some knowledge (expressed as a list of formulae) depends directly on the number of theories in which the formulae from this list are inferred (not necessarily together), depends inversely on the numbers of theories in which the consequence from this knowledge (Y) is inferred from various lists, and does not depend at all on the total number (N) of theories.

The N strict theories can readily be taken, in the author's opinion, for N probability models. In the general case, N is infinite.

Construction of strict theories and a logical account of probability based on these is how both the intellectual ambition of logicians and the practical need of geoscientists in probabilistic estimation of knowledge-related uncertainty (see above) can be satisfied. Moreover, this may substantially alter (and optimize, in the author's opinion) 'science's own understanding of the nature of science', as put by Frodeman (1995). Also, a minor yet important benefit is that elaboration of a logical account of probability frees us from a delicate mission of measuring experts' metabolism to estimate the speed of passage of experts through an array of probability models.

To summarize the account of logic, its main virtue is the relation of logical inference, or deducibility. Logic *sensu stricto* consists of two major parts: propositional logic and predicate logic. Propositional logic is a simple tool useful to introduce the logical calculi that actualize the relation of inference in certain formal procedures. However, alone, it is insufficient to process knowledge. Predicate logic incorporates all laws of propositional logic but offers specific means for construction of strict theories, which have been successfully tested in exact and experimental sciences. Consideration of inference of a statement in a set of strict theories composed in similar language opens an opportunity to define and calculate logical probability, which meets general requirements to probability and, in particular, fits well the Bayesian approach.

Discussion: logic and existing approaches

Consideration of logic as a tool to obtain probabilistic estimates of knowledge-related uncertainty raises three principal questions:

(1) How good is logic in comparison with existing methods?
(2) Is it competitive or complimentary with them?
(3) What can be done to make it work best?

Addressing point (1) first, logic may provide a means to assess probability of knowledge just as well as the existing methods of expert judgment processing which lead to subjective probability. However, contrary to subjective probabilities, the logical probability is normalized to 1, complies with the condition of additivity, and is based on absolutely strict foundations. This definitely favours application of logical rather than subjective probability in the Bayesian approach and elsewhere. Nevertheless, it imposes very strict conditions on knowledge:

- it requires a formal language sufficiently well describing the field of interest;
- the theories should be formulated in this language, each satisfying the rigours of self-consistency, completeness, deducibility and independence of axioms;
- even among these formally correct theories it cannot be excluded that some will appear senseless and would be rejected by scientists, and this might revive the undesirable 'aleatory' component in the logical approach to probability. Though, in contrast with the existing methodology, even if present, here it will be formulated in crisp yes/no mode (to include or not to include a theory in the set N), instead of assigning unverifiable values.

However, perhaps the most palpable obstacle is that the geoscientific community is psychologically not ready to accept a formal treatment of knowledge, possibly because of impending consequence of changing, by Frodeman (1995), the 'self-understanding of science'. Only in some exceptional cases like EYDENET, an expert system supporting decisions on landslide hazard warning in northern Italy (Lazzari *et al.* 1997; see reference in Woo 1999), a kind of propositional logic is used.

Besides, even if we advance far in logical formalization of some domains of geo-knowledge, there will always be problems unreached by logic but requiring urgent decision-making. Moreover, even concerning the well-formalized fields, to relate new knowledge resulting from generalization of new data or from intuition of a new expert, the conventional methods of expert judgement processing will be required. Thus, logic is a theoretically better, but practically less feasible option than the existing methods of evaluating epistemic uncertainty, and in any case it cannot replace these methods.

Addressing point (2), the above notwithstanding, there is still room for logic in the framework of existing procedures. Logical deducibility is the criterion for correctness of formulation of judgements. However, along with deducibility, there are a number of other logical relations between statements and concepts (e.g. controversy, compatibility, subordination, incompatibility), elucidation of which can be useful at least in the reconciliation of views, in processing the results of collective brainstorming (e.g. Morgan & Henrion 1990), in producing collective scientific opinion by a decision-support system (Woo 1999), in various elicitation protocols (Baddeley *et al.* 2004; Curtis & Wood 2004) and in the compilation of test datasets and questions to qualify experts in Cooke's method (Aspinall & Woo 1994).

To identify these relations between the statements or concepts formulated in natural language, Classical logic described above is necessary but not sufficient. Psychological, linguistic and other aspects should be accounted for in order to extract crisp sense from the loose and 'woolly' record in natural language, be it in text or speech.

It should be recalled herewith that logic in its modern form was developed only a few decades ago, while for centuries, since Aristotle, it existed in semi-intuitive, verbal form, perhaps with the single exception of syllogistic deduction. Until now the so-called 'traditional', or Aristotelean logic (e.g. http://plato.stanford.edu/entries/aristotle-logic/) occupied the first half of textbooks on logic for first-year students, serving as the introduction to the strict study of reasoning. It offers rules, quite clear by intuition, to define and classify concepts, to identify the relations between statements and to conclude from premises. However, these rules give unambiguous results only in a limited context (i.e. the number of involved entities – objects and/or properties – is finite). Hence, it should be accompanied by means of distinguishing data from knowledge and relevant knowledge from noise, recognition of linguistic ambiguities, extraction of knowledge from texts and speech and its compact presentation in various form (textual, graphical etc.) forms. All this is the virtue of the new field rapidly developing in recent years, *knowledge engineering* (thousands of publications every year and hundreds of thousands of references on the Internet (e.g. the website of Knowledge Engineering Review, http://titles.cambridge.org/journals/journal_catalogue.asp?mnemonic=ker; or a comprehensive, though rather old, review of Lukose 1996), and information technologies that utilize it (e.g. Loudon 2000; Smyth 2003; http://www.jiscmail.ac.uk/files/GEO-REASONING/dist.ppt). It incorporates a variety of tools united by goal rather than by methodology, ranging from psychological approaches to interviewing experts to statistical methods of extracting relevant words from texts or speech. The possibilistic or fuzzy logic approach advocated by Bardossy & Fodor (2001) is likely to be related to knowledge engineering when used to evaluate opinions. Virrantaus (2003) pointed out the necessity to involve knowledge engineering to decrease the uncertainty ('imprecision') of knowledge. The work on building ontologies for various fields of geoscience reported by Smyth (2003) is one of the first examples of this.

Likewise, the methods of expert judgement processing developed *ad hoc* to estimate subjective probability and support decision making can also be considered a part of knowledge engineering. In terms of the approach of Pshenichny (2003), they all fall into the field of methods enabling the passage from loose knowledge to strict theory and are naturally connected with logic. Therefore, a strategy is needed for efficient co-application of existing approaches of expert judgement processing, knowledge engineering and Classical logic to obtain the best estimation of epistemic uncertainty by possibilistic measures or by subjective probability where necessary, and to proceed to estimation of logical probability with time.

Addressing point (3), the passage from a loose knowledge to a structured finite domain, then to formal language and on to strict theories, is the optimal way to implement logic to estimate uncertainty. This may count as the answer to Question 3 at the beginning of this section. However, the outlined succession generates three successive sub-questions:

(1) How do you confine a context?
(2) How do you make a language from it?
(3) How do you create N theories in this language?

A serious problem in limiting geoscientific contexts is the involvement of diverse sources of knowledge, some of which are, in addition, largely subjective. Observation in many cases can hardly be distinguished from interpretation. Modelling, especially at a regional and planetary scale, often relies on a good portion of imagination. Putting together, say, a field description of a sandstone, equations from mechanics and dialectical laws widen the context rapidly. To limit it, it is pertinent, in the author's view, to avoid 'global' (philosophical, natural-scientific, physical, chemical and even regional geological) terms whenever possible. Rather, one should focus on a set of features furnished by the object in question and akin to objects, recognized at a given scale.

However, even if feasible, this is only part of a solution because interpretation of an outcrop, specimen or thin section unavoidably bears personal or common bias (see above). For instance, when describing a thin section, the optical constants of minerals are determined by what Frodeman calls, after Heidegger, the 'fore-havings' and 'fore-structures' – properties of the microscope and of the eye of the petrographer. While the parameters of the microscope are at least objective, and it can definitely be said that they have nothing to do with the mineral in question, what the eye sees is not only subjective, but we cannot even firmly decide whether it is relevant to the context or not. In a psycho-physiological sense (the ability of the eye to see and discern that depends on the eye proper and on the personality of the researcher), it should not be relevant to the nature of the object studied, though certainly is influential on the result of the study! Just as irrelevant is the capacity of the microscope, while in an epistemiological sense the eye is in fact a combination of 'eye and mind', by the locution of Merleau-

Ponty quoted by Johnson (see reference in Frodeman 1995), and therefore is loaded with prior knowledge about what minerals are, and is obviously relevant. Hence, focusing on the set of features furnished by the object in question, one should abstract not only from too 'general' terms but also from the 'side' (technical, physiological) terms if possible, *or*, if these terms are crucial, involve them somehow in the context. The latter is again illustrated by Frodeman (1995, p. 962):

> If the energy crisis is defined as a problem of supply ('we need more oil'), we will find a different set of facts and a different range of possible solutions than if it is defined as a problem of demand ('we need to conserve').

This recalls the point made in the 'Sources and measures of uncertainty' section, that science has two alternatives: either to ignore bias or to accept, reveal, assess and manage, but not to decrease it. As is clear now, the latter case implies that bias is explicitly formulated and included in the context of consideration and, further, into the strict language and, ultimately, in the theories describing, for example, the (im)possibility of a new volcano forming in non-volcanic regions. This reflects the 'division of labour' acknowledged by Frodeman between Continental and Analytical philosophies in hermeneutic (interpretive) sciences like geology: like the humanities, geology is not bias-free, but like mathematics, it *should* be strict.

In the limited context where the number of involved entities (objects and/or properties) is finite, a strict language can be constructed. The main problem here, in the author's view, is minimization and strict formulation of terms, making them universal for expression of various standpoints.

For instance, a peculiar context of volcanology is the formation of rocks of transitional lava–pyroclastic outlook. In a few decades, a number of concepts about their formation were suggested. These include:

(1) pyroclastic flows (Smith 1960; Ross & Smith 1961; and others)
(2) lava flows (Abich 1882)
(3) two immiscible lavas in one flow (Steiner 1960)
(4) 'boiling' lava turning into foam (Boyd 1954; and others)
(5) hydrothermally altered lava (Naboko 1971 – see reference in Sheimovich 1979)
(6) products of subaqueous lava flows (Ivanov 1966 – see reference in Sheimovich 1979)
(7) product of redeposition of ash in lakes (Karolusova-Kočiščákova 1958 – see reference in Sheimovich 1979).

Despite the striking diversity of views, they form a united and long-lived context. In some periods, for example, in the 1980s to early 1990s, a particular concept (e.g. formation of all rocks of transitional outlook from pyroclastic flows) was popular (e.g. Fisher & Schmincke 1984). However, later some objects (Rooiberg Felsite in South Africa for instance) were reported that can hardly be explained by this concept (Twist & Elston 1989) as well as the cases of formation of patches of pyroclastic-looking material in the rocks confidently interpreted as lavas (Allen 1989; Fink 1989) and intrusive bodies filled with the material of the same outlook (Stasiuk *et al.* 1996). Still, these authors, contrary to early adepts of the 'effusive' concept, reconciled their points with the concept of pyroclastic flow. Another concept, that of rheomorphism, proclaimed by Smith (1960) and Ross & Smith (1961), was developed that actually showed a possibility of 'effusive' environment in a thick body of pyroclastic flow material (Milner *et al.* 1992; Streck & Grunder 1995) after, and (according to the most recent works) before and during the deposition (Branney & Kokelaar 2002). We cannot exclude the possibility of seemingly weird ideas of hydrothermal alteration of lava, subaqueous lava flows or redeposition of ash in lakes, reformulated in new terms, being revived and successfully used for interpretation of the same rocks in the future. Moreover, perhaps the maturity of a hermeneutic science might be understood as the moment when new concepts are no longer elaborated, and new data continue to support or slightly modify some of the previously suggested concepts, none of which can ever be totally rejected or totally adopted, so that further development of science is an endless 'championship' of concepts, in which no one wins the cup or leaves the league.

However, once the 'league' has been formed, the language should be able to describe the concepts from, and relationships between each 'player' – and to do that in as strict and concise form as possible in order to be translatable into predicate logic language. In this language, the standpoints are expected to become strict theories, but some of them, if they appear compliant, would merge to produce one theory, and new theories may emerge by automatic operation with variables and predicates. These 'artificial' theories should be examined by scientists and either adopted or rejected (herewith the 'aleatory component' may emerge again, as discussed above). Nevertheless, as the logical studies show, correct formulation of a large number of theories in a given language is a task requiring time and patience, with a high risk of human error, even for a simple case of Smullyan's puzzles (Moukhachov & Netchitailov 2001). At the same time, the degree of formality principally allows its complete automatization, and this work is actually ongoing (V. P. Moukhachov pers. comm.). Needless to say, for geoscience, such an automated 'theory generator' is perhaps the only way to make the assessment of logical probability practically achievable.

Summing up the discussion, logic is a theoretically better, but practically less feasible option than the existing methods of evaluating epistemic uncertainty, but in any case it can neither be a total substitute for these methods nor totally abolish the aleatory uncertainty. Therefore, a strategy is needed for efficient co-application of existing approaches of expert judgement processing, knowledge engineering and classical logic to assess epistemic uncertainty by possibilistic measures or by subjective probability where necessary, and to proceed to logical probability with time. This strategy should aim to confine the context of geoscientific study but this context may include, where relevant, the forestructures and bias. To advance the logical estimation of probability, an automated generator of logical tasks (theories) is a prerequisite.

Conclusions

(1) Based on the approach of Pshenichny (2003), uncertainty can be classified as being either that of knowledge or that of data. The uncertainty of data might be considered: (a) in terms of additivity and measured by probability in the frequentist sense, (b) in fuzzy terms and measured by a variety of fuzzy parameters, or (c) jointly. Reasoning and language contribute to the uncertainty of knowledge, or epistemic uncertainty, and its refinement, aleatory uncertainty. As for the measure of epistemic uncertainty, if the approach to knowledge is intuitive, it may only be fuzzy; if the approach is to treat knowledge objectively, in the tradition of analytical philosophy, this measure, if it exists, should be probabilistic in a logical but not in a subjective sense.

(2) The main virtue of logic is the relation of logical inference, or deducibility. Logic *sensu stricto* consists of two major parts: propositional logic and predicate logic. Propositional logic is a simple tool, useful to introduce the logical calculi that actualize the relationship of inference in certain formal procedures. However, it is insufficient alone to process knowledge. Predicate logic incorporates all of the laws of propositional logic but offers specific means for construction of strict theories, which have been tested successfully in exact and experimental sciences. Consideration of inference of a similar statement in a set of strict theories composed in similar language opens the opportunity to define and calculate logical probability, which, being a measure of epistemic uncertainty, meets the general requirements of probability and, in particular, fits well with the Bayesian approach.

(3) Logic is a theoretically better but practically less feasible option than the existing methods of evaluating epistemic uncertainty, but in any case it can neither be a total substitute for these methods, nor totally abolish the aleatory uncertainty.

(4) Urgent tasks for future research include, firstly, a strategy for co-application of existing approaches of expert judgement processing, knowledge engineering and classical logic, able to incorporate prejudgements and bias if necessary, and secondly, an automated generator of logical tasks (strict theories) in a given language.

The author is deeply obliged to A. Curtis and R. Wood and other participants of the RAS meeting in London, May 2003, for their interest, attention and enthusiasm in discussing this work, as well as their help in its preparation. Special gratitude should be expressed to S. Henley for his intellectual and moral support and encouragement over the years. The keen attitude of M. Baddeley and J. Fodor, who reviewed the paper, hopefully has helped to constructively reconsider it. P. Vaganov guided the author in the field of methodology of uncertainty analysis. V. Moukhachov instructed and consulted the author on logical matters. Y. Simonov made very important observations on the material.

The work was funded by grant PD02-1.5-369 from the Ministry of Higher Education (Ministerstvo Obrazovaniya) of the Russian Federation and grant PD03-5.0-206 from the City Administration of St Petersburg.

References

ABICH, H. 1882. *Geologische Forschungen in den Kaukasischen Ländern (Geologie des Armenischen Hochlandes)*. Vol. 2. Alfred Holder Wien.

ALLEN, R. L. 1989. False pyroclastic textures in silicic lavas: continental magmatism. *IAVCEI Abstracts. New Mexico Bureau of Mines and Mineral Resources Bulletin*, **131**, 3.

ASPINALL, W. & COOKE, R. M. 1998. Expert judgement and the Montserrat Volcano eruption. *In:* MOSLEH, A. & BARIS, R. A. (eds) *Proceedings of the 4th International Conference on Probabilistic Safety Assessment and Management (PSAM4), 13–18 September 1998, New York City*, **3**, 2113–2118.

ASPINALL, W. & WOO, G. 1994. An impartial decision-making procedure using expert judgement to assess volcanic hazards: large explosive eruptions. *Accademia Nazionale dei Lincei – British Council International Symposium, 24–25 May 1993, Rome. Atte dei Convegni Lincei*, **112**, 211–220.

BACON, F. 1620. *The Novum Organon*, Oxford University Press, Oxford.

BADDELEY, M. C., CURTIS, A. & WOOD, R. 2004. An introduction to prior information derived from probabilistic judgments: elicitation of knowledge, cognitive bias and herding. *In:* CURTIS, A. & WOOD, R. (eds) *Geological Prior Information: Informing Science and Engineering*. Geological Society, London, Special Publications, **239**, 15–27.

BARDOSSY, G. & FODOR, J. 2001. Traditional and new ways to handle uncertainty in geology. *Natural Resources Research*, **10**, 179–187.

BOWDEN, A. 2004. Building confidence in geological models. *In:* CURTIS, A. & WOOD, R. (eds) *Geological Prior Information: Informing Science and Engineering*. Geological Society, London, Special Publications, **239**, 157–173.

BOYD, F. R. 1954. The Yellowstone rhyolite plateau. *Carnegie Institute, Washington, Yearbook*. **53**, 39–140.

BRANNEY, M. J. & KOKELAAR, P. 2002. *Pyroclastic Density Currents and Sedimentation of Ignimbrites*. Geological Society, London, Memoirs, **27**.

CARNAP, R. 1950. *Logical Foundations of Probability*. University of Chicago Press, Chicago.

COOKE, R. M. 1990. *Experts in Uncertainty: Expert Opinion and Subjective Probability in Science*. Oxford University Press, Oxford.

CURTIS, A. & WOOD, R. 2004. Optimal elicitation of probabilistic information from experts. *In:* CURTIS, A. & WOOD, R. (eds) *Geological Prior Information: Informing Science and Engineering*. Geological Society, London, Special Publications, **239**, 127–145.

DIRAC, P. A. M. 1964. *Lectures on Quantum Mechanics*. Belfer Graduate School of Science, Yeshiva University, New York.

FINK, J. H. 1989. Morphologic characteristics of silicic lava flows: continental magmatism. *IAVCEI Abstracts. New Mexico Bureau of Mines and Mineral Resources Bulletin*, **131**, 90.

FISHER, R. V. & SCHMINCKE, H.-U. 1984. *Pyroclastic Rocks*. Springer-Verlag, Berlin and Heidelberg.

FITELSON, B. 2003. Inductive logic: philosophy of science. *In:* PFEIFER, J. & SARKER, S. (eds) *An Encyclopedia*. Routledge Press. Worldwide web address: http://www.fitelson.org/il.pdf.

FREGE, G. 1896. Über die Begriffsschrift des Herrn Peano und meine eigene. *Berl. d. math. Gl. d. Kgl. Sächs. d. Wiss. zu Leipzig*, vol. 48 (1897), 361–378.

FRODEMAN, R. 1995. Geological reasoning: geology as an interpretive and historical science. *Bulletin of the Geological Society of America*, **107**, 960–968.

GENTZEN, G. 1934. Untersuchungen über das logische Schliessen, I-II. *"Math. Z."*, Bd. 39, H. 2, 3.

HALPERN, D. F. 1996. *Thought and Knowledge: An Introduction to Critical Thinking*. Lawrence Erlbaum Associates, Mahwah, New Jersey.

HILBERT, D. & BERNAYS, P. 1956. *Crundlagen der Mathematik*, I. Springer-Verlag.

KLEENE, S. C. 1952. *Introduction to Mathematics*. Van Nostrand Co. Inc., New York and Toronto.

LOUDON, T. V. 2000. *Geoscience After IT: A View of the Present and Future Impact of Information Technology on Geoscience*. Elsevier, Oxford [See also *Computers & Geosciences, Special Issue*, **26**, A1-A142.]

LUKOSE, D. 1996. *Knowledge Engineering*. University of New England Press, Armidale. Worldwide web address: http://www.pages.cpsc.ucalgary.ca/~kremer/courses/CG/

MILNER, S. C., DUNCAN, A. R. & EWART, A. 1992. Quartz latite rheoignimbrite flows of the Etendeka Formation, north-western Namibia. *Bulletin of Volcanology*, **54**, 200–219.

MORGAN, M. G. & HENRION, M. 1990. *Uncertainty: A Guide to Dealing with Uncertainty in Quantitative Risk and Policy Analysis*. Cambridge University Press, New York. [Reprinted 1998]

MOUKHACHOV, V. P. & NETCHITAILOV, I. V. 2001. An improvement of Brodsky's coding method for the sequent calculus of First Order Logic. *In:* GORE, R., LEITSCH, A. & NIPKOV, T. (eds) *International Joint Conference on Automated Reasoning Short Papers, Siena*, 113–121.

NAM, S. 1995. The semantics of locative prepositional phrases in English. PhD thesis, University of California. Worldwide web address: http://www.cl.snu.ac.kr/nam/papers/

PSHENICHNY, C. A. 2003. A draft for complex formal approach in geoscience. *In:* CUBITT, J., WHALLEY, J. & HENLEY, S. (eds) *Modelling Geohazards*. Proceedings of the 10th Annual Conference of the International Association for Mathematical Geology, 7–12 September 2003, Portsmouth. Worldwide web address: http://www.jiscmail.ac.uk/files/GEO-REASONING/Pshenichny_paper.doc

PSHENICHNY, C. A., MOUKHACHYOV, V. P. & KHRABRYKH, Z. V. 2003. Logical assessment of observational knowledge in volcanology. *Journal of Volcanology and Geothermal Research*, **128**, 287–298.

ROSS, C. S. & SMITH, R. L. 1961. *Ash-Flow Tuffs: Their Origin, Geologic Relations and Identification*. US Geological Survey, Professional Papers, **336**.

ROWE, W. D. 1988. *An Anatomy of Risk*. Malabar.

SHEIMOVICH, V. S. 1979. [*Ignimbrites of Kamchatka*]. Nedra, Moscow. [In Russian]

SMITH, R. L. 1960. *Zones and Zonal Variations in Welded Ash Flows*. US Geological Survey, Professional Papers **354–F**.

SMULLYAN, R. 1978. *What is the Name of this Book?* Prentice-Hall.

SMYTH, C. 2003. Geotechnical hazard modelling and geological terminology. *In:* CUBITT, J., WHALLEY, J. & HENLEY, S. (eds) *Modelling Geohazards*. Proceedings of the 10th Annual Conference of the International Association for Mathematical Geology, 7–12 September 2003, Portsmouth.

STASIUK, M. V., BARCLAY, J., CARROL, M. R., JAUPART, C., RATTE, J. C., SPARKS, R. S. J. & TAIT, S. R. 1996. Degassing during magma ascent in the Mule Creek vent (USA). *Bulletin of Volcanology*, **58**, 117–130.

STEINER, A. 1960. Origin of Ignimbrites of the North Island, New Zealand. A New Petrologic Concern. *New Zealand Geological Survey Bulletin*, **68**.

STRECK, M. J. & GRUNDER, A. L. 1995. Crystallization and welding variations in a widespread ignimbrite sheet; the Rattlesnake Tuff, eastern Oregon, USA. *Bulletin of Volcanology*, **57**, 151–169.

TAKEUTI, G. 1975. *Proof Theory*. North-Holland, Amsterdam and London, American Elsevier Publishing Co., New York, Studies in Logic and the Foundations of Mathematics, **81**.

TARSKI, A. 1959. What is elementary geometry? *In:* HENKIN, L. (eds).

TWIST, D. & ELSTON, W. E. 1989. The Rooiberg felsite (Bushveld complex): textural evidence pertaining to emplacement mechanisms for high-temperature siliceous flows: continental magmatism. *IAVCEI Abstracts. New Mexico Bureau of Mines and Mineral Resources Bulletin*, **131**, 273.

VIRRANTAUS, K. 2003. Analysis of the uncertainty and imprecision of the source data sets for a military terrain analysis application: *Proceedings of the International Symposium on Spatial Data Quality* Worldwide web address: www.hut.fi/Units/Cartography/research/ materials/kirsivirrantaus04032003.pdf

WHITEHEAD, A. N. & RUSSELL, B. 1910. *Principia Mathematica*. Cambridge University Press, London.

WOO, G. 1999. *The Mathematics of Natural Catastrophes*. Imperial College Press, London.

WOOD, R. & CURTIS, A. 2004. Geological prior information and its application to geoscientific problems. *In:* CURTIS, A. & WOOD, R. (eds) *Geological Prior Information: Informing Science and Engineering*. Geological Society, London, Special Publications, **239**, 1–14.

Optimal elicitation of probabilistic information from experts

ANDREW CURTIS[1,2] & RACHEL WOOD[1,3]

[1]*Schlumberger Cambridge Research, High Cross, Madingley Road, Cambridge CB3 0EL, UK (e-mail: curtis@cambridge.oilfield.slb.com)*

[2]*Grant Institute of Earth Science, School of GeoSciences, University of Edinburgh, West Mains Road, Edinburgh EH9 3JW, UK*

[3]*Department of Earth Sciences, University of Cambridge, Downing Street, Cambridge CB2 3EQ, UK*

Abstract: It is often desirable to describe information derived from the cumulative experience of human experts in a quantitative and probabilistic form. Pertinent examples include assessing the reliability of alternative models or methods of data analysis, estimating the reliability of data in cases where this cannot be measured, and estimating ranges and probable distributions of rock properties and architectures in complex geological settings.

This paper presents a method to design an optimized process of elicitation (interrogation of experts for information) in real time, using all available information elicited previously to help in designing future elicitation trials. The method maximizes expected information during each trial using experimental design theory. We demonstrate this method in a simple experiment in which the conditional probability distribution or relative likelihood of a suite of nine possible 3-D models of fluvial-deltaic geologies was elicited from a geographically remote expert. Although a geological example is used, the method is general and can be applied in any situation in which estimates of expected probabilities of occurrence of a set of discrete models are desired.

The generic task of gathering quantitative, probabilistic information on any topic usually requires the interrogation of experts in that field. Usually such an interrogation revolves around some model or representation of the topic of interest. Expert judgement and intuition is used to place constraints both on the range of complexity that should be included in the model and on the range of model parameter values to be considered. Everyone can recall situations in which experts (including ourselves) have poorly estimated such information, or in which information from multiple experts was inadequately assimilated, resulting in inefficiency or errors.

Poor results usually occur because all humans – experts included – are subject to natural biases when trying to estimate probabilities or risks mentally. Some of these are shared with other mammals, others only with primates, and some appear to be exclusively human. These biases, and techniques to ameliorate their effects, have been the subject of much study and debate in psychological and statistical research literature, and form the field known as *elicitation theory*, i.e. the theory concerning the interrogation of subjects for information. Surprisingly however, few elicitation techniques appear to be used in the geosciences, and so one aim of this paper is to review some of the most pertinent elicitation literature.

In order to develop the theory further, we here consider explicitly a generic situation often encountered in the hydrocarbon exploration and production industry, where estimates of the detailed stratigraphic architecture and properties of reservoir rocks in 3-D (henceforth, the reservoir *geology* in *3-D*) are required, using only remote sensing methods which have either lateral or vertical spatial resolution that is poorly matched to this task (e.g. seismic and well log data). Wood & Curtis (2004) demonstrate how quantitative geological information that is independent of the remotely sensed data, called *prior information*, can be used to constrain probabilistically the architecture of the geology some distance from a single vertical well at which (in that study) the only available data was collected. How best to elicit the prior information that is used implicitly by experts in placing such constraints is the subject of this study. Once

From: CURTIS, A. & WOOD, R. (eds) 2004. *Geological Prior Information: Informing Science and Engineering*. Geological Society, London, Special Publications, **239**, 127–145. 186239-171-8/04/$15.00

captured, such information might be re-used in multiple similar situations.

Consider a case where an expert geologist knows the typical characteristics of the geology in an area of interest, for example, the likely range of either sinuosity of river channels in a fluvial system, or different reef types within carbonate strata. Assume also that a computer program exists that creates synthetic geological cross-sections given a set of input parameters (e.g. Griffiths *et al.* 2001; Tetzlaff & Priddy 2001; Burgess 2004; Tetzlaff 2004). Boschetti & Moresi (2001) demonstrated how a genetic algorithm (GA) might be used to find values of the computer program's input parameters that produce the desired characteristics in the modelled geology (see also Wijns *et al.* 2004). The expert simply ranks sets of cross-sections from different modelling runs in order of how well they match the desired characteristics; the GA iterates through further generations of modelling runs, requesting further rankings from the expert, until sets of input parameters are found such that geologies with the desired characteristics are modelled.

Both the strength and weakness of the method of Boschetti & Moresi (2001) is that the interpreter never needs to produce quantitative estimates of *how* well or poorly any model run matches the desired characteristics – only relative ranks of the models are requested. This property renders the method easy to use and widely applicable, but has the potential drawback that it cannot deliver quantitative, probabilistic information about the range of model parameters that fit the desired characteristics. It also implies that only optimization algorithms that do not require probabilistic inputs (such as the GA) can be used.

We extend these methods by asking experts to provide a minimal set of probabilistic information from their prior experience. This allows the construction of fully normalized probability distributions representing geological prior information.

We begin this paper with an overview of background material. This includes a discussion of the general nature of geological prior information that one might collect and use quantitatively, and a description of a particular geological system about which we require information (the problem addressed later in this paper). Elicitation theory has a large associated literature in engineering, statistical and psychological literature, so we also provide a brief overview of some of this work in the background material section. In the subsequent section we present a new method for collecting the required information. This method employs techniques from the field of statistical experimental design, and these are described in a separate section. We then demonstrate the benefits of this approach in a simple geological elicitation experiment, and discuss how to apply the method to solve the problem posed in the overview of background material below.

Overview of background material

The nature of geological prior information

Geological prior information is often qualitative. This is usually because it is based on interpretations, best guesses from past individual experiences, and intuition. Although useful to obtain other qualitative results, such information cannot be used directly in any other probabilistic methodology without implicitly or explicitly quantifying this qualitative knowledge (Wood & Curtis 2004).

Additionally, almost all geological prior information is partly subjective. This is a direct consequence of the fact that prior information usually comprises generalizations of previous experience. There are exceptions to this: for example, in a seismic survey, pertinent prior information might include previously acquired surface gravity measurements, or measurements derived from wells, and that information may be objective. Such objective information is relatively easy to capture; however, less easy to capture is geological prior information derived from tacit experience. For example, information about formations and rock characteristics expected in any particular type of geology usually consists of patterns observed previously in outcrops, which have been generalized by experts. Each expert may generalize their various experiences of different outcrops in markedly different ways. Therefore, estimates of prior probabilities are themselves necessarily inaccurate because of the uncertainty and biases implicit within each expert from whom prior information was elicited. Hence, it is necessary to understand the kinds of biases and uncertainties that humans tend to exhibit when estimating probabilities in order that the effects of these might be mitigated. Later we summarize some of the psychological and statistical literature examining such behavioural tendencies.

Problem description

We now describe the particular type of problem for which prior information is elicited in this paper: define the rock in two layers intersected by a vertical well to be of types T_1 and T_2, respectively (Fig. 1a). We consider the most general case where the range of rock types for T_1 is continuous rather than discrete (e.g. T_1 = % carbonate minerals), and similarly for T_2.

Let us describe the Earth model by a vector **m** containing the unknown model parameters: $\mathbf{m} = [T_1, T_2]$. Vector **m** is clearly 2-D, and a value for **m** corresponds to a single point on the 2-D plane in Figure 1a. If we now consider a well that intersects three unknown rock types then $\mathbf{m} = [T_1, T_2, T_3]$, corresponding to a single point in the 3-D space shown in Figure 1b. By analogy, if four rock types were intersected then $\mathbf{m} = [T_1, T_2, T_3, T_4]$ corresponds to a single point in 4-D hyperspace (hyperspace is simply space in more than 3-D; Fig. 1c). Generally, if we have an unknown model that is described by N parameters (N rock types in this example) then $\mathbf{m} = [T_1, T_2, \ldots, T_N]$ corresponding to a point in N-dimensional hyperspace. We call the plane, space, or hyperspace the *model space* M.

Each time the number of parameters (layers) is increased by one, the cost of exploring the model space M increases. Consider the number of points required to sample the model space to some level of detail. Assuming that 10 values of each model parameter would provide a sufficiently detailed exploration, then, for the 2-D problem in Figure 1, we require $10^2 = 100$ samples, for the 3-D problem we require $10^3 = 1000$ samples etc. If a well in the field intersected even only 20 rock types, the 20-dimensional hyperspace spanned by the unknown model parameters is surely incomprehensibly large to all humans, and exploration to the required level of detail would require 10^{20} samples.

Estimating the true Earth model (the true rock types defined by a single point in hyperspace) requires that we have sufficient information to be sure that all other points represent untrue models. As the number of other points increases, so the amount of information required increases.

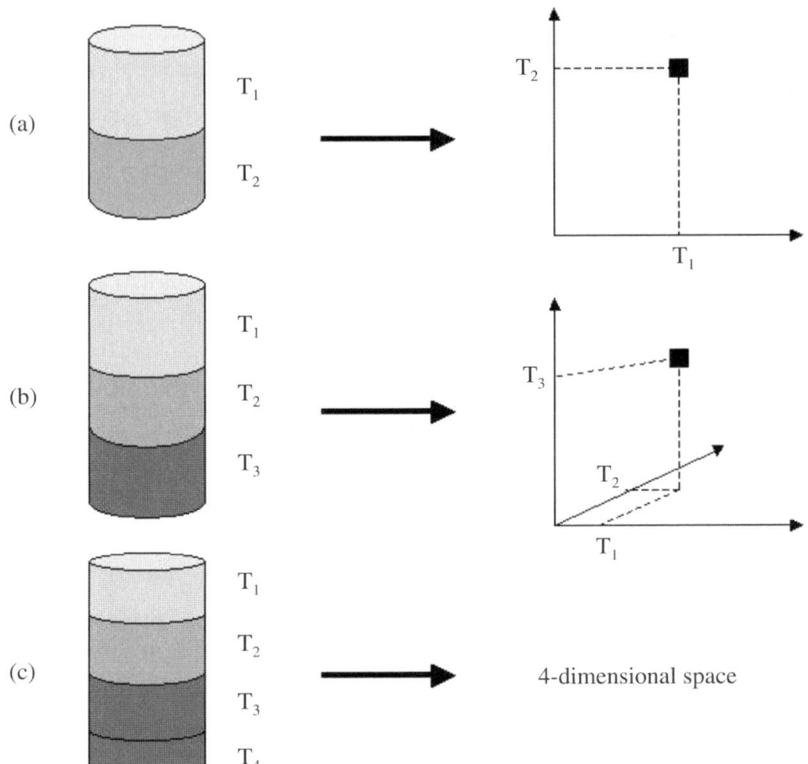

Fig. 1. Representation of rock types intersected by a well in a Cartesian model space. (**a**), (**b**) and (**c**) show a well intersecting 2, 3 and 4 rock types respectively on the left, with the corresponding model space representation on the right.

The property of increasing hyperspace volume and the corresponding increase in information required to estimate the true model as its dimensionality increases can be defined formally and is referred to as the 'curse of dimensionality' (Scales 1996). Curtis & Lomax (2001) show that, in such large hyperspaces, intuitive or innate notions of probabilities derived from 1-D–3-D space no longer hold. In turn this implies that simply asking a subject to estimate probability distributions in such spaces will surely lead to poor results.

Consider estimating rock types and architectures in a 2-D cross-section through two vertical wells (Fig. 2a); a similar problem is addressed in Wood & Curtis (2004). Given no information at all about the rocks intersected by the wells, any point in space on the cross-section could contain any rock type. One method of parameterizing the model space is to discretize the cross-section as shown in Figure 2b. Each discrete cell might contain only one model parameter, 'rock type', that we wish to estimate. This results in an N-dimensional model vector $\mathbf{m}$, and even if only 10 cells are used both horizontally and vertically, then $N = 100$ and the resulting 100-dimensional model space (Fig. 2c) is unimaginably large. If the model was 3-D and only 10 cells were used in the third dimension then the model space would be 1000-dimensional.

When we include all parameters in which we might be interested in a reservoir situation, it is always the case that, even after constraints have been imposed from the most comprehensive geophysical data, there will remain lines, surfaces or hypersurfaces in M on which any model is equally likely. These lines or surfaces span what is called the *null space* in M: the part of M about which we have no information that differentiates between models that are more or less likely to occur in the Earth. Thus, the null space represents fundamental uncertainty in estimated Earth models. No amount of geophysical data can remove all null space directions, so other information is always necessary to constrain our estimates. Since such information is

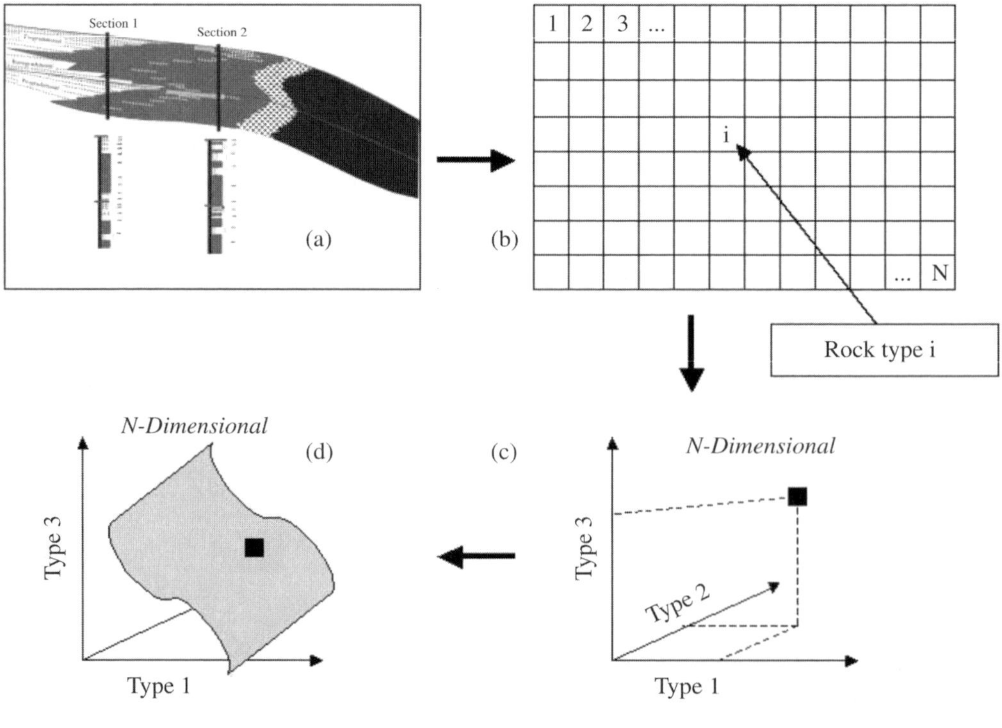

Fig. 2. (**a**) Geological cross-section with logs of rock type through two intersecting wells, (**b**) Possible parameterization of rock type in the 2-D section comprising cells within which any rock type might occur in the Earth. This model space has dimension N. (**c**) Representation of the N-dimensional model space in Cartesian form. However, if prior information (e.g. from a geological modelling package) exists, the possible model space can be restricted to a manifold shown in (**d**).

independent of the geophysical survey data, it is called *prior* information.

The problem that we address in this paper is that of collecting and usefully representing reliable geological prior information that reduces the portion of model space M that represents feasible Earth models (Fig. 2d), even before any (current) geophysical data is collected. This has two principal, tangible benefits. Firstly, this information will reduce the null space post-data acquisition, thereby improving Earth model estimates and reducing uncertainty. Secondly, as this information is available prior to acquiring data it can be used to design an optimal data acquisition survey. Data need only be collected if it fills gaps or reduces large uncertainties in the prior information; the optimal dataset to collect will be that which fulfils as many of these requirements as possible, and at either minimal or acceptable cost.

Elicitation theory

The problem tackled in the field of elicitation theory is to design the best way to interrogate experts or lay people (henceforth, *subjects*) in order to obtain accurate information about a topic in question. Extensive psychological research has shown that this is a difficult problem: simply asking even expert subjects to provide a (numerical) probability estimate results in poor probability assessments. The reason is that people find providing such estimates difficult and hence tend to use heuristics to help themselves; these in turn introduce biases and poor calibration (Kahneman *et al.* 1982). Baddeley *et al.* (2004) review various biases that are observed commonly in humans. Below we briefly summarize some of these and highlight various key works from elicitation theory.

Individuals' biases and heuristics

At least two main types of bias can be distinguished (Skinner 1999): *motivational bias* (caused by the subject having some vested interest in biasing the results) and *cognitive bias* (caused by incorrect processing of information available to the subject). The former can often be removed by explaining that unbiased information is critical, and by using all possible means to remove the particular motivations for biasing the results. The latter is typically the result of the use of heuristics, and suitable elicitation methods can sometimes correct part of such biases.

At least four types of heuristics causing cognitive biases are commonly encountered: availability, anchoring and adjustment, representativeness, and control (Kahneman *et al.* 1982). *Availability* is the heuristic of assessing the probability of an event by the ease with which occurrences of the event are brought to mind and is biased by prominence rather than frequency of different information (e.g. bicycle accidents are more frequent, incur a greater number of fatalities overall, but are less easy to recall than aeroplane crashes). *Anchoring and adjustment* is a single heuristic that involves making an initial estimate of a probability called an *anchor*, and then revising it up or down in the light of new information (e.g. information about opinions of others on the matter in question). This typically results in assessments that are biased towards the initial anchor value. The *representativeness* heuristic is where people use the similarity between two events to estimate the probability of one from the other (e.g. we may use knowledge of an individual's personality traits to estimate their political persuasions). The *control* heuristic is the tendency of people to act as if they can influence a situation over which they have no control (e.g. buying lottery tickets with personally chosen rather than random numbers).

Well-known consequences of these heuristics are the gambler's fallacy, the conjunction fallacy, base-rate neglect, probability matching and over-confidence (for definitions see Baddeley *et al.* 2004). Of all of these biases, the most prevalent may be over-confidence and base-rate neglect (Baecher 1988). Over-confidence is particularly a problem for extreme probabilities (close to 0% and 100%), which people find hard to assess.

The biases described above imply that the results of elicitation from each individual expert may need to be calibrated. This in turn requires some statistical model of the elicitation process. The model to which most references are made in the literature is that of Lindley *et al.* (1979). This model requires that there be an objective assessor who will consolidate the results derived from subjective experts. We will not make explicit use of this model, however, as it is not clear why an assessor should be any more objective than the experts.

Other work that attempts to recalibrate experts' judgement includes that of Lau & Leong (1999), who created a user-friendly Java interface for elicitation that includes graphical illustrations of possible biases and any inconsistencies in elicited probability estimates (see the review of graphical methods by Renooij

2001). The interface then enters into dialogue with the expert until consistency is achieved.

It should be noted that some of the heuristics described above can perform extraordinarily well in some situations (e.g. Gigerenzer & Goldstein 1996; Juslin & Persson 2002). In practical situations, however, it is never clear from the results alone whether or not this is the case since there is no objective answer with which the results of the use of heuristics can be compared. Hence, one role of the elicitor is to try to reduce the use of heuristics as much as possible.

Elicitation protocols and strategies

There are no universally accepted protocols for probability elicitation and there is relatively little formal, empirical evaluation of alternative approaches. There are, however, three common assessment protocols (Morgan & Henrion 1990). These generally include five phases: motivating the experts with the aims of the elicitation process, structuring the uncertain quantities in an unambiguous way, conditioning the expert's judgement to avoid cognitive biases, encoding the probability distributions, and verifying the consistency of the elicited distributions.

Within each protocol the elicitor must decide exactly what problems to tackle or what questions to ask in order to maximize information about the topic of interest. Coupé & van der Gaag (1997) showed how a sensitivity analysis might sometimes be carried out in order to see which elicited probabilities would have most influence on the output of a Bayesian belief network. Using their terminology, later in this paper we extend their '1-way sensitivity analysis' by introducing a new method of analysing the sensitivity test results that allows optimal decisions to be made regarding which probabilities should be elicited.

We have discussed only a few of the many references in the field of elicitation theory (see also Baddeley *et al.* 2004). It is clear that eliciting prior information is not a trivial problem. Below we make the first attempt (to our knowledge) to optimize the elicitation process in real time using all information available. This must be an optimal strategy in principle. However, in practice the details of our particular method may be improved in future: for example, we make no attempt to correct biases in the subject's judgement, but optimize the elicitation process within the framework of the biases described above. It is likely, therefore, that our method can be improved by combining it with bias-reducing techniques such as those of Lindley *et al.* (1971) or Lau & Leong (1999).

Methodology

We now present a method to derive, parameterize and describe geological prior information. We distinguish two main types of geological prior information: *static* information, about reservoir architectures and their properties as they exist today; and *dynamic* information, about the various processes that created the present characteristics of the rock (Wood & Curtis 2004). Such a classification implicitly incorporates any information about past geological events as these can only be estimated using a combination of static and dynamic information. The method proposed includes both types of information.

Geological information is introduced in two steps, each described separately below. Firstly, geological modelling software that encapsulates dynamic prior information is used to remove large portions of model space M that are geologically infeasible (Fig. 2d). Secondly, within the remaining portions of M, a new method uses the experiences of interpreters or geologists (henceforth called 'experts', or 'subjects' [of an experiment]), obtained from previously observed static information, to assess the relative likelihood of occurrence of different models and to convert the experts' knowledge into prior probabilities. In order to estimate these probabilities we set up an elicitation procedure that is described in part of this section. In the following section we describe how this elicitation procedure was optimized. The result of these steps is a normalized probability distribution representing geological prior information.

Geological modelling software

Software exists that models the architecture of existing rock structures based either on statistics from existing, observed outcrops (static information, e.g. Wood & Curtis 2004), or on geological processes of deposition, erosion, transport, re-sedimentation and subsequent diagenesis acting over geological time scales (dynamic information e.g. Griffiths *et al.* 2001; Tetzlaff & Priddy 2001; Burgess 2004; Tetzlaff 2004). In this study we will use an example of the latter type of package, the Geologic Process Model (GPM; Tetzlaff & Priddy 2001), but any other preferred software may be used to replace this without any other change in our methodology.

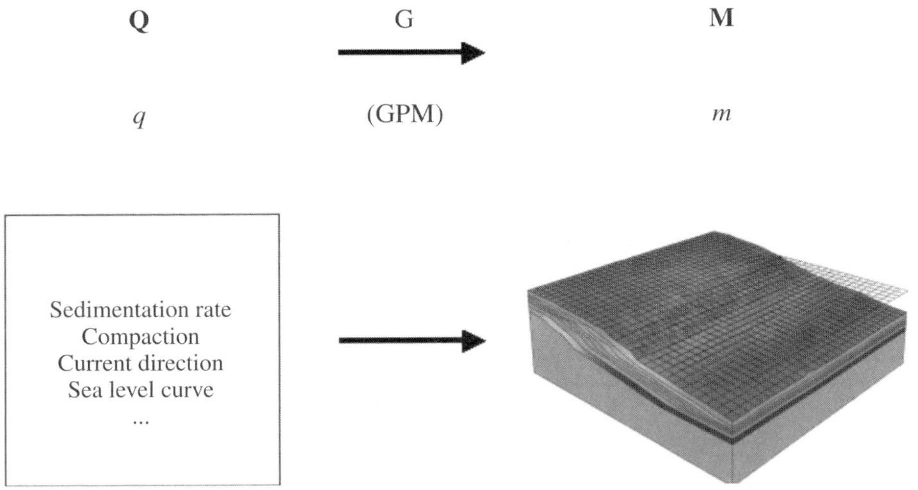

Fig. 3. Representation of the relationship between parameters **q** in parameter space **Q** that are input to the Geological Process Model (GPM), and the output geological sections parameterized by models **m** in model space **M**.

The GPM software has been developed by Tetzlaff of Schlumberger, based on initial research conducted at Stanford (Tetzlaff & Harbaugh 1989; Griffiths *et al.* 2001; Tetzlaff & Priddy 2001) but with many recent additions and enhancements. GPM currently models at least: (1) depositional sedimentation, (2) compaction and diagenesis, and (3) erosion and subsequent redeposition within siliciclastic settings.

Assumption 1. We make the assumption that whatever software is used embodies at least sufficient geological processes to model the range of possible geologies that are likely to be encountered within the true geology in question – at least to the level of detail relevant to the problem of interest.

The validity of results obtained will be conditional on this assumption. The software may, additionally, be able to model some geologies that are infeasible in the Earth (due either to infeasible parameter values input to the software or to errors in the software), or to model some geologies that are feasible in the Earth but are impossible in the true geological domain of interest, without damaging the principles of our method.

Since the components of GPM used naturally model only deposition and some diagenetic processes, our use of this package combined with Assumption 1 implicitly implies the following, second assumption:

Assumption 2. Either (a) the geology has not been significantly modified by unmodelled tectonic or diagenetic processes, or (b) we are only interested in the non-tectonic components of the architecture and rock types and will compensate for the tectonic modifications in a separate modelling or data decomposition step.

Any deviation from these assumptions will introduce errors into the prior probability distributions derived and may require a different package to be used. In such cases the rest of our elicitation method would be unchanged.

GPM requires various input parameters in order to create 3-D geological scenarios. Denote these parameters by **q**, where **q** is K-dimensional if there are K parameters, and let their range of values define parameter space Q. Then GPM can be regarded in purely functional terms as a black box that translates parameters **q** into Earth models **m** (Fig. 3). This black box embodies much dynamic prior information about depositional and diagenetic processes. Hence, if we allow parameters **q** to vary over their entire feasible ranges, the range of models produced by GPM represents a range of possible 3-D geological models. By Assumption 1, this range spans the range of possible real geologies, at least to the desired level of detail.

Although there may be huge variation in the models produced, this range is still very much smaller than the N-dimensional model space produced by simply discretizing the reservoir into N cells (as described earlier, $N = 1000$ in a relatively coarse model) with no additional information about the distribution of rocks within the cells (compare Fig. 2c and d). In the latter context, both credible and incredible models are allowed, for example, a model where every consecutive cell has alternately siliciclastic and carbonate rock types in a checkerboard pattern is feasible. In contrast, GPM would not be able to reproduce such a model; the model is therefore deemed impossible and removed from further consideration. Thus, if we consider only models that can be produced by GPM we have automatically greatly reduced the possible model space simply by the introduction of dynamic geological prior information. Such information is difficult to represent without using either GPM or another modelling package explicitly.

Defining prior probabilities

GPM reduces our range of possibilities from an N-dimensional hypervolume to some manifold that is at most K-dimensional, where K is usually much lower than N (Fig. 3). In this context a manifold can simply be thought of as a lower dimensional structure embedded within a higher dimensional space (Fig. 2d). An example of a manifold is the familiar Swiss roll structure; intrinsically this is a 2-D plane ($K = 2$) that is 'rolled up' within a 3-D space ($N = 3$).

By Assumption 1, the manifold defined by GPM spans at least the range of all possible geological models. For reasons described earlier, some additional models spanned by the manifold may be impossible. Even within the range of models that are possible, some will be more or less likely to occur in practice. We would therefore like to define over this manifold a probability distribution that describes the likelihood of occurrence of each of the models spanned, in order to further augment the prior information injected into subsequent investigations.

In principle there are at least two methods that can be used to create this distribution. Firstly, we could define the prior probability distribution across parameters $\mathbf{q}$ directly. Since GPM merely translates $\mathbf{q}$ into the model space M, we can also use GPM to translate this distribution into the model space. The resulting probability distribution in model space M would represent both prior information embodied within GPM and information from the prior distribution of parameters $\mathbf{q}$. In practice this method is problematical: parameters $\mathbf{q}$ include sedimentation rates, loci of sediment origination, pre-existing topography, compaction heterogeneity etc. Such information can only be derived from observations of existing geological architectures, properties and observable processes that occur on Earth today. Parameter values are then inferred by geologists who, in effect, carry out an 'inverse GPM' in their heads: they unravel the complex dynamic sedimentary and diagenetic processes, usually non-quantitatively, to infer rough ranges on some parameters in $\mathbf{q}$. For other parameters in $\mathbf{q}$ their relationship to observations in the present day will be so complex that no geologist can infer sensible estimates for their ranges, other than by intuition. In turn, it is not clear from where such intuition would be derived since no geologist can have observed true processes that take place over geological time scales, and hence geologists cannot intuitively estimate their effects quantitatively. The only alternative would be to observe present-day values for parameters $\mathbf{q}$ and describe these by a probability distribution, but these values may well not be pertinent over geological time scales.

Here, we opt for a second approach: we directly interrogate experts about the likelihood of occurrence of models produced by GPM. Our approach is to take sample models from the manifold and interrogate experts about the likelihood of their occurrence in the reservoir or geology of interest; we can then interpolate between these models along the manifold, to define the probability of occurrence over the entire section of the manifold bounded by the samples. Interrogating people for information requires an elicitation process; we now describe a new elicitation method suitable for this purpose.

Method to elicit geological information

In light of the brief summary of heuristics and biases given above and described in Baddeley *et al.* (2004), we need a strategy that does not ask for absolute probabilities, that allows the reliability of probabilistic estimates to be assessed, and that also mediates the effects of individuals' biases and heuristics by sensibly consolidating the input of multiple experts. We therefore include the following elements in our elicitation method:

(1) We ask only that experts provide *relative* probabilities of occurrence of sets of a small number of models at a time. This has the advantage that the subject need only have knowledge about the *relative* abundance of Earth structures that are similar to the models in question, rather than knowledge of *all* such structures

existing in the Earth. The small number also ensures that they can keep all examples in memory at once, rather than allowing one or more to be displaced by the availability heuristic.

(2) We ensure that 'loops' exist in the models presented within the sets. Loops can be illustrated if we consider only three models, A, B and C. Denote the relative probability of A and B by

$$\Pr(A, B) = \frac{P(A)}{P(B)}, \quad (1)$$

where $P(A)$ denotes the probability of model A being true. If we ask a subject to estimate Pr(A, B) and Pr(B, C) then we can estimate

$$\Pr(\mathrm{A}, \mathrm{C}) = \Pr(\mathrm{A}, \mathrm{B}) \times \Pr(\mathrm{B}, \mathrm{C}). \quad (2)$$

If we also ask the subject to estimate Pr(A, C) then the match between this estimate and the calculated value provides a consistency or reliability check on their probability estimates. Obviously loops can involve many more models than three. The requirements that probabilities around each loop are consistent, and that other required relationships between valid probabilities are satisfied, are referred to as *coherency conditions* (Lindley *et al.* 1979).

In practice we ensure that loops exist by presenting each subject with a set of L models ($L = 5$ in the current study) and asking them first to rank them in order, and second to estimate their relative likelihoods of occurrence. They do this by first assigning the most probable model in their view an (arbitrary) likelihood value of 10, then by assigning the other models (possibly non-integral) likelihoods relative to that model. Simply by dividing the different likelihood estimates thus obtained in all possible ways, we obtain estimates of the required relative probabilities around all possible loops that can be created with L models, similarly to Equations 1 and 2. All probabilities around loops within each set of L models are automatically consistent, but loops between different sets remain as coherency conditions.

(3) We ensure that each set incorporates models that are also members of other sets, and that at least one loop can be constructed between a set of models that span all other sets, such that relative probabilities around that loop have been constrained. Without this condition it is possible that pairs of models exist which occur in different sets, and about the relative probabilities of which we obtain no information.

(4) In principle we would like absolute probabilities of occurrence of models, not relative probabilities. If the manifold is bounded, then this problem can be solved by introducing the fundamental property of all valid probability distributions, that

$$\int_{Man} P(\mathbf{m})d\mathbf{m} = 1 \quad (3)$$

where the domain of integration is the entire manifold. This condition can be used to provide a constant value that normalizes the integral on the left of Equation 3 to unity and thus provides absolute probabilities.

If the manifold is unbounded then it is only possible to estimate absolute probabilities given some assumptions about the behaviour of the probability distribution approximately outside of the minimum convex hull (on the manifold) containing all assessed models. If the models assessed do not contain all possible degrees of freedom (e.g. if not all parameters in **p** were varied so as to span their true possible ranges) then the normalization procedure above will provide only normalized probability distributions that are conditional on those parameters lying within the ranges across which they were varied.

The elicitation process is carried out by e-mail with subjects in remote locations from the person running the elicitation process, henceforth called the *elicitor*. Whenever possible, and certainly on the first two occasions when a particular subject was involved in trials, the subject is in contact by telephone with the elicitor during the entire trial. The elicitors ensure that the subjects understand what is being asked of them, and encourages the subjects to talk through exactly what they were doing during every stage of the trial.

Our method consists of repeatedly presenting subjects with sets of $L = 5$ models selected to fulfil the above criteria and asking each subject to estimate the relative likelihoods of models as described in element (2) above. The subjects are presented with each set of models as a zipped attachment to a standard e-mail that explains how to assess relative likelihoods. The five models are given in the output file format of

GPM; each subject requires a copy of the GPM viewing program, GS, to view each model. Subjects are asked to evaluate the likelihood of occurrence of the final architecture and sedimentary grain-size distributions produced in the model. However, they can also use GS to run backwards and forwards through geological time to see how each model was created. Trials continue to be presented to subjects until either sufficient information is obtained or an equilibrium level of residual uncertainty is reached.

The elicitors take notes of the subjects' description of what they are doing from the opening of the standard e-mail through to the final evaluation of all models. The notes focus on how the subjects view the models (what angles of view are used, and in what modes of view – both dynamic and map view, either with and without any water layer are possible) and how they assess the models' likelihoods (on what their assessment is based, how and whether they incorporate dynamic information about how each final model was created in their assessment), and any other details mentioned by the subjects.

The four strategic elements described above do not uniquely define the sets of models to be presented to subjects in each trial. Thus, we are left with an experimental design problem: given a particular modelling package (here, GPM), and assuming that we vary parameters $\mathbf{q}$ over ranges that include the correct values, exactly which models should we include within each set presented to experts, and how many experts should we use, in order to obtain maximum information about the relative probability of models at reasonable cost? This is a complex question due to the large number of often intricate loops that might be created between sets, and to the unknown degree of variability that might occur between different subjects. In the next section we show how to design such experiments optimally in real time using statistical experimental design theory. To our knowledge this has not been done previously.

Experimental design

We define the experimental design scenario as follows: let $\mathbf{p} = [P_1, \ldots, P_M]^T$ be the vector of probabilities that we would like to estimate by elicitation. Usually $P_i = P(\mathbf{m}_i)$, $(i = 1, \ldots, B)$, where $P(\mathbf{m}_i)$ denotes the probability of model $\mathbf{m}_i$ occurring in reality, and $\mathbf{m}_i$ is the ith of a total of B models, sampled from the manifold, over which we wish to estimate a prior probability distribution.

Constraints offered by trials

To elicit relative probabilities during each trial we present a subject with a subset of L of the model samples, $\{\mathbf{m}_{k_1}, \ldots, \mathbf{m}_{k_L}\}$, and request estimates of the relative likelihoods of these models as described in the previous section. Denote the elicited estimates thus obtained by $\hat{P}_{K_iK_j}$, where $P_{K_iK_j}$ is defined by

$$P_{K_iK_j} = \frac{P_{K_j}}{P_{K_i}}, \tag{4}$$

where $\hat{P}_{K_iK_j}$ is an estimate of $P_{K_iK_j}$, and where $i = 1, \ldots, L-1$ and $j = i+1, \ldots, L$ form a nested loop. Re-arranging we obtain linear constraints of the form

$$\hat{P}_{K_iK_j} P_{K_i} = P_{K_j} \quad \Leftrightarrow \quad \hat{P}_{K_iK_j} P_{K_i} - P_{K_j} = 0.$$

Let matrix $\mathbf{A}_1$ and vector $\mathbf{d}_1$ be defined such that all equations of the form (5) that are *expected to be derived from the next trial* may be expressed as

$$\mathbf{A}_1\mathbf{p} = \mathbf{d}_1. \tag{6a}$$

If we also have estimates of the uncertainties that might be expected of the vector $\mathbf{d}_1$, expressed as a covariance matrix $\mathbf{C}_1$, then the equations describing constraints available after the next trial become

$$\mathbf{A}_1^T\mathbf{C}_1^{-1}\mathbf{A}_1\mathbf{p} = \mathbf{A}_1^T\mathbf{C}_1^{-1}\mathbf{d}_1 \tag{6b}$$

In practice we estimate a diagonal $\mathbf{C}_1$ based on results of previous trials (see below). Note that Equation 6a is simply a particular case of the more general Equation 6b, with $\mathbf{C}_1 = \mathbf{I}$ where $\mathbf{I}$ is the appropriate identity matrix, and where Equation 6a has been multiplied by $\mathbf{A}_1^T$. We therefore consider Equation 6b only.

We assume that several estimates of relative probabilities may have been elicited in previous trials, providing a set of equations similar to

$$\hat{P}_{ij} = \frac{P_j}{P_i} \quad \Leftrightarrow \quad \hat{P}_{ij}P_i - P_j = 0, \tag{7}$$

where i and j run over all indices for which estimates of $\hat{P}_{ij}$ already exist. Let matrix $\mathbf{A}_2$ and vector $\mathbf{d}_2$ be defined such that all equations derived from previous trials may be expressed in the linear form of Equation 7 as

$$\mathbf{A}_2\mathbf{p} = \mathbf{d}_2. \tag{8a}$$

If uncertainties in $\mathbf{d}_2$ are known and expressed as a covariance matrix $\mathbf{C}_2$ then the corresponding equations derived from previous trials become

$$\mathbf{A}_2^T\mathbf{C}_2^{-1}\mathbf{A}_2\mathbf{p} = \mathbf{A}_2^T\mathbf{C}_2^{-1}\mathbf{d}_2 \quad (8b)$$

with solution

$$\mathbf{p} = \left[\mathbf{A}_2^T\mathbf{C}_2^{-1}\mathbf{A}_2\right]^{-1}\mathbf{A}_2^T\mathbf{C}_2^{-1}\mathbf{d}_2 \quad (9a)$$

and post-inversion covariance matrix

$$\mathbf{C}_{post} = [\mathbf{A}_2^T\mathbf{C}_2^{-1}\mathbf{A}_2]^{-1}, \quad (9b)$$

assuming that sufficient constraints exist from previous experiments so that Equations 8a or 8b are strictly over-determined and hence that matrix inverses in Equations 9a or 9b exist. In practice we estimate a diagonal matrix for $\mathbf{C}_2$ by assuming that relative likelihood estimates provided are uncertain with a standard deviation of 50% of each estimate. This would seem to be a conservative estimate.

Note that Equation 8a is simply a particular case of the more general Equation 8b under the substitution $\mathbf{C}_2 = \mathbf{I}$, where $\mathbf{I}$ is the appropriate identity matrix and where Equation 8a has been multiplied by $\mathbf{A}_2^T$, and the solution to Equation 8a is given by making the same substitution in Equations 9a and 9b. We therefore only consider Equations 8b, 9a and 9b.

If it is the case that Equation 8b is under-determined then the system must be regularized to remove non-uniqueness in the solution (Equation 9a). This is often achieved either by damping out zero or near-zero eigenvalues, or by smoothing the solution according to some chosen kernel (Menke 1989). However, both of these strategies effectively add additional prior information to the system – information that we do not have. Rather than pursue this line of discussion further, we note that, if a single, non-normalized probability value is fixed at an arbitrary value, at least in principle it is relatively easy to carry out sufficient trials that systems 8a or 8b become over-determined. This is achieved by fixing an arbitrary (non-zero) probability value to 1, by including every individual probability in $\mathbf{p}$ within at least one trial, and by ensuring that at least one particular individual probability in $\mathbf{p}$ is included in every trial conducted. We therefore assume that previous trials have been carried out sensibly and that there is no need to give further consideration to under-determined systems.

Design problem definition

The experimental design problem to be solved in order to design each trial given all previous information can now be defined precisely.

Let

$$\mathbf{A} = \begin{bmatrix} \mathbf{A}_1 \\ \mathbf{A}_2 \end{bmatrix}, \quad \mathbf{d} = \begin{bmatrix} \mathbf{d}_1 \\ \mathbf{d}_2 \end{bmatrix}, \quad \mathbf{C} = \begin{bmatrix} \mathbf{C}_1 & \mathbf{0} \\ \mathbf{0} & \mathbf{C}_2 \end{bmatrix}. \quad (10)$$

The design problem is to select the subset of L models $\{\mathbf{m}_{k_1}, \ldots, \mathbf{m}_{k_L}\}$ that results in a matrix $\mathbf{A}_1$ and vector $\mathbf{d}_1$ such that the set of equations

$$\mathbf{A}^T\mathbf{C}^{-1}\mathbf{A}\ \mathbf{p} = \mathbf{A}^T\mathbf{C}^{-1}\mathbf{d} \quad (11)$$

is expected to place maximum constraint on (result in minimum residual uncertainty in) the probability vector of probabilities $\mathbf{p}$ of interest.

Optimal design algorithm

Since Equation 11 is linear, this problem can be solved using linear experimental design theory. Curtis *et al.* (2004) present a deterministic algorithm that approximately solves this kind of problem. The only limitation is that the covariance matrices must be diagonal (i.e. measured or estimated correlations between probability estimates are neglected). This limitation is not severe since in practical situations it is unlikely that sufficient trials will be conducted that correlations between individual probability assessments can be measured robustly, and estimates of expected correlations should in principle be based on such measurements. Hence, although interprobability correlations may exist it is unlikely that accurate descriptions of these will be accessible, so diagonal covariance matrices should suffice for practical purposes.

The algorithm of Curtis *et al.* (2004) initially includes *all M* models within the set that would (hypothetically) be presented to the subject during the next trial (i.e. set $L = M$ above). This results in matrices $\mathbf{A}_1$ and $\mathbf{C}_1$ and vector $\mathbf{d}_1$ defined such that Equation 6b contains *all possible* equations of the form (5). The design problem is then solved by deleting *M-L* models from this set, so that the remaining L models form a subset that ensures that Equation 11 results in low residual uncertainty in $\mathbf{p}$.

In essence, residual uncertainty in $\mathbf{p}$ occurs because matrix $\mathbf{A}^T\mathbf{C}^{-1}\mathbf{A}$ has small eigenvalues: uncertainties in the vector $\mathbf{d}$ propagate into vector **p** with a magnification in proportion to the inverse square root of these eigenvalues (e.g. Menke 1989). Hence, uncertainty in **p** will be minimized when matrix $\mathbf{A}^T\mathbf{C}^{-1}\mathbf{A}$ has eigenvalues with the largest possible magnitudes. Small eigenvalues generally occur because one or more rows of **A** are nearly linearly dependent on the other rows. Since $\mathbf{C}^{-1}$ is diagonal it merely weights rows of **A** relative to each other. Complete linear dependence results in a zero eigenvalue and singularity of the matrix $\mathbf{A}^T\mathbf{A}$ and $\mathbf{A}^T\mathbf{C}^{-1}\mathbf{A}$.

Each equation of the form (5) results in a single row of matrix $\mathbf{A}_1$ in Equation 6b. Rows of matrix $\mathbf{A}_1$ that are linearly dependent on other rows of $\mathbf{A}_1$ correspond to redundant constraints resulting from the current trial. Rows of $\mathbf{A}_1$ that are linearly dependent on rows of $\mathbf{A}_2$ correspond to constraints that effectively repeat those offered from previous trials.

The algorithm developed by Curtis *et al.* (2004) removes small eigenvalues from the spectrum by deleting those equations that correspond to rows of $\mathbf{A}_1$ that are most linearly dependent on other rows of matrix **A**. It does this by removing one model at a time from the set that will be presented to the subject. Each removal of a model, $\mathbf{m}_q$ say, results in several rows being deleted from **A**, namely all rows corresponding to equations concerning relative probability estimates involving P_q. The algorithm iterates, and at each iteration that model is deleted for which the corresponding rows are most linearly dependent on other rows in **A**. The measure of linear dependence employed by Curtis *et al.* (2004) also accounts for the relative weighting imposed by $\mathbf{C}^{-1}$. The algorithm stops iterating when the number of models to be presented to the subject reaches the required number, always five in our experiments.

While the remaining set of equations result in a matrix **A** that has rows that may not involve the least linear dependency possible, experiments in various domains of application have shown that the algorithm gives designs which significantly increase the smallest eigenvalues in the spectrum relative to those that would pertain if a random experiment was performed (Curtis *et al.* 2004). Alternative methods include the methods of Curtis (1999*a*, *b*), which tend to increase the largest eigenvalues of the spectrum, and those of Rabinowitz & Steinberg (1990), Steinberg *et al.* (1995), Maurer & Boerner (1998) and Maurer *et al.* (2000), which roughly, focus on increasing all eigenvalues equally. The design problem above is over-determined so all small eigenvalues will usually be included, unregularized, within the solution to Equation 11. Since small eigenvalues are principally responsible for large residual uncertainties in the probabilities estimated, the method of Curtis *et al.* (2004) is well suited to the design problem at hand.

An elicitation experiment

We carried out a simple elicitation experiment in order to demonstrate the methodology presented above. In this experiment only a single subject was used, and the manifold over which we estimated the prior probability distribution was 1-dimensional. Our aim in the current experiment was to demonstrate the elicitation technique and to show that the optimal design method presented above increases information obtained during the elicitation procedure. Our results are not sufficiently prolific to estimate statistically the percentage improvement that might be expected by using the design algorithm rather than conducting random trials. Instead our results simply demonstrate that the algorithm does provide improved estimates of elicited probabilities and hence encourage more extensive tests of the elicitation methodology in future.

During the experiment we assessed the probability distribution representing the prior likelihood of occurrence of different fluvial-deltaic geological models produced by GPM by varying only a single parameter, the 'diffusion coefficient' (explained below). The resulting elicited probability distributions will be conditional both on all the other parameters required to run GPM being fixed at those values used in this study, and on the diffusion coefficient best representing true sedimentary erosion and redeposition in the fluvial-deltaic systems modelled being within the range of diffusion coefficients spanned by this experiment (the range 0.7–30 $\mathrm{m^2/a}$ was used).

When modelling sedimentary erosion and redeposition, the cumulative effects of various erosional processes that transport sediment approximately (locally) downdip is modelled by applying a diffusion equation to simulate redistribution of existing sediment (e.g. Tetzlaff & Harbaugh 1989). A simple example would be

$$\frac{dh}{dt} = D\nabla^2 h, \tag{12}$$

where h is the sediment top-surface height, D is the diffusion coefficient and

$$\nabla^2 h = \left(\frac{\partial^2 h}{\partial x^2} + \frac{\partial^2 h}{\partial y^2}\right), \tag{13}$$

where x and y are orthogonal horizontal

coordinates, is the Laplace operator or second spatial derivative of h. In this equation, the rate of diffusion is moderated both by the local curvature in top sediment topography and by the diffusion coefficient.

Increasing the diffusion coefficient makes sediment redistribution more efficient. For a given set of initial conditions and a fixed period of geological time over which the diffusion process acts, the effects of increasing the diffusion coefficient are to transport more sediment. Since lower diffusion coefficients acting over a longer time period could produce a somewhat similar effect, increasing the diffusion coefficient creates final models where the geological landscapes appear to be more 'mature' to a geologist (compare the plots on the left and right of Fig. 4). Table 1 shows the diffusion coefficient values used in each model $\mathbf{m}_i$ used in this study and for which estimates of $P_i = P(\mathbf{m}_i)$ are required (see below for details).

Experimental description

In this study we elicit information about fluvial-deltaic environments. The single subject was geologically expert, but was not a specialist in this depositional setting. Hence, the knowledge elicited in this paper is the general knowledge of a good geologist about a fairly standard sedimentary environment.

Table 1 shows indices by which each model simulated is referred to in the text, the parameter values used in the GPM simulation for each model, and those models (and their ordering) that were included within each of nine trials conducted on the subject. Trials 1–5 were designed randomly (i.e. models to be included in each of these trials were selected at random,

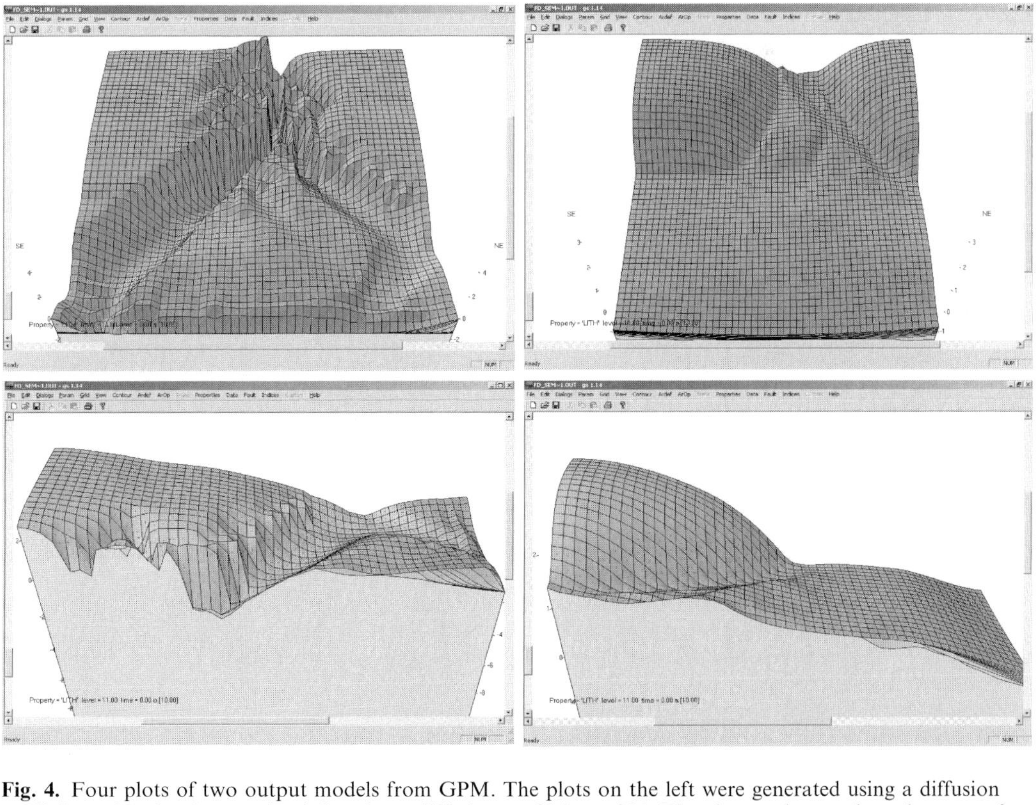

Fig. 4. Four plots of two output models from GPM. The plots on the left were generated using a diffusion coefficient of 1, the plots on the right using a diffusion coefficient of 30. The plots at the top show the top surface with the water layer removed; those at the bottom show cross-sections through the centre of each model showing the prograding sequence developed in the sub-surface. Darker grey, deposited rocks of different grain sizes; plain light grey, basement rock.

Table 1. *Parts of the framework for Geological Prior Information addressed by each paper in the current volume*

Model Indices[a]	Models Included[b]									Diffusion Coefficient[c]	Transport Coefficient[c]	Basement erodability[c]
	T1	T2	T3	T4	T5	T6	T7	T8	T9			
1		b		b			b	b		0.7	1	1
2				a		a	a	a	a	1	1	1
3					a			c	b	1.3	1	1
4				c		b		d		2	1	1
5					b	c		e	c	3	1	1
6					c	d	c			5	1	1
7	d	c	c	d	d		d		d	10	1	1
8		d		e						20	1	1
9	e				e	e	e		e	30	1	1
10	a									1	1	0.7
11			a							0.9	1	1
12	b									1.5	1	1
13	c	a	b							2	1	1
14		e	d							1	0.3	1
15			e							1	2	1

The letters in each column show those five models that were included within each trial
[a] Model index numbers referred to in text
[b] Models included in each trial
[c] Values used in each model simulation
T1, etc, trial numbers; letters show the five models included in each trial
a–e, models presented to subject in each trial

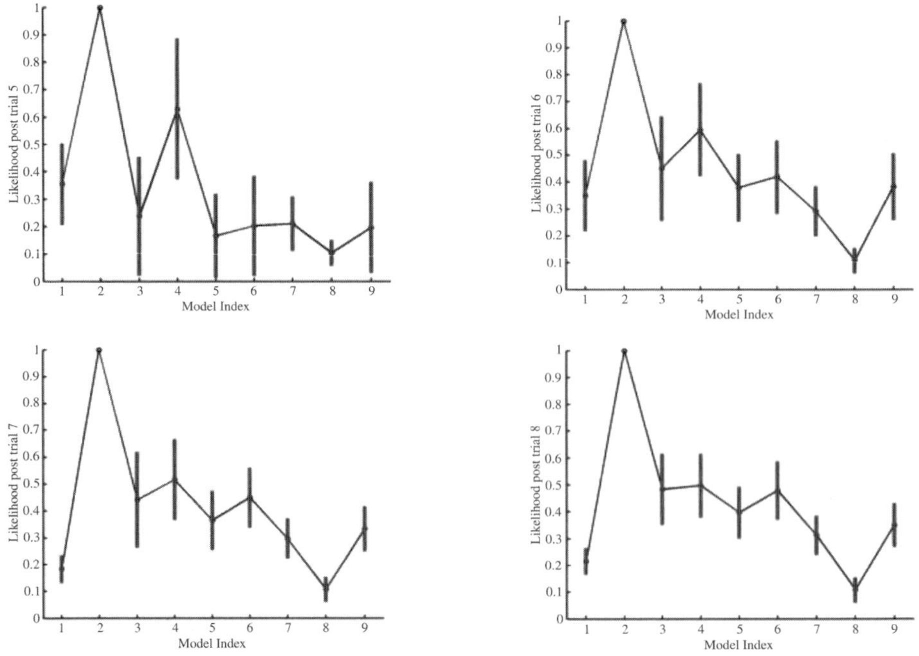

Fig. 5. Non-normalized prior distribution (black line) with one standard deviation uncertainties (vertical bars) after likelihoods had been elicited for models. Trials 1–5 (top left), 1–6 (top right), 1–7 (bottom left) and 1–8 (bottom right).

Table 2. *Parameters used in the model in Fig. 1 and the range of each given prior knowledge. This is sufficient information to define a uniform prior probability distribution across all parameters*

Model indices[a]	Likelihood estimates[b]								
	T1	T2	T3	T4	T5	T6	T7	T8	T9
1		0		5			2	6	
2				10		9	10	7	6
3					10			10	10
4				8		10		9	
5					7	8.5		8.5	9
6					8.5	8	7		
7	2	3		3	8		4		9.5
8		4		1					
9	1				9	7	3		2

[a] Model index numbers referred to in text. (Only models 1–9 are shown since these are examples of our desired conditional distribution; see main text.)
[b]Likelihood estimates assigned by each subject during each trial. (Only relative likelihoods—ratios between different models—should be interpreted; see main text.)
T1 etc., trial numbers (Trial 3 provided no useful information)

without using the experimental design procedure); trials 6–8 were designed by running the design procedure before each trial, including all knowledge gained from previous trials within matrix $\mathbf{A}_2$ in Equations 6 and 7. Trial 9 was again designed randomly and was used as a control trial as shown below, since the set of models used in this trial did not appear in any of the first five trials or in the three designed trials.

Several points should be noted from Table 1. Firstly, we are only interested in estimating prior probabilities conditional on all parameters other than the diffusion coefficient being fixed. Table 1 also shows that in some models, other parameters were varied (models 10, 14 and 15). Models 10–13 were additionally performed with boundary conditions that were all closed to sediment and fluid flow; in all other simulations boundaries were open towards the basin, open sideways below sea level, and closed sideways above sea level and landward (see Fig. 4). Hence, all models below the dashed line (10–15) do not constitute samples of our conditional distribution, and can be regarded as 'nuisance' models. These simulations were made in order to test the effects of outlying models on elicited probability estimates within the conditional range; the results are not central to this paper and will be analysed elsewhere. Here we only consider relative likelihoods from each trial that pertained to models 1–9. As a consequence, trial 3 is immediately rendered useless for this particular study since it includes only a single model from within the conditional range and hence results in no relative likelihood estimates.

The time simulated in each run was 1 ka. Sediment input had four different grain sizes, 1, 0.1, 0.01 and 0.001 mm, in equal proportions. Each GPM run was produced by a single sediment source point at the centre of the landward boundary of the model, with identical initial topography, and in models 1–9 with flow boundary conditions open below sea level and closed above sea level. Sea level was constant throughout each run.

Results

Table 2 shows likelihood estimates of probabilities sampled from the conditional distribution described earlier, from trials involving our single subject. In each trial the subject was asked to order the models in terms of their likelihood of occurrence and to assign a likelihood value of 10 to the most likely model. Other models were then to be assigned (not necessarily integral) likelihoods relative to this maximum value. Hence, only relative likelihoods (ratios of the values in columns of Table 2) should be interpreted because the absolute values are arbitrarily scaled.

Many inconsistencies are evident in the data. For example, the relative likelihood of models 1 and 2 was estimated three times, in trials 4, 7 and 8, with estimates of $P_{12} = P_2/P_1$ assuming values 2, 5 and 1.17 respectively. Similarly estimates of the relative likelihood $P_{92} = P_2/P_9$ assume values 1.29, 3.33 and 3 in trials 6, 7 and 9 respectively. However, some consistencies are also evident in the data. Whenever model 3 is included within a trial it is assigned the maximum likelihood value of 10. When model 3 is not included, the maximum likelihood value is always assigned to models 2 or 4 (when either

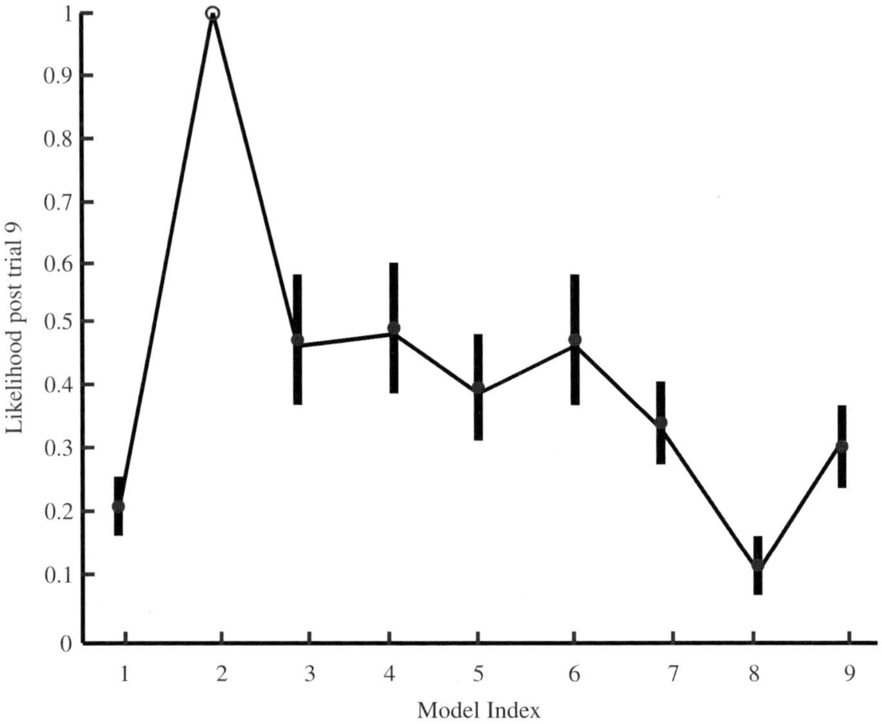

Fig. 6. Final non-normalized prior distribution (black line) with one standard deviation uncertainties (vertical bars) after likelihoods had been elicited for models 1–9 (bottom right).

or both are included). Hence, the conditional distribution appears to assume a maximum somewhere around the diffusion coefficient $1.3\,m^2/a$, and this maximum almost certainly occurs between coefficients 1 and 2 (a maximum at $D = 1$ is confirmed in Fig. 6, which shows the final estimate of the conditional distribution). Additionally, notice that likelihoods lower than 5 are all confined to models 1, 7, 8 or 9, i.e. towards the two extremes of the range of diffusion coefficient values considered. Therefore, assuming that the trend of diminishing likelihoods continues outside of the range of coefficients included in these trials, we appear to span a range of coefficient values over which the probability distribution reduces to at most one half of its maximum value.

Inconsistencies such as those observed above are not unexpected. The subject in this case did not have a strong quantitative background and had only a general knowledge of probability theory. General knowledge usually includes only an inkling of the concept of consistency of relative probabilities, so the subject would not have been checking to ensure such consistency. Additionally, it has been shown in numerous previous studies that subjects have most difficulty in assessing extremes in probability, i.e. values closest to zero or to one. Probability ratios between high and low probabilities such as P_{12} and P_{92} described above are therefore expected to exhibit maximum inter-trial variation. It is likely, therefore, that the current method could be enhanced by illustrating such inconsistencies graphically during the elicitation procedure in order that the subject could reconcile their relative probability estimates (e.g., Lau & Leong 1999; Renooij 2001).

Trials 6–8 were designed using the algorithm presented above, using information from previous trials to construct the design of each subsequent trial. Trial 9 on the other hand was designed randomly. We used this as a single control trial: after each optimally designed trial we compare results obtained with those results that would have been obtained if the control trial had been conducted instead, assuming that the same results would have been obtained during the control trial whenever it was carried out. While the percentage improvement (if any) in results observed in this test will certainly not be statistically significant, we simply test whether an improvement is consistently observed.

We first compare the standard deviation (s.d.) of the probability estimates, calculated using Equation 9b, using optimal and control trials for each designed trial. The difference in the mean s.d. (control minus optimal) expressed as a percentage of the mean s.d. estimate obtained in each trial was 3.8%, 10% and 2.6% in trials 6, 7 and 8 respectively where a positive percentage represents an improvement. While these improvements in the post-trial s.d.'s are small, they are consistently positive. Also, the total improvement due to the optimal strategy is compounded in each successive trial and hence is 15.7%. The actual percentage values are statistically unconstrained since they depend both on the (random) control trial used and on the estimates obtained from the subject during the control trial. Hence, positivity of the percentages and hence consistent improvement is probably the limit of interpretation possible given only these results.

In order to search for any improvement in the mean probability estimates, we assume that the probability estimates obtained using all information from trials 1–9 in Equation 9a provide the best estimates available, and these are shown in Figure 6. We calculate the mean absolute difference between these best probabilities and those estimated after each of trials 6–8. We compare these with the differences that are obtained if each of trials 6–8 is, respectively, replaced with the control trial 9. The results expressed as percentage of the probabilities in Figure 6 are 15%, 6% and 2% after trials 6, 7 and 8 respectively. Again, the improvements are all positive, indicating improvement in all cases. The decrease in magnitude of the improvement over successive trials is expected: as trials progress we approach an equilibrium level of uncertainty: that which is inherent within our subject's general knowledge. As we approach that equilibrium, additional trials should on average show diminishing improvement in the mean estimate.

Discussion

The results presented above confirm that the optimal elicitation strategy described earlier improves the information content expected in conducted trials beyond that expected in random trials, at least in the simple experiment that we performed and describe herein. Clearly more detailed and extensive trials are required in order to constrain exactly how much improvement one might reasonably expect using such a strategy over other methods in particular elicitation experiments. However, the strategy makes use of real-time experimental design between elicitation trials, including all pertinent information available; this must be strategically optimal, although the details of the algorithm can probably be improved in future (e.g. by including graphical illustrations of probabilistic inconsistencies in real time as described above).

The assignment of uncertainties in matrix **C** is probably the least satisfactory part of the elicitation algorithm described herein. If uncertainties in the P_{K_i} are symmetric then uncertainties in the $P_{K_iK_j}$ are usually not symmetric. This fact is responsible for most of the approximations employed: the requirement imposed by most experimental design algorithms (including that of Curtis *et al.* 2004) to represent uncertainties by symmetric standard deviations implies that uncertainties in equations involving terms like $P_{K_iK_j}$ are necessarily only crude approximations. Although the approximations currently employed are adequate to provide more accurate results than would be provided by random trials, improving this aspect of the algorithm should be the subject of future research.

As in many other studies, our strategy would attempt to compensate for various natural human biases (see review in Baddeley *et al.* 2004) by including multiple experts. If possible, trials should be continued until an equilibrium level of residual uncertainty is attained. Once this equilibrium has been reached, the addition of further randomly selected experts to those already interrogated should have little effect on the probability estimates and their uncertainties. This test will diminish some of the effects of over-confidence of individual experts and also in the base-rate neglect heuristic. However, if all or most experts suffer from identical biases in either of these respects, these effects will not be removed.

By solving Equation 8b using Equations 9a and 9b using data from multiple experts, the strategy also creates a consensus distribution by taking a least-squares fit to the set of individual probability distributions. By differentiating estimates from different experts explicitly within the probability vector and design matrices, we might also allow the design algorithm to select the expert from whom information should be elicited in each trial. Thus, the algorithm itself might be designed to ameliorate the other common problem of anchoring. Lastly, by careful choice of the set of experts, selected so that their ranges of experience overlap minimally, we can also diminish the availability bias. Additionally, this would presumably reduce off-diagonal correlation terms in matrix **C** (which are ignored by the current algorithm), since inter-expert correlation is likely to be reinforced if experts have similar background knowledge or experience. Hence, the strategy presented should help to diminish many of the principal biases that commonly affect elicitation procedures.

We return finally to the situation of a well intersecting two rock types as described in the Introduction and in Figure 1a. Assume the geology was known from prior information to have been formed in a siliciclastic, fluvial-deltaic environment. A method to quantify an approximate conditional prior distribution for the rock types intersected by the well is as follows:

- Define several sedimentary rock types and for each of models 1–9 assign rock types throughout the model in 3-D.
- Calculate the histogram of rock types that could be intersected by a well drilled vertically through that environment at any horizontal location.
- Calculate the mean weighted histogram of rock types across all models, where each individual model histogram is weighted by: (a) the best estimate of the prior probability of that model occurring given in Figure 6 and, if desired, by (b) the proportion of the diffusion parameter space represented by that model given the sampling of this space shown in Table 1 (i.e. by the inverse sampling density).
- Normalizing this final histogram to have unit area results in a prior probability distribution of the rock types intersected by the well, conditional on all of the same conditions as those pertaining to the prior distribution in Figure 6.

Conclusions

In many situations central to academic and business research, quantitative information must be elicited from experts. Such information is often valuable, uncertain and difficult to obtain, and it must be derived by directly interrogating people residing in many different geographical locations. In such situations, poor results often occur due to expert over-confidence and other natural human biases. This paper presents a new, optimized method to mediate the effects of such biases in order to gather more accurate and probabilistic information. The information thus obtained forms prior probability distributions for further analysis or decision making.

The new elicitation strategy involves real-time design of elicitation trials based on all available information. The proposed strategy allows more information to be gained from expert elicitation than would be gained through non-optimally designed interrogation. We demonstrate this improvement in a simple experiment in which the conditional probability distribution (or relative likelihood) of a suite of nine possible models of fluvial-deltaic geologies was elicited, both optimally and randomly, from a single expert in a location remote from the elicitor. These results will be augmented in future by a more extensive application involving multiple experts in multiple locations and increased diversity of geological settings.

References

BAECHER, G. B. 1988. *Judgemental Probability in Geotechnical Risk Assessment*. Technical Report, Prepared for The Office of the Chief, US Army Corps of Engineers. World Wide Web Address: http://www.ence.umd.edu/~gbaecher/papers.d/judge_prob.d/judge_prob.html.

BADDELEY, M., CURTIS, A. & WOOD, R. 2004. An introduction to prior information derived from probabilistic judgements: elicitation of knowledge, cognitive bias and herding. *In:* CURTIS, A. & WOOD, R. (eds) *Geological Prior Information: Informing Science and Engineering*. Geological Society, London, Special Publications, **239**, 15–27.

BOSCHETTI, F. & MORESI, L. 2001. Interactive inversion in geosciences. *Geophysics*, **66**, 1226–1234.

BURGESS, P. M. & EMERY, D. J. 2004. Sensitive dependence, divergence and unpredictable behaviour in a stratigraphic forward model of a carbonate system. *In:* CURTIS, A. & WOOD, R. (eds) *Geological Prior Information: Informing Science and Engineering*. Geological Society, London, Special Publications, **239**, 77–93.

COUPÉ, V. M. H. & van der GAAG, L. C. 1997. Supporting probability elicitation by sensitivity analysis: *Knowledge Acquisition, Modeling and Management, Lecture Notes in Computer Science*, **1319**, 335–340.

CURTIS, A. 1999*a*. Optimal experiment design: cross-borehole tomographic examples. *Geophysical Journal International*, **136**, 637–650.

CURTIS, A. 1999*b*. Optimal design of focussed experiments and surveys. *Geophysical Journal International*, **139**, 205–215.

CURTIS, A. & LOMAX, A. 2001. Prior information, sampling distributions and the curse of dimensionality. *Geophysics*, **66**, 372–378.

CURTIS, A., MICHELINI, A., LESLIE, D. & LOMAX, A. (2004). A deterministic algorithm applied to experimental design applied to tomographic and microseismic monitoring surveys. *Geophysical Journal International*, **157**, 595–606.

GRIFFITHS, C. M., DYT, C., PARASCHIVOIU, E. & LIU, K. 2001. SEDSIM in hydrocarbon exploration. *In:* MERRIAM, D. F. & DAVIS, J. C. (eds) *Geological Modeling and Simulation: Sedimentary Systems*. Kluwer Academic/Plenum Publishers, New York.

JUSLIN, P. & PERSSON, M. 2002. Probabilities from examplars (PROBEX): a 'lazy' algorithm for probabilistic inference from generic knowledge. *Cognitive Science*, **26**, 563–607.

KAHNEMAN, D., SLOVIC, P. & TVERSKY, A. (eds) 1982. *Judgement Under Uncertainty: Heuristics and Biases*. Cambridge University Press, Cambridge.

LAU, A.-H. & LEONG, T.-Y. 1999. Probes: a framework for probability elicitation from experts. *American Medical Informatics Association Annual Symposium*, 301–305.

LINDLEY, D. V., TVERSKY, A. & BROWN, R. V. 1979. On the reconciliation of probability assessments (with discussion). *Journal of the Royal Statistical Society*, **142**, 146–180.

MAURER, H. & BOERNER, D. 1998. Optimized and robust experimental design: a nonlinear application to EM sounding. *Geophysical Journal International*, **132**, 458–468.

MAURER, H., & BOERNER, D. & CURTIS A. 2000. Design strategies for electromagnetic geophysical surveys. *Inverse Problems*, **16**, 1097–1117.

MENKE, W. 1989. *Geophysical Data Analysis: Discrete Inverse Theory*. Revised edition. Academic Press, Orlando, International Geophysics Series.

MORGAN, M. G. & HENRION, M. 1990. *Uncertainty: A Guide to Dealing with Uncertainty in Quantitative Risk and Policy Analysis*. Cambridge University Press, Cambridge.

RABINOWITZ, N. & STEINBERG, D. M. 1990. Optimal configuration of a seismographic network: a statistical approach. *Bulletin of the Seismological Society of America*, **80**, 187–196.

RENOOIJ, S. 2001. Probability elicitation for belief networks: issues to consider. *The Knowledge Engineering Review*, **16**, 255–269.

SCALES, J. 1996. Uncertainties in seismic inverse calculation. *In:* JACOBSON, B. H., MOSEGAARD, K. & SIBANI, P. (eds.) *Inverse Methods*. Springer-Verlag, 79–97.

SKINNER, D. C. 1999. *Introduction to Decision Analysis*. Probabilistic Publishing.

STEINBERG, D. M., RABINOWITZ, N., SHIMSHONI, Y. & MIZRACHI, D. 1995. Configuring a seismographic network for optimal monitoring of fault lines and multiple sources. *Bulletin of the Seismological Society of America*, **85**, 1847–1857.

TETZLAFF, D. 2004. Input uncertainty and conditioning in siliciclastic process modelling. *In:* CURTIS, A. & WOOD, R. (eds) *Geological Prior Information: Informing Science and Engineering*. Geological Society, London, Special Publications, **239**, 95–109.

TETZLAFF, D. & HARBAUGH, J. W. 1989. *Simulating Clastic Sedimentation: Computer Methods in the Geosciences*. Van Nostrand Reinhold, New York.

TETZLAFF, D. & PRIDDY, G. 2001. Sedimentary process modeling: from academia to industry. *In:* MERRIAM, D. F. & DAVIS, J. C. (eds) *Geological Modeling and Simulation: Sedimentary Systems*. Kluwer Academic/Plenum, New York, 45–69.

WIJNS, C., POULET, T., BOSCHETTI, F., DYT, C. & GRIFFITHS, C. M. 2004. Interactive inverse methodology applied to stratigraphic forward modelling. *In:* CURTIS, A. & WOOD, R. (eds) *Geological Prior Information: Informing Science and Engineering*. Geological Society, London, Special Publications, **239**, 147–156.

WOOD, R. & CURTIS, A. 2004. Geological prior information and its application to geoscientific problems. *In:* CURTIS, A. & WOOD, R. (eds) *Geological Prior Information: Informing Science and Engineering*. Geological Society, London, Special Publications, **239**, 1–14.

Interactive inverse methodology applied to stratigraphic forward modelling

C. WIJNS[1,2], T. POULET[1], F. BOSCHETTI[1], C. DYT[1] & C. M. GRIFFITHS[1]

[1]*Commonwealth Scientific and Industrial Research Organization, PO Box 1130, Bentley, WA 6102, Australia*

[2]*School of Earth and Geographical Sciences, University of Western Australia, Crawley, WA 6009, Australia*

Abstract: An effective inverse scheme that can be applied to complex 3-D hydrodynamic forward models has so far proved elusive. In this paper we investigate an interactive inverse methodology that may offer a possible way forward. The scheme builds on previous work in linking expert review of alternate output to rapid modification of input variables. This was tested using the SEDSIM 3-D stratigraphic forward-modelling program, varying nine input variables in a synthetic example. Ten SEDSIM simulations were generated, with subtle differences in input, and five dip sections (fences) were displayed for each simulation. A geoscientist ranked the lithological distribution in order of similarity to the true sections (the true input values were not disclosed during the experiment). The two or three highest ranked simulations then acted as seed for the next round of ten simulations, which were compared in turn. After 90 simulations a satisfactory match between the target and the model was found and the experiment was terminated. Subsequent analysis showed that the estimated input values were 'close' to the true values.

Underlying most forward-modelling exercises in geosciences is an implicit inverse question. A formal inverse approach to geological modelling has only recently started to develop, due mostly to two reasons: first, standard geological modelling is extremely computationally intensive and, second, it is hard to develop numerical cost functions, in order to drive the inverse search, that can capture enough geological knowledge to make the search meaningful. Creating a cost function is difficult because it involves distilling into a single value an assessment of the geometries, volumes and positions, as well as determining how geologically reasonable a solution may be.

We employ interactive inversion to tackle the second problem. The concept is of such simplicity as to appear little more than a novelty: replace the numerical evaluation of solution quality with a subjective value provided by an expert user. Underneath the simplicity there is a very powerful concept. We replace what in most cases is an artificial numerical function with the computational power of an expert brain. This provides not only the best 'geological processing power' we currently have, but also intuition and expertise, in the form of theoretical knowledge, experience and *a priori* information. Here, the notion of 'expert user' may take different meanings. At the most superficial level, a geologist may have enough expertise reliably to evaluate basic similarities in different geological fields. This is the level we use in this work. At another level, a user who is an expert in some specific forward modelling can employ his or her experience to better interpret the modelling output in terms of the software functionality. As a result, for example, convergence may be improved and geological understanding may benefit. Yet another level involves a user who is an expert in the optimization tool used (a genetic algorithm in this work) and who may be able to perceive when the inversion itself needs 'help' and tune inversion parameters on the run. Boschetti & Moresi (2001) demonstrate applications of these last two cases of expertise. It would be possible to employ all these levels jointly by asking users with different expertise to rank models together.

Attempts to formalize the geological processing power of an expert have been presented in the literature but, to our knowledge, are incomplete. Our own experience in attempting to develop a geological image similarity module, accounting for human input, is encouraging (Kaltwasser *et al.* in press) but still far from being applicable to real-world problems. Extensive reference to interactive inversion and

From: CURTIS, A. & WOOD, R. (eds) 2004. *Geological Prior Information: Informing Science and Engineering*. Geological Society, London, Special Publications, **239**, 147–156. 186239-171-8/04/$15.00

standard implementations, and a review of several artistic and engineering applications, can be found in Takagi (2001). Boschetti & Moresi (2001) and Wijns *et al.* (2003) present applications to geoscientific problems.

More recently, we have focused on visualizing the results of an interactive inversion. The purpose is to enable our inverse strategy to return information to the user in a 'language' similar to the one the user employs to feed information to the code. We expect the user to be an expert geoscientist, not an expert in inverse theory. We want the user to understand visually what the inversion has achieved, to obtain further insights into the problem, and, by so doing, to provide more information to the inversion. Such information can be used as further *a priori* input in a subsequent inversion run (e.g. Boschetti *et al.* 2003).

We extend this approach by employing more than one visualization method. Complicated information can be better represented by combining different tools. This method allows the user to provide more information to the inversion than a single measure of quality. The information, i.e. quality measure, that the user may decide to provide does not need to be determined before the inversion, but may arise spontaneously during the procedure because of new insight into the problem.

The problem

The inversion has been applied to a simple sedimentation problem where the initial conditions are known exactly. A genetic algorithm (GA) provides an engine for the inversion. Since geological forward modelling is time consuming, the GA population is kept fairly small. Boschetti *et al.* (2003) discuss convergence issues related to interactive GAs, while a more general discussion of GAs can be found in Wright (1991). Here we use populations of ten individuals. At the end of the exercise, we have collected 90 individuals ranked by a user in terms of similarity to a target image. Intragenerational rankings are simultaneously accumulated in a global ranking order for post-inversion analysis.

The sedimentary model was deliberately chosen to be quite simple, and represents a ramp setting over an area of 19 × 19 km, with a 1-km grid spacing. The topography slopes down from the south at 0.1 degree. Neither tectonic movement nor sea-level change was incorporated. This scenario is more complex than it may initially seem, as the below sea-level accommodation was rapidly filled and a more complex meandering stream and delta formed. Likewise, the low angle of the initial topography leads to a greater variation in possible sediment profiles than would be expected from a steeper gradient. The model incorporates four different classes of sediments, including a medium sand (0.5 mm), a fine sand (0.2 mm), a finer sand (0.07 mm) and a clay (0.0004 mm). The only parameters permitted to vary in these simulations are those affecting the input conditions of the fluid laden with sediment. These are the initial east–west and north–south locations of the fluid source, the initial east–west and north–south velocities of the fluid source, the flow rate of the fluid source and the initial sediment concentrations of each of the four different grain sizes.

The true values used in the SEDSIM simulation are given in Table 1, along with the allowed search ranges. The model was run for 50 ka. The initial parameters and the step sizes outlined in Table 1 were deliberately chosen so that the target simulation could not be exactly reproduced. The goal of the inverse model is, of course, to retrieve the true (or closest) input parameters from the observations (the output of a stratigraphic simulation). The initial conditions for the trial simulation (Table 1) were chosen by a domain expert as being likely 'reasonable' priors. A more formal approach to

Table 1. *Parameter values used in the SEDSIM simulations*

Variable	True	Step size	Min.	Max.
E–W source location (m)	9500	200	8000	12,000
N–S source location (m)	600	90	20	1100
E–W source velocity ($m\,s^{-1}$)	0	0.25	−1.0	1.0
N–S source velocity ($m\,s^{-1}$)	0.5	0.2	0.4	2.2
Flow rate ($m^3\,s^{-1}$)	9	1.5	1	13
Conc. (0.5 mm) ($kg\,m^{-3}$)	0.025	0.01	0.01	0.1
Conc. (0.2 mm) ($kg\,m^{-3}$)	0.10	0.03	0.0	0.3
Conc. (0.07 mm) ($kg\,m^{-3}$)	0.175	0.06	0.0	0.3
Conc. (0.0004 mm) ($kg\,m^{-3}$)	0.20	0.06	0.0	0.3

prior selection may involve analogue studies in a real case, or the complementary interactive methods of Curtis & Wood (2004).

Each of the variables was randomly initialized by a genetic algorithm. This is a standard strategy in GA inversion, the influence of which, in the final GA result, is discussed in Wright (1991). Ten simulations were run at each generation, and two output sections, displaying lithology and sediment ages, were used to rank the results in order of similarity to the true sections (Fig. 1), while the true input values were hidden. Ranking was carried out by evaluating models according to the classes contained in Table 2, which include both specific characteristics and overall resemblance to the target. The two or three highest ranked simulations then acted as seed for the next round of ten simulations, which were compared in turn. Lithology was used exclusively to rank the first six generations, after which it became necessary to analyse the alternating sediment age distribution in order to distinguish between some lithologically similar models. No separate class was adopted for age characteristics (cf. Table 2); this comparator was used only for a relative rank between two good models that otherwise shared all the same lithological class rankings.

Lithology was chosen as the primary comparator for several reasons. SEDSIM's main purpose is to predict lithological variations away from well holes, as this is one of the key features that controls the formation and behaviour of a petroleum system, affecting the source rock, migration of hydrocarbons, reservoir formation and seal quality. It is also one of the most variable characteristics, especially in near-shore and fluvial environments, where local variations in flow or sediment volume can have a marked effect on the lithology. It may have been more representative of the typical SEDSIM usage had the lithology been evaluated at several vertical locations representing the lithology identified at a well location, rather than along an entire dip section. The secondary measure of simulation quality was taken to be a display of age layers, with successive ages alternately coloured red and blue. These give a good idea of the volume of sediment deposited at each layer as well as the shape of each deposit, roughly akin to what may be seen on a seismic section.

Table 2. *Definitions of different classes for ranking*

Class	Ranking
1	Best models
2	Good models ('acceptable')
3	Acceptable sediment topography
4	Acceptable 0.5 to 0.2 mm grain distribution
5	Acceptable 0.5 mm grain distribution
6	Acceptable 0.07 mm grain distribution
7	Acceptable clay distribution

SEDSIM stratigraphic forward modelling

Three-dimensional stratigraphic forward modelling enables the combined influence of a variety of interdependent basin processes to be studied at geological time scales. The results reflect possible changes in sediment distribution over time as a function of the changing depositional environment. Past studies have demonstrated the value of this approach (Koltermann & Gorelick 1992; Martinez & Harbaugh 1993; Griffiths & Paraschivoiu 1998; Griffiths & Dyt 2001; Griffiths *et al.* 2001). Any computer modelling is only as good as the validity of the input data and the algorithms used in the program. The selection of prior information for computer models should be no more onerous than that required for a conceptual geological model of an area. However, the need for quantitative data forces the geologist to a greater degree of commitment than may otherwise be the case.

SEDSIM is a 3-D stratigraphic forward-modelling program developed originally at Stanford University in the 1980s and extensively modified and extended in Australia since 1994. SEDSIM flow and sedimentation programs are linked to modules including subsidence, sea-level change, wave, storm and geostrophic current transport, compaction, slope failure and carbonates. The program models sediment erosion, transport and deposition, and predicts elastic and carbonate sediment distribution on a given bathymetric surface. The conceptual background to SEDSIM is described by Tetzlaff & Harbaugh (1989). SEDSIM is controlled by a parameter input file, and files describing relative sea-level change, initial topography/bathymetry and tectonic movement for each grid cell over time.

Basic principles of SEDSIM operation

SEDSIM uses an approximation to the Navier–Stokes equations in 3-D. The full Navier–Stokes equations describing fluid flow in 3-D are currently impossible to solve due to limitations in computer speed (it would take longer to simulate a flow than the real event). SEDSIM instead simplifies the flow by using isolated fluid elements to represent continuous flow (Tetzlaff & Harbaugh 1989, Ch. 2). This Lagrangian approach to the hydrodynamics allows for a massive increase in speed of computation and simplification of the fluid flow equations. At best, simulations over geological periods can only hope to capture the mean conditions and create a general pattern of sediment distribution, rather than capture the exact timing of each individual pulse of material.

Fluid elements travel over an orthogonal grid describing the topographical surface, reacting to the local topography and conditions such as the flow density and the density of the medium through which the element is passing (e.g. air, sea water or fresh water). Fluid elements are treated as discrete points with a fixed volume, an approach known as 'marker-in-cell'. Several simplifications are made to the Navier-Stokes equations (Tetzlaff & Harbaugh 1989). The most important of these is that the flow is expected to be uniform in the vertical direction (i.e. the whole of the fluid element has the same velocity), and that the friction experienced by the fluid element is controlled by Manning coefficients. The net result of these simplifications is that the Navier–Stokes equations are modified into non-linear ordinary differential equations. These equations are now solved using a modified Cash–Karp Runge Kutta scheme (Press *et al.* 1986) that ensures stable and accurate fourth-order-in-time solutions.

Although SEDSIM has the capability to model many aspects of the depositional process, the current experiments used only a small subset of the modules available. Modules switched off were: internal nested grids, slope failure, wave, storm and geostrophic current, sea-level change, tectonic subsidence or uplift, syn-depositional compaction, isostasy and carbonates. Even though the model used here was an extremely simplified version of a typical SEDSIM simulation, the depositional environment modelled is typical of a river-dominated delta in a cool, temperate, shallow lacustrine setting with no significant subsidence over a 50 ka period.

The simulations

Apart from the variables discussed above and listed in Table 1, all other input values remained the same for all 90 simulations. Each simulation took 4 hours under MS-DOS on an Intel computer (Fujitsu LifeBook©) running at 850 MHz with 256 Mb RAM.

Visualization and analysis of the solution space

The goal of the inversion is to optimize nine parameters controlling a sedimentation process. This results in a 9-D search space. A number of tools have been proposed in the literature to visualize high-dimensional (nD) spaces (Buja *et al.* 1996). In this work we employ multi-dimensional scaling (MDS) and a self-organizing map (SOM). Both methods attempt to plot n-D vectors on a 2-D surface in such a way that topology is best respected, i.e. in such a way that points close to one another in the original n-D space are plotted in close proximity in the 2-D map. Obviously, with the exception of trivial cases, respecting the topology perfectly is impossible. MDS and SOM differ in the way they approximate such an impossible mapping. In a nutshell, MDS yields a better overall picture of the distance relationships between the n-D points, while SOM provides a better picture of local relationships, acting more like a clustering algorithm. Accordingly, different kinds of information can be extracted by these two tools. Using the accumulated global rankings of simulation outputs, we visualize and analyse, *a posteriori*, the effectiveness of the inversion.

Multi-dimensional scaling

MDS works by calculating all the pair-wise distances between points in the nD space. Points are then positioned on a 2-D surface in such a way that pair-wise distances in the 2-D map match the original distances as much as possible. Among possible MDS implementations, we have adopted Sammon's mapping (Sammon 1969), which emphasizes the preservation of small distances.

As explained above, we cannot expect this mapping to be optimal. Consequently, we cannot expect the map to be always easily interpretable. The highly non-linear mapping inherent in the MDS approach may result in a series of projections, rotations and stretches that may hinder the interpretation of data relationships. Fortunately, this is not the case in our test. Figure 2 shows an MDS map containing all 90 points collected during the inversion. Different symbols indicate the simulation quality. The 'best' points (ranks 1–3) are coloured red, while other 'good' points (ranks 4–20) are blue, 'mediocre' points are grey and 'very bad' points are magenta. Points belonging to these first two sets plot close to one another, which may suggest the existence of an approximately convex area in

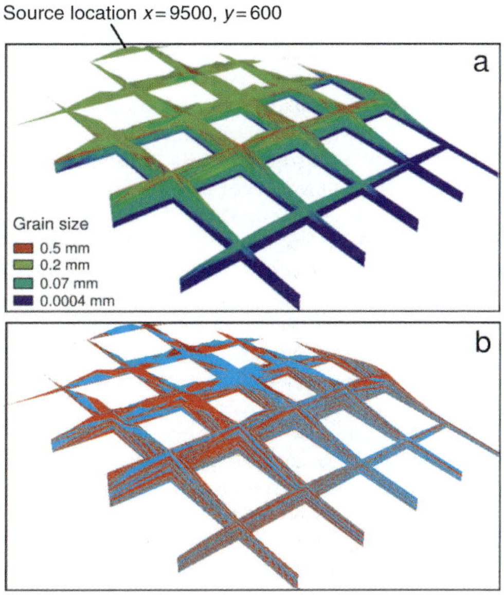

Fig. 1. Target simulation with every fourth fence (section) shown for (**a**) lithology (grain size) and (**b**) 'alternating age'.

the original n-D search space containing good models. It is tempting to infer something about the uniqueness of the solution. However, the limited number of points, as well as the approximate mapping we are using, compels us to use caution in drawing such conclusions. This map should be used as a tool to suggest patterns that need to be properly verified by further analysis.

An obvious (though not sufficient) test to verify convexity is to generate an input by the weighted average of all good models and to judge whether the resulting geological model is also good. The simulation output from the weighted average model, shown in Figure 2 as the 'combined best estimate', weakly confirms the convexity hypothesis. Since an MDS map is not reversible, it is not possible to take a generic point on the 2-D map and transform it into the original n-D space.

The above analysis refers only to the subjective ranking of image similarity. In producing this ranking, the user found it useful to assign the solutions to one or more of the classes in Table 2. The first two classes represent the good models, as mentioned above. The other classes represent models that are not good overall, but that possess a good feature. Because of the non-linearity inherent in the physics of the model,

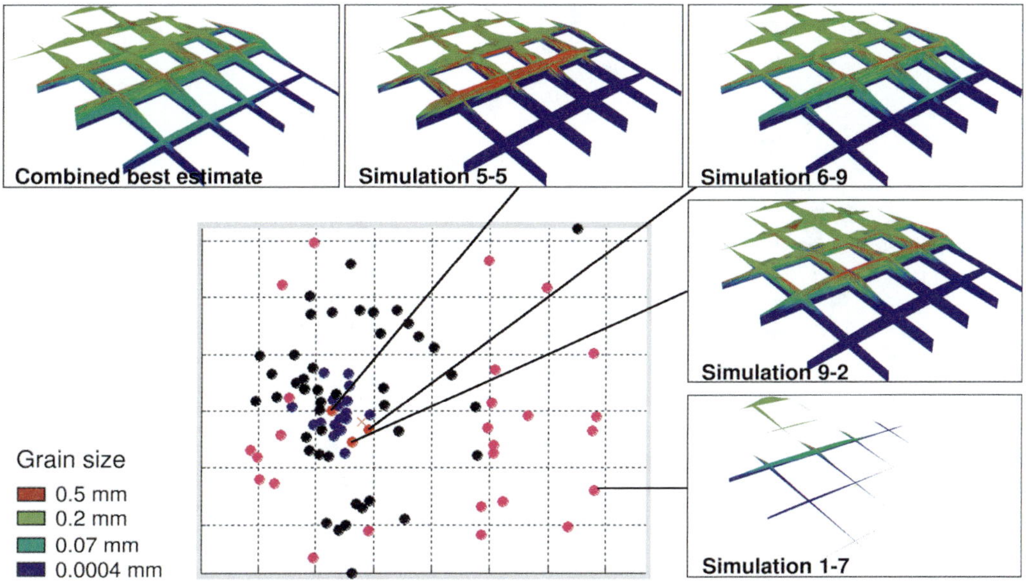

Fig. 2. Sammon's map, with each point assigned a colour according to its quality ranking. Axes represent distance, but there is no physical scale. Red, the three 'best' points (simulations very similar to the target); blue, 'good' points (somewhat similar); grey, 'mediocre' points; magenta, 'very bad' points. Simulation results for the three best models are mapped onto their corresponding points, and the true (target) solution is plotted on the map by an ×. The 'combined best estimate' is obtained by averaging the three best (red) points.

relationships between input parameters and classes are not obvious beforehand. This is information we hope to obtain from the inversion process.

The MDS map can be used to visualize the comparative locations of the different classes within the parameter space. Figures 3 and 4 contain the points belonging to, respectively, classes 4 and 6, plotted with grey circles. Figure 3 suggests that '0.5 to 0.2 mm distribution' is an important factor in judging the quality of a model, as it is rare to have a bad simulation within this class. On the other hand, Figure 4 suggests that models are not very sensitive to '0.07 mm grain distribution', since both good and bad models fit this criterion. Simulations such as these also indicate the sensitivity of simulation results to the various input parameters. The Sammon's map in Figure 3 shows that the quality of the simulation is strongly dependent on the accuracy of predicting the correct concentration of coarser grain material, with class 4 points strongly matching with classes 1 and 2 (best and good models). This makes sense in several ways. The coarse material drops out of suspension from a fluid source first, and hence is typically deposited in either a fluvial or near-shore environment. Its presence has a strong effect on subsequent physical processes in the area. It is also capable of forming steep sediment profiles, again altering the hydraulic surface significantly. Fine-grained materials tend to form lower slope angles and drop out of suspension typically in deeper, calmer water. This has less effect on subsequent flows, which is why we see in Figure 4 that the quality of a solution is very poorly correlated with class 6, the concentration of fine-grained material.

The MDS visualization provides some other information. First, models closely resembling the target seem to be located within a relatively small sub-domain of the initial search space. Second, this domain seems to be approximately convex. Third, the visual analysis of the models,

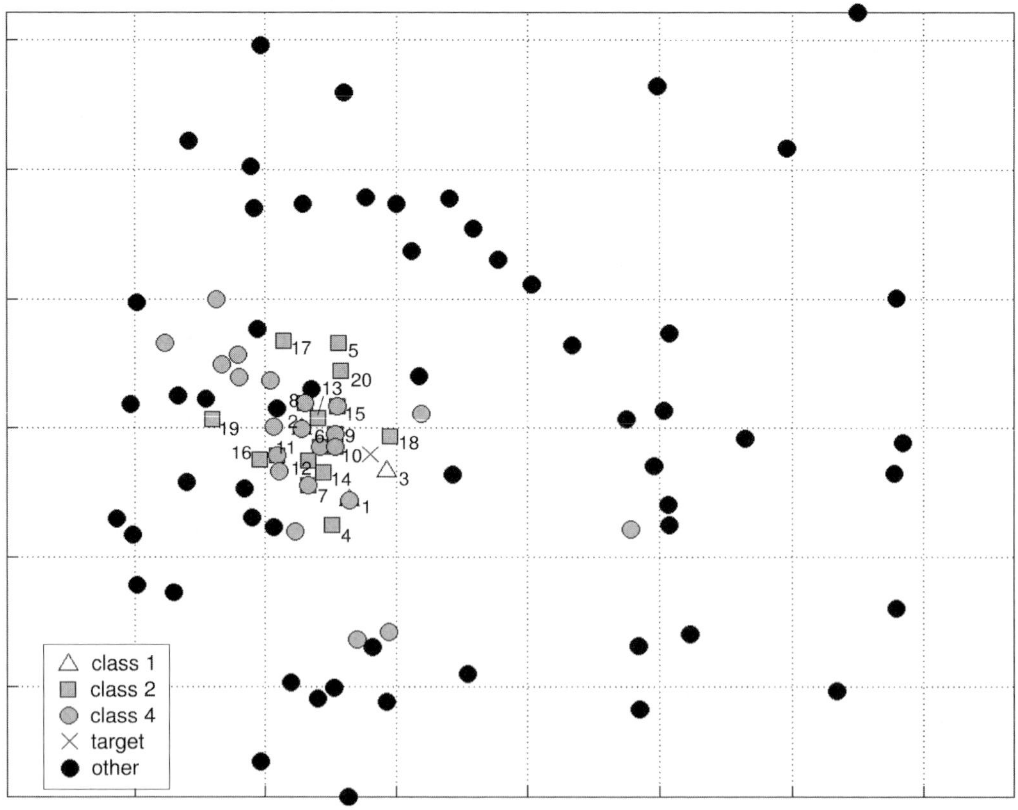

Fig. 3. Sammon's map. Points are labelled with a quality ranking according to similarity to the target, and 1 to 20 refer to good points. Classes 1 and 2 contain the best simulation results, but class 4 points are close to the good points.

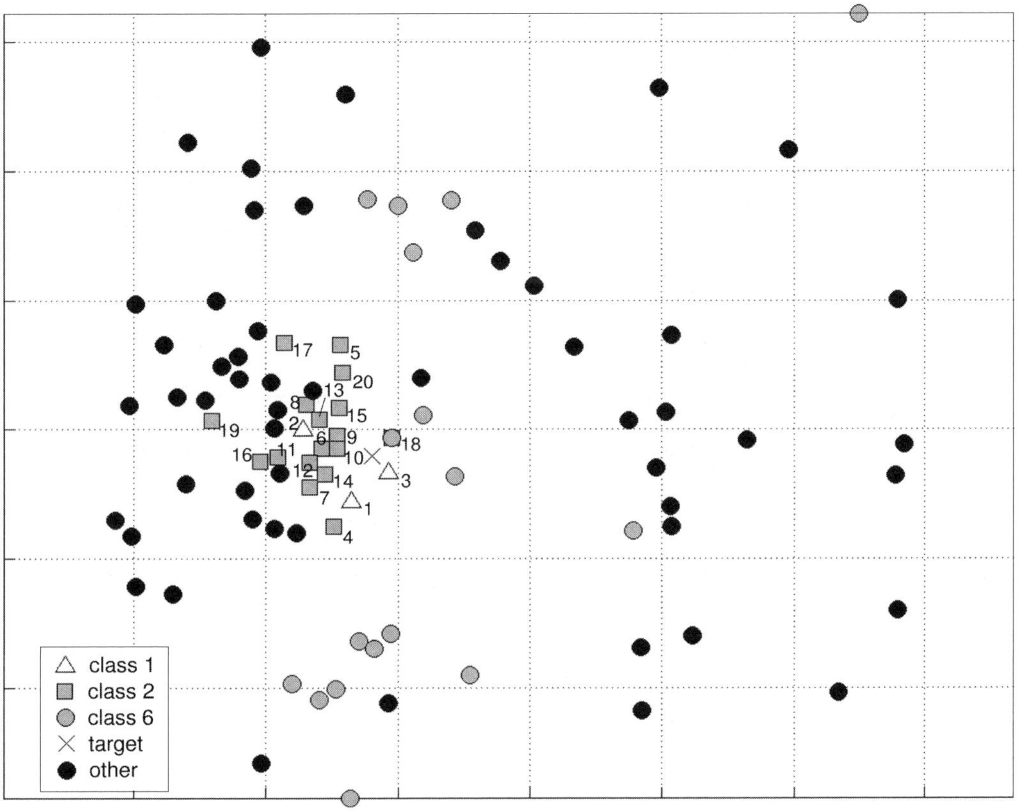

Fig. 4. Sammon's map. Points are labelled with a quality ranking according to similarity to the target, and 1 to 20 refer to good points. Classes 1 and 2 contain the best simulation results, and class 6 points are relatively far from the good points.

underlying the interactive inversion approach, apparently captures important aspects of the geological interpretation, since we cannot expect the previous two features to arise completely by chance. Although we cannot exclude the existence of other areas in the parameter space containing good models, it is natural to focus the search on the smaller sub-domain identified above, should we wish to further improve our analysis.

The self-organizing map

The last, and probably most important aspect of our analysis, is to understand which input parameters most control the variability in the geological process. We explore this important element via the use of Kohonen's SOM (Kohonen 2001). This algorithm spreads n-D points over a 2-D plane in such away that neighbourhood topology is respected, i.e. two points lying close to one another in the higher-dimensional space should lie close in the 2-D plot. In doing so, it acts as a classification algorithm that separates all the input data into clusters according to similarity. Depending on the topology of the n-D points, more than one data point may be assigned to a 2-D cell. We prevent this from happening by choosing a map containing many more cells than data points.

The SOM is usually displayed by assigning a colour to each cell, which represents the distance between neighbouring cells in the original n-D space. These distances show how the map needs to stretch in order to accommodate the complex n-D topology. By doing so, it implicitly displays basic data clusters, i.e. sets of data points that are all in close proximity. An example is given in the top left image of Figure 5. Unfortunately, a SOM is more reliable in its representation of local cluster relations. For global relations, the MDS described above is better suited.

An attractive feature of a SOM is that it allows a display of the magnitude of each

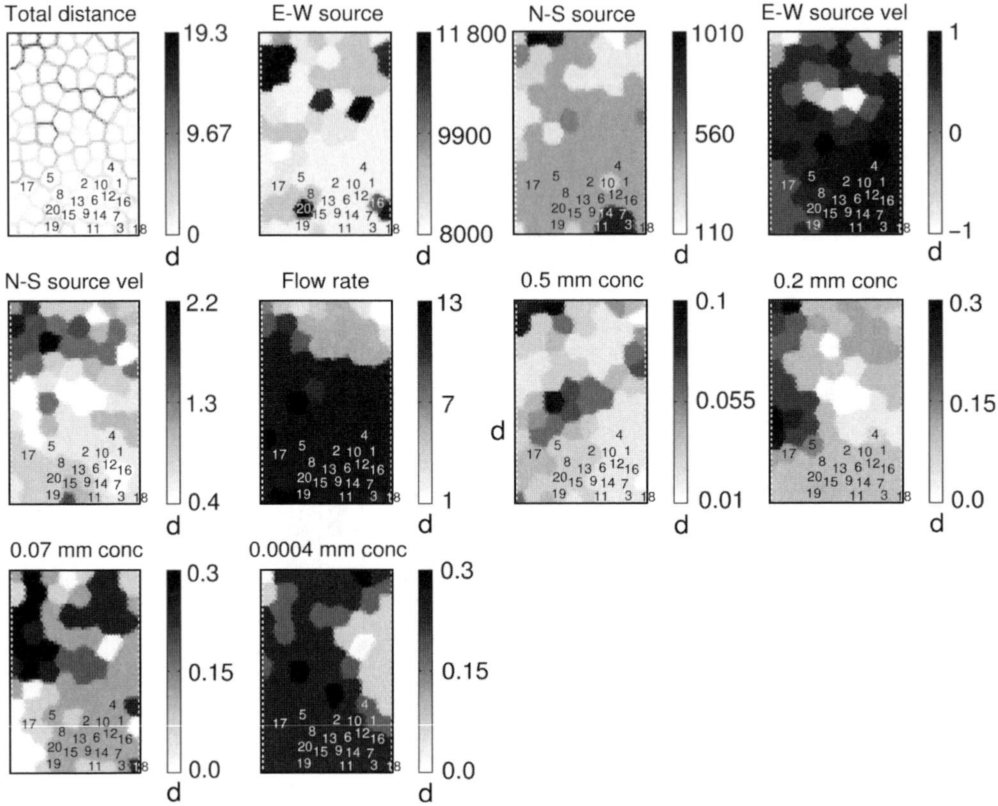

Fig. 5. SOM representing the clustering of model outputs (total distance) and the distribution of the magnitudes of the nine variables. Variables correspond to those in Table 1. Only the good points (quality ranking 1 to 20) are labelled on the map.

original dimension (i.e. each model parameter) at each cell location. This is shown in the remaining images in Figure 5. The labels indicate the location on the SOM of the 20 best models generated during the inversion. Here also, they cluster close to one another. By analysing the variability (and values) of each dimension at the locations corresponding to the best models, we can estimate which input parameters most control the geological process. When good models show a wide range of values for a specific variable, such as for the north–south source velocity, we conclude that the solution is not very sensitive to this variable. On the other hand, if all good models share a fairly constant value of a specific parameter, such as for the flow rate or 0.2 mm grain concentration, we cannot ascertain whether the solution is extremely sensitive to this parameter or whether the genetic algorithm has fixed upon one value. Our further work will focus on the choice of sample points to avoid the latter scenario.

Discussion

The results show that, at least in this case, convergence to visually similar sections occurred within very few generations. This offers hope that such a technique may be a practical approach to the inversion of complex 3-D stratigraphic forward models. The danger of getting trapped in local minima is always present, in GA approaches as much as in other inverse strategies. One approach to testing for the existence of these in a GA environment is to introduce larger value 'jumps' every *n*th generation and test for survivability.

There are obviously many different ways one could extract final value estimates from the resulting predictions. One could take modal values from the ten most successful of all generations. Figure 2 shows that, even without using a weighted average of predicted values, the best results from simulations in the sixth and ninth generations (6–9 and 9–2) are visually

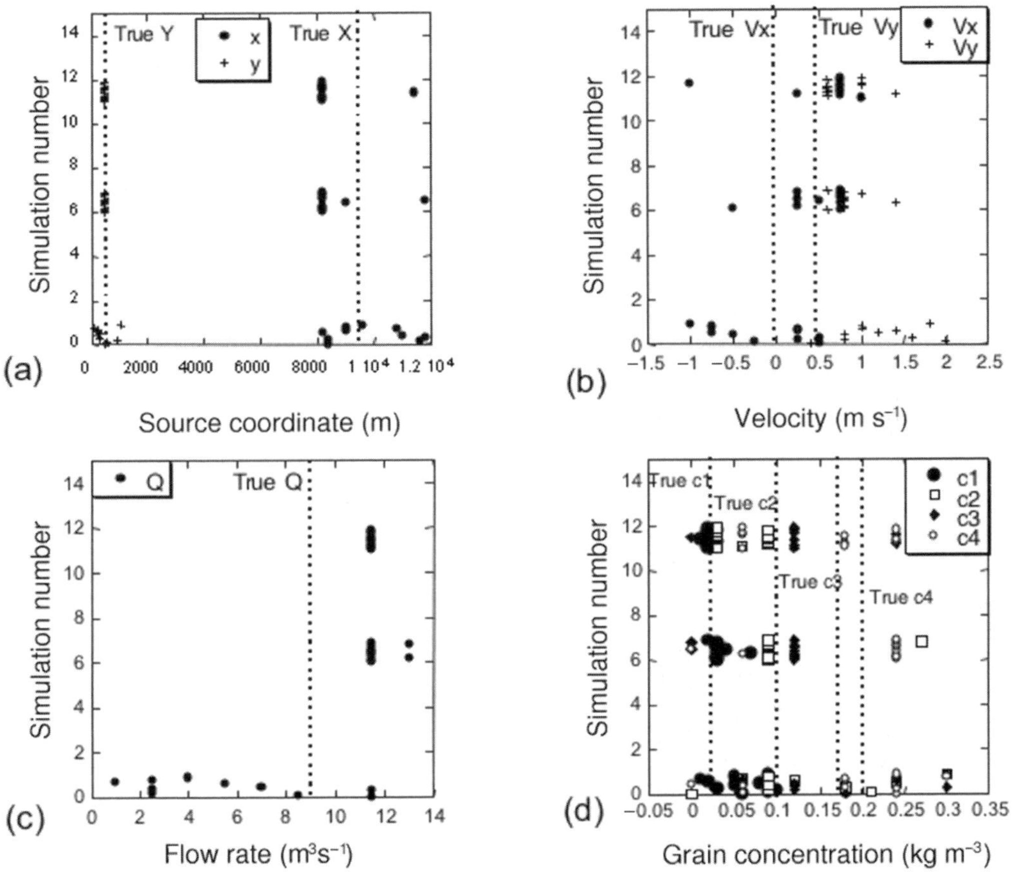

Fig. 6. Convergence graphs of the nine variables, for selected generations, with the underlying true values shown as vertical lines. (**a**) Source location in x and y coordinates, (**b**) initial velocities in the x and y directions, (**c**) flow rate, and (**d**) grain concentrations.

similar to the target simulation. In the case of the combined best-estimate model shown in Figure 2, a study carried out (before the model was run) using between 3 and 20 best simulations, with weights corresponding to their quality, suggested that the optimal choice was three. Thus the combined best estimate was a weighted average of the input values from the three models most visually similar to the target, all shown in Figure 2.

Some interesting features emerge from the predicted input values associated with these visually similar sections. First, the simulations do not improve markedly after the sixth generation. Second, there is a convergence to source (X) values, source velocity (V_x) values, and flow rate (Q) values that are different from the target values. The stable estimated values are not randomly distributed, but are systematically offset by up to 20% (Fig. 6). The most obvious explanation for this is that the GA has settled on a local numerical minimum that gives a result that is visually similar to the target, i.e. this minimum is beyond the resolution of a human interpreter. This may indicate that the suggested convexity of the solution space of good models is valid only at the visual level, which is a limitation of the visual ranking method.

It is clear that the comparison component could be automated given a suitable section comparator, as described by Griffiths *et al.* (1996) and Griffiths & Duan (2000), and the process speeded up considerably. This may be desirable considering the observation that the convergence slows after the sixth generation. Obviously, visual comparison is very effective for quick convergence in the initial generations. In fact, it is probably more effective than numerical comparisons at this first stage, where a human can quickly evaluate multiple criteria with ease when outputs are very different. However, as outputs become more similar, it is harder for a human to rank them. For this reason, age characteristics were used to distinguish between lithologically similar results. More types of representation could have been used, for example, water depth facies and porosity, but the challenge for a human evaluator increases accordingly. Thus, at this stage, when model results become similar, it may be desirable to substitute, or otherwise integrate, a numerical evaluation. Another approach could involve capturing the decision-making process of the modeller and using that to guide the inversion (e.g. Kaltwasser *et al.* in press). Nevertheless, this has been a valuable demonstration of the possibility of combining a relatively simple, subjective comparison scheme with a GA approach to taking geological priors and testing their validity.

Conclusion

An interactive inversion scheme, which combines human evaluation of model outputs with a genetic algorithm for the exploration of parameter space, is effective in converging towards a target section when using 3-D hydrodynamic modelling. The use of human interaction may have a particular advantage in ranking model outputs near the start of the process, where multiple criteria can be quickly evaluated. At the end of the inversion, we visualize the parameter space for all results, using, in this case, multidimensional scaling and SOMs, to draw conclusions about the relative importance of different physical parameters that control the sedimentation process. The visualization may also be used to delimit areas in parameter space where further investigations should be concentrated.

References

BOSCHETTI, F. & MORESI, L. 2001. Interactive inversion in geosciences. *Geophysics*, **66**, 1226–1234.

BOSCHETTI, F., WIJNS, C. & MORESI, L. 2003. Effective exploration and visualisation of geological parameter space. *Geochemistry, Geophysics, Geosystems*, **4**, 1086. [DOI:10.1029/2002GC000503]

BUJA, A., COOK, D. & SWAYNE, D. 1996. Interactive high-dimensional data visualization. *Journal of Computational and Graphical Statistics*, **5**, 78–99.

CURTIS, A. & WOOD, R. 2004. Optimal elicitation of probabilistic information from experts. *In:* CURTIS, A. & WOOD, R. (eds) *Geological Prior Information: Informing Science and Engineering.* Geological Society, London, Special Publications, **239**, 127–145.

GRIFFITHS, C. & DUAN, T. 2000. Quantitative comparison of observed stratigraphy and that predicted from forward modelling. *Proceedings of the 31st International Geological Congress, 6–17 August 2000, Rio de Janeiro*, CD-ROM.

GRIFFITHS, C. & DYT, C. 2001. Six years of SEDSIM exploration applications. American Association of Petroleum Geologists, Bulletin, **85**, 13. [Abstract]

GRIFFITHS, C. & PARASCHIVOIU, E. 1998. Three-dimensional forward stratigraphic modelling of Early Cretaceous sedimentation on the Leveque and Yampi Shelves, Browse Basin. *Australian Petroleum Production and Exploration Association Journal*, **38**, 147–158.

GRIFFITHS, C., DUAN, T. & MITCHELL, A. 1996. How to know when you get it right: a solution to the section comparison problem in forward modelling. Proceedings Numerical Experiments in Sedimentology. May 1996, University of Kansas, CD-ROM.

GRIFFITHS, C., DYT, C., PARASCHIVOIU, E. & LIU, K. 2001. SEDSIM in hydrocarbon exploration. *In:* MERRIAM, D. & DAVIS, J. (eds) *Geologic Modeling and Simulation.* Kluwer Academic, New York, 71–97.

KALTWASSER, P., BOSCHETTI, F. & HORNBY, P. (in press). Measure of similarity between geological sections accounting for subjective criteria. *Geophysical Research Letters.*

KOHONEN, T. 2001. *Self-organizing Maps.* 3rd edition. Springer, *Series in Information Sciences*, **30**, New York.

KOLTERMANN, C. & GORELICK, S. 1992. Palaeoclimatic signature in terrestrial flood deposits. *Science*, **256**, 1775–1782.

MARTINEZ, P. & HARBAUGH, J. 1993. *Simulating Nearshore Environments.* Pergamon Press, New York, *Computer Methods in the Geosciences*, **12**.

PRESS, W., FLANNERY, B., TEUKOLSKY, S. & VETTERLING, W. 1986. *Numerical Recipes: The Art of Scientific Computing.* Cambridge University Press, New York.

SAMMON, J. 1969. A nonlinear mapping for data structure analysis. *Institute of Electrical and Electronic Engineers, Transactions on Computers*, **C-18**, 401–409.

TAKAGI, H. 2001. Interactive evolutionary computation: fusion of the capacities of EC optimization and human evaluation. *Proceedings of the Institute of Electrical and Electronic Engineers*, **89**, 1275–1296.

TETZLAFF, D. & HARBAUGH, J. 1989. *Simulating Clastic Sedimentation: Computer Methods in the Geosciences.* Van Nostrand Reinhold, New York.

WIJNS, C., BOSCHETTI, F. & MORESI, L. 2003. Inverse modelling in geology by interactive evolutionary computation. *Journal of Structural Geology*, **25**, 1615–1621. [DOI:10.1016/S0191-8141(03)00010-5]

WRIGHT, A. H. 1991. Genetic algorithms for real parameter optimization. *In:* RAWLINS, G. (ed) *Foundations of Genetic Algorithms.* Morgan Kaufmann, San Mateo, 205–218.

Building confidence in geological models

R. ANDREW BOWDEN*

Quintessa Limited, Dalton House, Newtown Road, Henley-on-Thames, RG9 1HG, UK
(e-mail: andrewbowden@quintessa.org)
**Present address: 2928 Moggill Road, Pinjarra Hills, Brisbane, QLD 4069, Australia*
(e-mail: geodec@ozemail.com.au)

Abstract: Geological models are required because we do not have complete knowledge, in time or space, of the system of interest. Models are constructed in an attempt to represent the system and its behaviour based on interpretation of observations and measurements (samples) of the system, combined with informed judgement (expert opinion) and, generally, constrained for convenience by the limitations of the modelling medium. Geological models are inherently uncertain; broadly those uncertainties can be classified as epistemic (knowledge based) or aleatory (reflecting the variability of the system). Traditional, quantitative methods of uncertainty analysis address the aleatory uncertainties but do not recognize incompleteness and ignorance. Evidence-based uncertainty analysis provides a framework within which both types of uncertainty can be represented explicitly and the dependability of models tested through rigorous examination, not just of data and data quality, but also of the modelling process and the quality of that process. The inclusion of human judgement in the interpretation and modelling process means that there will be frequent differences of opinion and the possibility of alternative, inconsistent and/or conflicting interpretations. The analysis presented here uses evidence-based uncertainty analysis to formulate a complete expression of the geological model, including presentation of the supporting and the conflicting or refuting evidences, representation of the remaining uncertainties and an audit trail to the observations and measurements that underpin the currently preferred and the possible alternative hypotheses.

Geological models provide the basic framework as prior input to the development and parameterization of higher level process and risk analysis models in all areas of geoscience. Such geological models are usually deterministic, representing one preferred interpretation from many possible, and some equiprobable, alternatives. Application of these alternative models could result in significantly different results, which affect perceived project viability and decision making. Nowhere is this more important than in the radioactive waste disposal sector, where the selection and characterization of a site for deep underground waste disposal is required to demonstrate that the chosen site will remain geologically stable, ensuring the integrity of waste packages, for tens to hundred of thousand of years. Geological processes and events that threaten the integrity of a waste repository are greatest for those countries that happen to be located close to plate tectonic boundaries. Japan is a good example, being located in the circum-Pacific orogenic belt close to the boundaries of four tectonic plates and in one of the most tectonically active areas in the world. Japan has also recently (2002) begun a volunteer process for the selection of a site for high-level radioactive waste disposal (Kitayama 2001; NUMO 2002*a*, *b*) and therefore its use here, in illustration of the geological modelling issues, is topical. The integrity of an underground waste repository in this environment is threatened by a number of geological processes and events, including earthquake activity and active faulting, active volcanism and geothermal activity, uplift, erosion and other natural phenomena, and human intrusion, for example, in the search for mineral resources.

Models

In general terms, models are required because we do not have complete knowledge, in time or in space, of the system of interest. Models are therefore constructed in an attempt to represent the system and its behaviour based on interpretation of observations and measurements (samples) of the system, combined with

From: CURTIS, A. & WOOD, R. (eds) 2004. *Geological Prior Information: Informing Science and Engineering*. Geological Society, London, Special Publications, **239**, 157–173. 186239-171-8/04/$15.00

informed judgement (expert opinion) and, generally, constrained for convenience by the limitations of the modelling medium. Distinction is made between interpretation to provide a basic understanding of the geological framework within which a system operates (conceptual models and hypotheses), numerical models that allow the properties and behaviour of the system to be analysed quantifiably, and 3-D computer visualizations that allow the features and processes to be represented spatially and dynamically. Problems arise because the database on which geological models are constructed is incomplete, the data are sparse and possibly inconsistent and the quality of data and interpretation is variable. The models are gross simplifications and, for those geological models required to look into the future over thousands and even tens of thousands of years, extrapolations are little more than crude speculation. Inevitably, in such circumstances, geological models contain large elements of uncertainty. Geologists are used to dealing with this uncertainty and adept in making the imaginative leaps required in order to translate incomplete information into fully formed conceptual models. They are less adept at documenting the logic and justification behind their reasoning or in providing a well-documented audit trail from model back to the raw data. Dependability of a model, in the sense used here, is a measure of the ability of a model to deliver the function required of it, i.e. fitness for purpose. Since it is generally not possible to verify geological models, building confidence should not be a search to prove truth or reality but should be aimed at demonstrating their dependability with emphasis on the quality of the process involved in their construction, clarity of understanding, and transparency with respect to exposure of any inherent uncertainties. A complete expression of the model would ideally include the observed/measured 'facts', definitions of the currently preferred and possible alternative hypotheses, presentations of the supporting and the conflicting or refuting evidence, and some form of representation of the remaining uncertainties (Thury 1994).

Interpretation

No definitive, rule-based methodology exists for data integration and interpretation. Geological models are to a large extent based on 'implicit interpretation' and the degree of support, or confidence, for each conceptual step cannot be assessed easily. Any specific interpretation, even if based on all the available information, may be highly individualistic and reflect the personal views of the interpreter. It is generally accepted that individual, equally qualified geologists, geophysicists or geochemists provided with the same basic data will arrive at different interpretations. This is not to say that any one interpretation is any more or less valid than another; all may be equally valid and may be recognized as possible alternatives. However, individual interpreters will have their own personal, subjective views based on their prior experience and knowledge, from which they, consciously or subconsciously draw inferences and preferences with respect to possible alternative interpretations. Nevertheless, in order to provide a justified, if not completely objective interpretation that can be audit-tracked from start to finish, we need to examine and make judgements on both the quality of the data and the quality of the interpretation and modelling process. Commentary on how well the process has been carried out (e.g. whether interpretation was carried out under financial or time constraints, or the extent to which the interpreters' individual experience and personal biases may have influenced the outcome) should be recognized and recorded, and some qualitative measure of the associated uncertainty should be taken into account.

Uncertainty

Fundamentally uncertainty can be defined as 'anything we are not sure of'; as such it is a function of our belief in our understanding of a system and its behaviour. We will always have uncertainties because we do not and cannot know everything. Given the broad definition of uncertainty as 'anything we are not sure of', we can ask the question:

> In what ways can we be unsure?

for which we might come up with a list as follows:

- incomplete knowledge of process – we do not understand all the processes involved;
- incomplete knowledge of the system – we do not have all the data;
- uncertain quality – we have the data but we are not sure of their derivation;
- uncertain meaning – we have the data but we do not know what they mean;
- conflict – we have data from different sources but they do not agree;
- variability – we have data but it does not give us a clear answer.

Different researchers have come up with different classifications of uncertainty (e.g. Funtowicz & Ravetz 1990; Hoffman & Hammonds 1994; Ferson & Ginzberg 1996; Foley *et al.* 1997; Hoffman-Reim & Wynn 2002) but in general they can be mapped onto a similar framework (Fig. 1) that distinguishes uncertainties relating to system variability (aleatory) from the uncertainties to do with incompleteness and lack of knowledge (epistemic).

Traditionally, the established methods of uncertainty analysis in the geosciences are quantitative methods for treating the aleatory uncertainties identified in Figure 1. The methods are, in general, 'closed world' approaches requiring uncertainties to be 'captured', for example, as a frequency distribution or a probability density function (p.d.f.). A simple example is the probabilistic uncertainty analysis of the possible outcomes from a throw of dice. Likelihood probabilities can be precisely calculated because all possible outcomes of a throw of dice are known completely. This arises because the characteristics of the dice are precisely known, i.e. each die is an identical unbiased cube with six sides, equal in shape and with precisely known quantities on each face. However, what do we do if none of these parameters is known precisely (i.e. if our dice are not identical, they have an unknown number of sides of possible irregular shape and with unknown quantities on each face)? Under these conditions it is not possible to have prior knowledge of all the possible outcomes of a throw of dice, let alone assign probabilities to them. In the geosciences, where high levels of uncertainty

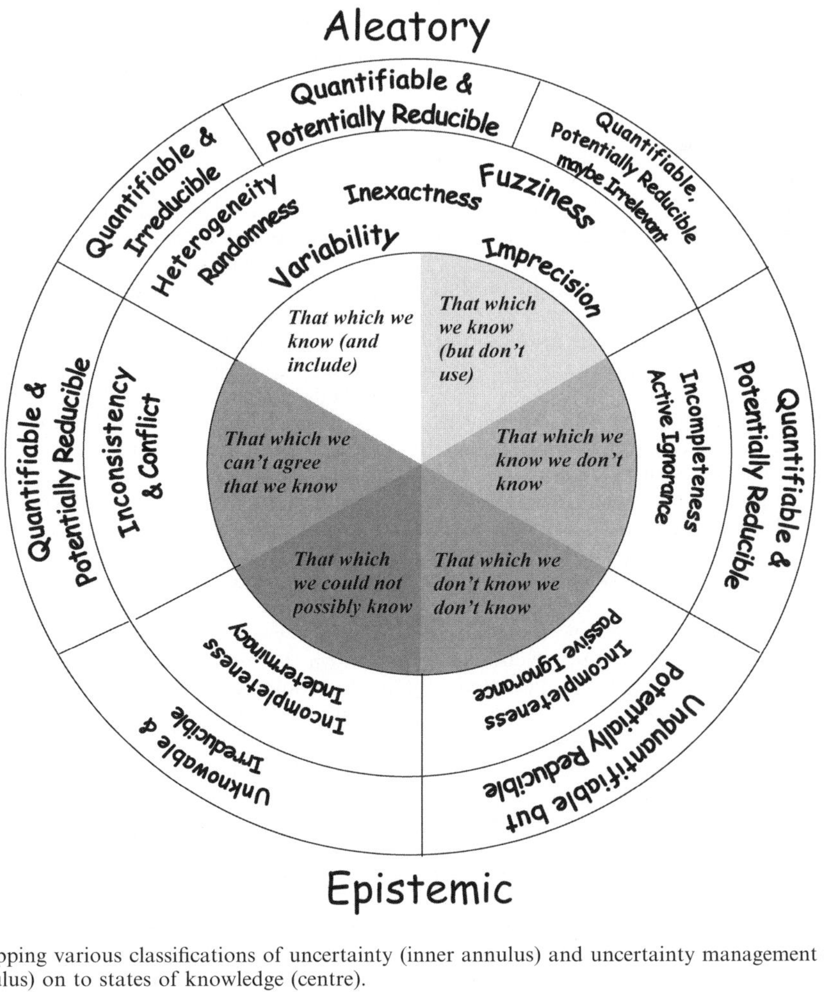

Fig. 1. Mapping various classifications of uncertainty (inner annulus) and uncertainty management capability (outer annulus) on to states of knowledge (centre).

are the norm, traditional probabilistic analysis may not be appropriate because insufficient information is available for complete assessment of all possible outcomes. Commonly, 'expert' judgement via a panel of experts is used in an attempt to elicit uncertain parameters and to define the shape of a p.d.f. intended to capture the full range of possible outcomes and the likelihood of their occurrence. There are several difficulties in this approach, not least of which are over-confidence and bias from experts in providing estimates (Capen 1976; Kahneman *et al.* 1982; Halpern 1996; Baddeley *et al.* 2004; Curtis & Wood 2004). Research has shown that people have a natural tendency to introduce bias into judgements and are more confident about probabilistic estimates than they should be. Some common examples of bias include:

- attaching undue weight to evidence that supports our beliefs and ignoring or downplaying contrary evidence (motivational or confirmation bias);
- a tendency to believe what we want to believe (motivational bias or confirmation emphasis);
- remembering cases involving extreme outcomes of success and failure (subjective emphasis or availability bias);
- weighting heavily our most recent experiences (recall or availability bias);
- resistance to changing our minds in the face of new evidence (anchoring);
- vulnerability to peer and institutional pressures (herding).

Curtis & Wood (2004) and Baddeley *et al.* (2004) provide a more complete and detailed account of the various sources of bias.

For many applications expert elicitation of uncertain geological parameters may be sufficient to cover the expected range of possible outcomes. However, in the face of very high uncertainty, for example, in long-term future geological processes and events, and particularly in apparently random catastrophic events such as volcanic eruptions and earthquakes, dependence on expert judgement leading to a quantitative calculation of a probability of occurrence may be at best misleading and, at worst, could lead to harmful consequences.

In the radioactive waste industry this problem has been addressed by use of deterministic evaluation of a discrete number of potential scenarios that, it is hoped, bound the full range of possible alternative futures (NEA 1999). Monte-Carlo type simulations try to expand the range of possibilities at the expense of increased modelling complexity but both methods suffer from the charge of 'incompleteness' since it is not possible to claim that all possible futures have been taken into account. It is acceptance of this epistemic uncertainty (i.e. the ability to say 'We do not know') that underpins the 'open world' philosophy and which is contrary to the notion of a probabilistic assessment that tries to capture and quantify uncertainties by restricting the possible future outcomes to only those that are currently thought to be possible.

Building confidence

If it is not possible to establish a model as a true representation of reality and, as in the case of extrapolations far into the future, it is not possible to test the dependability of different scenarios through predictive testing, a fundamental question is:

> When will we know when we have reduced uncertainty in a particular model sufficiently to make a dependable, defensible decision based on that model?

Without some decision framework that sets out what the model is trying to achieve, it will be difficult to demonstrate that sufficient confidence in the model has been generated. In such circumstances, despite the limitations on lack of objectivity noted earlier, the use of expert judgement is inevitable. It is also, in my view, justified, particularly if the evaluation provides robust procedures for formally documenting the evidence and the reasoning underpinning the judgements expressed. What is required is the establishment of a framework which includes a number of confidence-building steps and which recognizes that uncertainty cannot be eliminated completely. Such a framework should include explicit uncertainty analysis together with a 'quality' measure that incorporates expert judgement on the model-building process in terms of quality indicators such as, for example, the use of auditable interpretation procedures, adherence to scientific method, calibration, validation and verification, and peer review.

Evidence-based uncertainty analysis

The method of evidence-based uncertainty analysis described here is developed from a methodology described by researchers at the University of Bristol, UK (Cui & Blockley 1990; Foley *et al.* 1997; Hall *et al.* 1998; Blockley &

Godfrey 2000; Davis & Hall 2003), and adapted by the author for application to geological interpretation and modelling. The method uses a hierarchically structured process tree (e.g. Fig. 2) to represent the features, events or processes (FEPs) contributing to the development of the model or hypothesis. The tree structure is used to capture and propagate the evidence (both in support of and against) and the uncertainties associated with the individual FEPs at the lowest level of the tree in order to provide an overall outcome of support for the model. In practice, this is a top-down approach. The process tree links the top-level process to the raw data at the lowest level via a number of intermediate steps. If we consider the example given in Figure 2, the hypothesis that a specific site is suitable for deep geological disposal of radioactive waste is under evaluation. The corresponding top-level process is: 'Establishing confidence in the suitability of a site for deep geological disposal of radioactive waste'.

Development of the process tree, in terms of description of the contributory factors for site selection, is focused on those site-specific factors associated with evaluating whether a specific site should go forwards for detailed investigation or be excluded from further consideration. The top-level issue is too large and vague to be answered directly and it has been broken down into eight contributing processes, each of which considers a separate site selection issue allowing greater definition of the problem. The framework shown in Figure 2 is neither unique nor definitive. The definition of the structure of the process tree is subjective and, ideally, a more

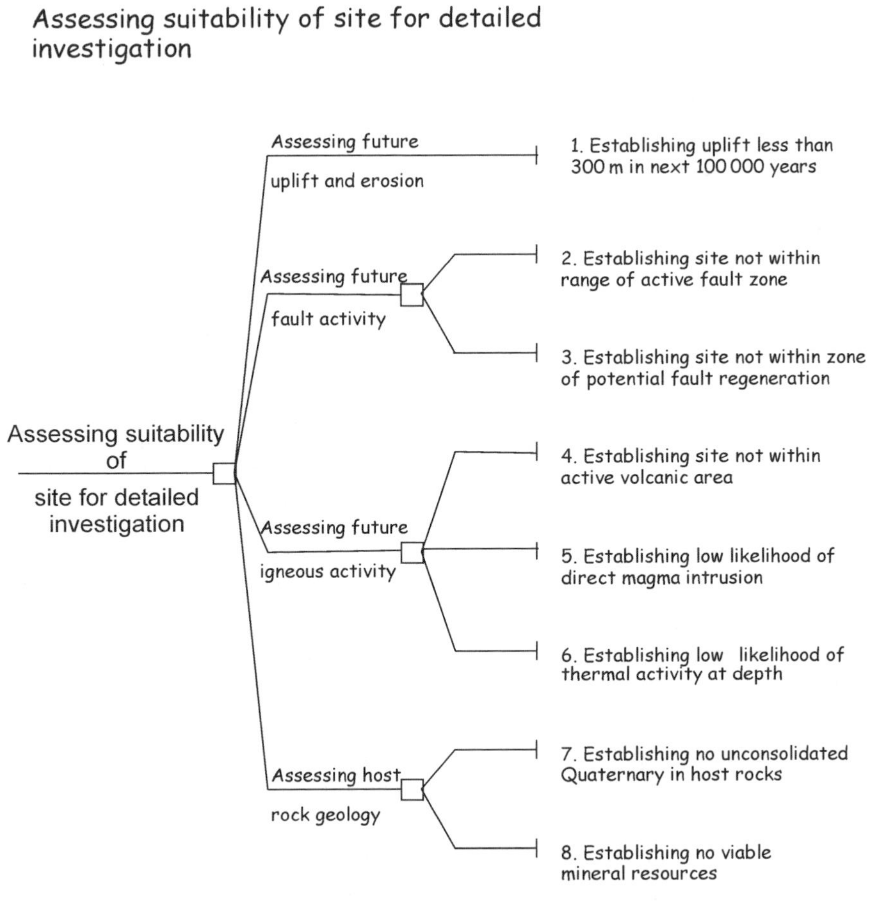

Fig. 2. Part of an evidence-based process model showing site-specific factors for the assessment of the suitability of a site for inclusion in a detailed investigation programme. Each of the lowest level process boxes is underpinned by a 'nested' model that develops the factor to be assessed to a level at which evidence, supported by data, can be provided.

definitive, defensible (and consensually acceptable) model will result from formal elicitation by a team of qualified experts.

The full process model for the generic example given in Figure 2 is complex and is not described fully in this paper. Instead one of the eight third-level processes has been developed further in Figure 3. This branch of the model considers the process 'Establishing low likelihood of thermal activity at depth'. This process is decomposed further until a level is reached at which the experts have confidence in being able to provide judgement on the dependability of the lower level process. For each process on which a judgement of evidence is required, a description of the process and some criteria for the success of the process are required. Thus, in Figure 3 (inset text-box), the process 'Establishing low value of Anion Index' is described and the success criteria for the process is given (Noda 1987; JNC 2000).

Evidence and belief

The dependability of each process is expressed using a three-way logic based on Dempster–Shafer evidence theory (Shafer 1976; Waltz & Llinas 1990) subsequently developed as interval

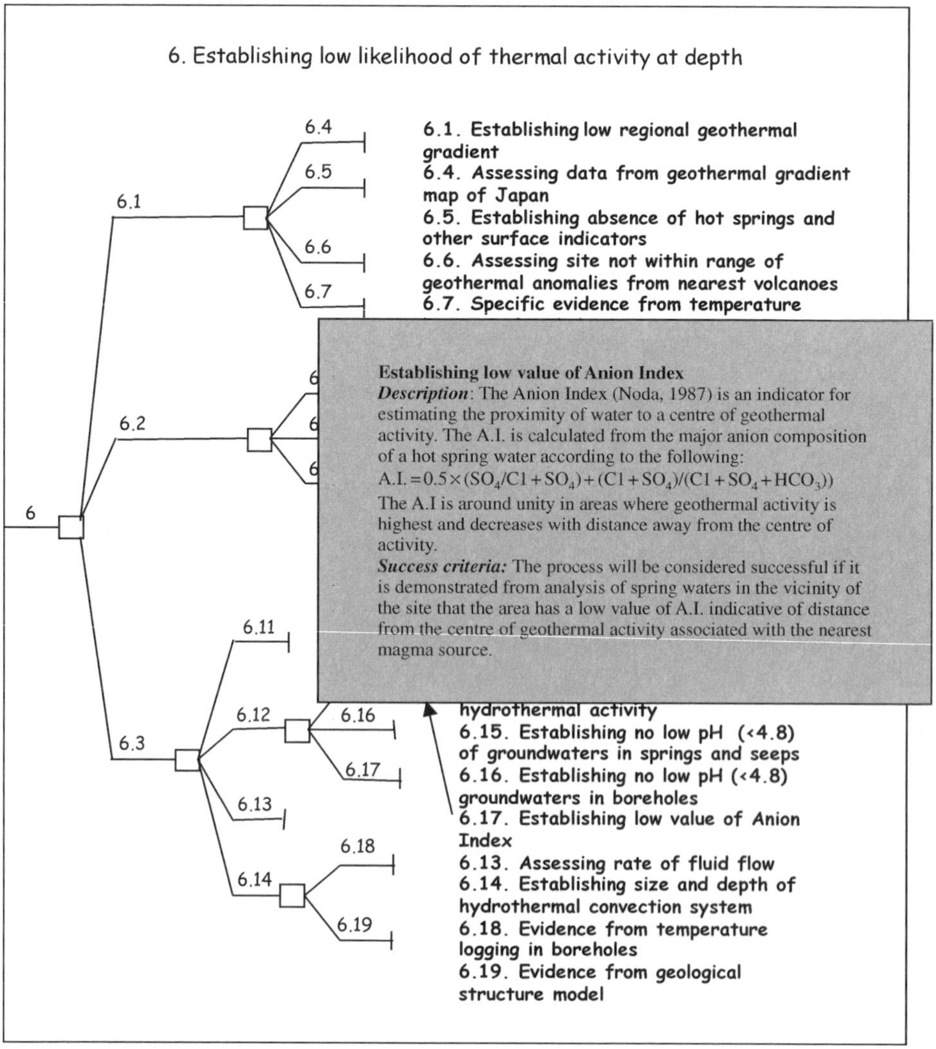

Fig. 3. Nested process model showing lower level of the tree for which evidence of support for, evidence against and remaining uncertainty is required. Inset shows an example of the attribute information for process 6.17.

probability theory by researchers at Bristol University, UK (Cui & Blockley 1990; Hall *et al.* 1998). Much of the value of the method and what makes it unique is the way in which uncertainty is explicitly identified and incorporated formally into the analysis. This uniqueness derives from its specialized handling of uncertainties as intervals (Fig. 4). In a major advance over traditional probability theory the interval representation allows us to admit to a general level of uncertainty in addition to support for and against a proposition (Waltz & Llinas 1990). The inadequacy of probabilistic two-value logic compared with the three-value logic approach to decision support in the field of radioactive waste disposal has been emphasised, for example, by Shrader-Frechette in her book *Burying Uncertainty* (Shrader-Frechette 1993). It is important to recognize that in this 'open world' view of uncertainty one of the principal axioms of probability theory, i.e. that $p(\mathrm{A}) + p(\mathrm{Not\ A}) = 1$, does not hold true. Since we are dealing here with 'beliefs and evidence' not 'probability and frequency' it is possible to recognize that belief in a proposition may be only partial and that some level of belief concerning the meaning of the evidence may be assigned to an uncommitted state. The equivalent axiom for assignment of belief is: $b(\mathrm{A}) + b(\mathrm{not\ A}) \leqslant 1$. In the 'Italian flag' representation of the three-value logic (Fig. 4) 'evidence for' is represented as green, 'evidence against' is red and 'remaining uncertainty' is white (Blockley & Godfrey 2000). The summation of evidence is expressed in belief terms and therefore is normalized to a frame of reference of unity (as above). However, the evidences in support of our belief for or against a proposition may be independent and in conflict, and/or may be more than sufficient for absolute commitment to the belief. It may be useful in such cases to consider the absolute weight of evidence underpinning the belief distribution in order, for example, to recognize cases where support for or against the proposition may be overwhelming, or where uncertainty due to conflict is dominant over uncertainty due to lack of knowledge. In this event evidence defined as 'the data or information used to support a belief' may be less than, equal to, or more than sufficient to fully support that belief. In evidence terms the equivalent axiom might be written as: $e(\mathrm{A}) + e(\mathrm{Not\ A}) \Leftrightarrow 1$.

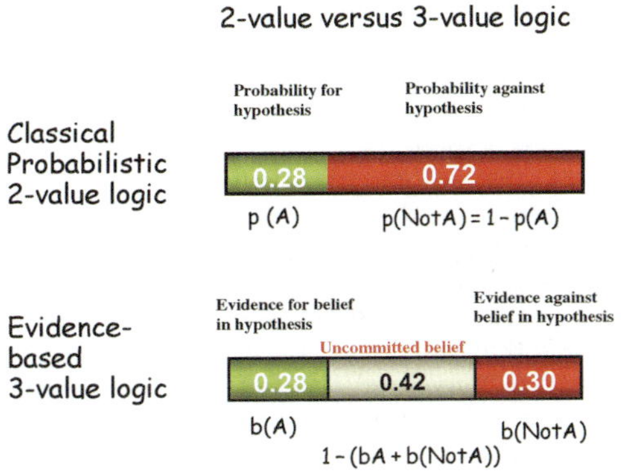

Fig. 4. Classical two-value probabilistic analysis compared with three-value interval analysis

Propagation of evidence

In the Bristol University application of the method, evidence is propagated using an approach to evidential reasoning called interval probability theory (IPT). Readers interested in the full mathematical formalism of IPT and its inference logic are referred to the original research (Cui & Blockley 1990). The following represents an attempt by a non-mathematician to present an account of the way in which values for the logical operators are elicited and to demonstrate the influence of the logic operators in propagating evidence through the process model.

At a basic level the summation of evidence using IPT provides a result similar to that arising from the fuzzy algebraic sum of the contributing elements where

$$\mu_{\mathrm{combination}} = 1 - \prod_{i=1}^{n} (1 - \mu_{\mathrm{i}}), \tag{1}$$

i.e. as with the fuzzy algebraic sum, the result is always greater than (or equal to) the largest contributing evidence value. The effect is additive in that two lines of evidence that both support (or refute) a hypothesis reinforce each other so that their combined evidence is more supportive than either piece of evidence taken individually (Bonham-Carter 1994). Although we may have some element of uncertainty with respect to each individual contributing piece of evidence, the application of the fragments of knowledge from each element allows us to increase our confidence in the hypothesis and hence reduce uncertainty overall.

In practice, the addition of evidence is modified in IPT by three factors: *sufficiency, dependency* and *necessity*. A weighting factor, *sufficiency*, represents the relative significance or relevance of each piece of evidence. In eliciting *sufficiency* we ask the question:

> Given that there is complete dependability of this process how much would it contribute to the dependability of the parent?

In the context of geological uncertainty, responses to this question may be too ill-defined or complex to be characterized precisely by numbers. Linguistic variables provide a means of approximate characterization using terms such as 'very high', 'high', 'intermediate', 'low', 'very low' that are subsequently converted to numerical form for computational purposes using a utility function. As with all linguistic to numerical conversions it is important that all the contributing 'experts' understand the conversion factors used and have the opportunity to modify, revise or refine their initial judgement if the conversion appears to misrepresent their intention. If necessary the utility function can be elicited as part of the evaluation.

The second factor, *dependence*, or *dependency* (Cui & Blockley 1990), describes the degree of overlap or commonality between the contributing processes and is introduced to avoid double counting of support from any mutually dependent pieces of evidence. Dependency therefore provides a subtractive element to the additive propagation of evidence. In eliciting *dependency* we ask the question:

> How much shared information do the processes use in contributing evidence to the dependability of the parent?

Dependency is assessed for each pair of processes. As with *sufficiency* the values assigned to *dependency* are elicited as linguistic responses.

A Boolean operator '*necessity*' is used to indicate a *necessary* process, i.e. a process whose dependability or success is essential to the success of the parent process. The *necessity* operator acts as a 'red flag', over-riding the normal propagation rules when, in the Bristol University application of the method, the 'evidence against' is greater than the combined values of 'evidence for' and 'remaining uncertainty'. In the writer's modification, *necessity* operates when the combined values of 'evidence against' and 'remaining uncertainty' are greater than the value of 'evidence for'. This modification is introduced to ensure that not only evidence of failure but also the potential for failure of an essential process is highlighted as an issue of concern. In the event that the *necessity* operator is brought into action the unfavourable outcome is copied directly to the parent process irrespective of the outcomes from other, non-essential, processes contributing to the same parent. If there is more than one necessary process contributing to the same parent then the outcome from the process with the highest level of combined uncertainty and evidence against is propagated, providing that their combined outcome is greater than the value of evidence in favour.

The following simple examples illustrate the impact that the logic operators for propagation of evidence have on the top-level outcome. Consider three processes contributing to a parent process (Fig. 5). The evidence is the same for all cases. In the first example (case a), *sufficiencies* are all 1 (i.e. each process is totally sufficient to satisfy the parent), *dependency* is 0 (all processes are independent), no processes are considered to be *necessary*. In this case the calculation of 'evidence for', using the fuzzy algebraic sum method of combining the contributing elements, is given by $1-[1-(1-0.36)*(1-0.24)*(1-0.12)]$ which equals 0.57 and, similarly, for 'evidence against', the fuzzy algebraic sum of $(0.21, 0.14, 0.07)$ is $1-[1-(1-0.21)*\ (1-0.14)*(1-0.07)]$, which equals 0.36. The 'evidence for' and 'evidence against' values represent the committed belief. In a belief system normalized to a frame of reference of unity, the uncommitted belief (i.e. remaining uncertainty) is given by $1-(0.57+0.36)=0.07$.

In the second example (case b), *sufficiencies* are 1.0 (total), 0.667 (high) and 0.167 (very low) respectively, *dependency* is 0, no processes are *necessary*. In this case the standard calculation of the fuzzy algebraic sum is modified by the sufficiency weighting factors. For the same evidence input as in case (a), the equivalent sum for 'evidence for' is $1-[1-(1-(0.36*1))*(1-(0.24*0.667))*(1-(0.12*0.167))]=0.47$ and for 'evidence against' is $1-[1-(1-(0.21*1))*(1-(0.14^*0.667))*(1-(0.07*0.167))]=0.29$. Remaining uncertainty is $1-(0.47+0.29)=0.24$. Thus, since two of the contributing processes supply only partial support to the parent, the uncommitted belief is increased at the expense of reduced accumulation of evidence.

In the third example (case c), *sufficiencies* are total, high, and very low respectively, pairwise *dependencies* are high ($D_{1,2}=0.667$), moderate ($D_{1,3}=0.5$) and low $D_{2,3}=(0.333)$, no processes are considered *necessary*. In this case the standard calculation is modified by both the weighting factors and by the double-counting *dependency* operator. The dependency factor introduced as a development of IPT by Cui and Blockley (1990) is given, for a combination

Fig. 5. Case examples (a–d) showing the different outcomes of the propagation algorithm resulting from different values of the inference operators *sufficiency*, *dependency* and *necessity* applied to the same evidence in a simple three-process tree. See text for details.

of two evidences e_1 and e_2, by

$$\rho_{12} = \frac{(1 - D_{12})(w_1 e_1 w_2 e_2)}{Min(w_1 e_1, w_2 e_2)} + D_{12}, \qquad (2)$$

where D_{12} is the dependency between e_1 and e_2 on a scale of 0 to 1 with zero indicating mutually independent processes and 1 indicating total dependence between the processes; w_1 and w_2 are the respective *sufficiency* values.

Introducing 'dependency' into the calculation results in a value of 'evidence for' of 0.40 and for 'evidence against' of 0.24. The overall outcome is a further reduction in accumulated evidence and an increase in the level of uncommitted belief or remaining uncertainty. In effect, the dependency factor operates to discount redundant information.

In the fourth example (case d), *sufficiencies* and *dependencies* are unchanged from the previous example, but process 3 is identified as a *necessary* process. In this case, since the necessary process has 'evidence for' of less than 0.5 the normal calculation is over-ridden and the unfavourable outcome from the necessary process is propagated forward.

For Bayesian enthusiasts, the calculation described above reverts to the conventional Bayesian revision when:

(1) There is no uncommitted belief in the inputs i.e. $b(\text{A}) + b(\text{Not A}) = 1$.
(2) Sufficiencies are normalized and sum to 1.
(3) Dependencies are all 0 (i.e. the input evidences are mutually independent).

In this respect IPT may be viewed as a generalization of Bayes' rule, or put another way, Bayes' rule is a specific case of the interval analysis approach when all the contributing processes are mutually exclusive and there is no general level of uncertainty.

Eliciting evidence

Judgements of evidence can be elicited in several ways. The approach recommended here involves separate elicitation of 'evidence for' and 'evidence against' using qualitative linguistic judgements that are subsequently converted to numbers via a utility function. One of the problems in eliciting evidence concerning geological models, particularly in the case where alternative sites are being compared or where a site is being evaluated at different stages in an investigation programme, is that we generally do not have the same level of knowledge for each site or for each evaluation period.

Investigations at one site may have advanced much further than at another. Models may have been constructed for both sites but the knowledge base underpinning the models of each site will be quite different. One approach that is useful in 'levelling the playing field' is to establish, for each process for which evidence is to be elicited, a measure of the knowledge base underpinning the process and to compare this with an estimate of the knowledge base that would be required for the expert to be fully confident in providing the judgement. In order to accommodate differences in knowledge base, elicitation of evidence, as practised by the writer, is carried out as a three-stage process (Fig. 6). Initially, participants are asked about the adequacy of the knowledge base on which they are being asked to make judgements. In practical terms one might ask the following two questions:

> How much information would you ideally wish to have to be confident in providing a judgement of evidence for or against the hypothesis?
>
> In relation to your answer from the previous question how much information do you actually have on which to make the judgement?

The response to the second question may be given as a quantitative fraction or as a linguistic variable, depending on preference. The difference between the two quantifies 'uncertainty due to lack of knowledge' for the particular process. For the remainder, i.e. the information base from which evidence is drawn, judgement is required of the 'face' value of the evidence in support of the process in question, and this is subsequently modified by judgement of the quality of the evidence. Thus, in eliciting 'evidence for the proposition', we ask two questions:

> Assuming that the information is of high quality and trustworthy, what support does it give to the dependability of the process?
>
> How much faith do you have that the information on which you have based your judgement is of high quality and is trustworthy?

Evidence against the hypothesis is elicited separately but in the same way. Assuming a frame of reference of unity for assignment of

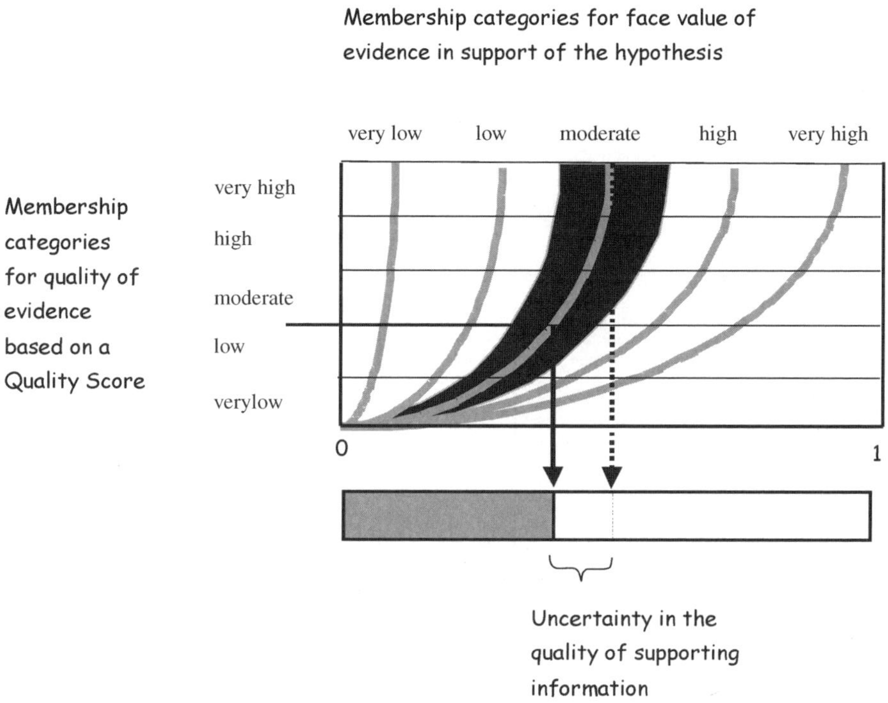

Fig. 6. Linguistic variables are used to define membership categories for 'face value' and 'quality' of evidence. The initial judgement of the 'face' value of the evidence is modified via a utility function linked to the quality score. The utility function is chosen so that the initial judgement of the face value of the evidence could be downgraded, if necessary all the way to zero, if, for example, the information is judged to be worthless.

belief, then uncommitted belief (remaining uncertainty) is given by $1 - (b(\text{A}) + b(\text{Not A}))$.

Assessing quality

The value that should be attached to information, whether quantitative or qualitative, can be developed to provide a more robust justification than is possible through a single linguistic response to the judgement of quality. As part of their NUSAP (**Nu**merical, **U**nit, **S**pread, **A**ssessment and **P**edigree) scheme for uncertainty analysis, Funtowicz & Ravetz (1990) introduced the concept of 'Pedigree', in which the origin and trustworthiness of knowledge is based on some measure of who has it, how it was derived and what went into its derivation. The form of the Pedigree evaluation is a rectangular array, the columns representing different quality indicators. The cells in each column describe the particular criteria against which judgements are made in rank order from top to bottom (Funtowicz & Ravetz 1990; van der Sluijs *et al.* 2002). The quality indicators used may vary depending upon the subject of the analysis. In the geological context, the indicators proposed here (Fig. 7) are familiar confidence-building measures appropriate to the peer review process, for example, theoretical basis, scientific method, auditability, calibration, validation and objectivity. For each process for which evidence is input, scores are given for the various quality indicators. A quality score is calculated from the cumulative, normalized scores for the individual indicators. The indicators themselves may, if desired, be weighted according to judgement of their relative importance. Separate scores are elicited for the supporting and refuting evidence.

Quality indicators

Theoretical basis

This indicator is used to indicate whether the evidence under consideration conforms to well-established theory at one extreme or is pure speculation at the other.

Scientific method

As a general rule the scientific method follows the sequence: Observation; Formulation of

Quality score	Theoretical basis	Scientific method	Auditability	Calibration	Validation	Objectivity
V high (1)	Well established theory	Best available practice; Large sample; direct measure	Well documented trace to data	An exact fit to data	Independent measurement of same variable	No discernible bias
High	Accepted theory; high consensus	Accepted reliable method; small sample; direct measure	Poorly documented but traceable to data	Good fit to data	Independent measurement of high correlation variable	Weak bias
Mod	Accepted theory; low consensus	Accepted method; derived data; analogue; limited reliability	Traceable to data with difficulty	Moderately well correlated with data	Validation measure not truly independent	Moderate bias
Low	Preliminary theory	Preliminary method; unknown reliability	Weak, obscure link to data	Weak correlation to data	Weak, indirect validation	Strong bias
V low (0)	Crude speculation	No discernible rigour	No link back to data	No apparent correlation with data	No validation presented	Obvious bias

Quality Indicators

Fig. 7. Judgement of the 'quality' of the information on which evidence is given is based on a number of quality indicators.

hypothesis; Test hypothesis; Reject/Fail to reject hypothesis. If rejected, a new hypothesis can be proposed and the process begun again. If not rejected, the hypothesis stands as a valid explanation of the observations. But it is not necessarily the only valid explanation; there may be other equally valid alternative explanations of the observations, and future observations or experiments could cause the hypothesis to be rejected.

This indicator provides a measure of how well the scientific method and best practice has been followed in the production of the information on which judgement of evidence has been made.

Auditability

No definitive, rule-based methodology exists for data integration, interpretation and modelling. However, in order to demonstrate confidence in the models, it is essential to document and justify the sequence of conceptual steps contributing to specific interpretations such that an audit trail is discernible from the model to the data from which it has been conceived. This indicator provides a measure of how well the information presented can be traced back to the raw observations.

Calibration and validation

Models are abstractions of reality. Calibration may be used to confirm the legitimacy of the model as 'consistent with all the observational data', and validation may establish the model as permissible through some form of prediction testing, but neither calibration nor validation can establish whether or not the model is correct. Nevertheless both calibration and validation are valuable and necessary confidence-building measures for demonstrating quality in the modelling process.

These indicators are used to make judgements on whether the information is: (1) calibrated to data; and (2) validated through independent measurement or prediction testing.

Objectivity

While the scientific method provides a logical framework for improving understanding it does not guarantee objectivity. The influence of entrenched values, motivational bias and peer and institutional pressures may obscure true objectivity. In order to maintain a check on the quality and objectivity of our interpretations we rely on peer review and exposure to critique through peer-reviewed publication.

This indicator is used to give a judgement on the extent to which the information can be said to be objective and free from bias.

Analysing the inputs

Once the process tree has been constructed and evidence supplied, it not only provides a top-level view of the confidence in the model but also stands as an audit record of the evaluation process and continues to be a tool for sensitivity analysis, value of information studies and ongoing evaluations and updates. For each process for which evidence is given, a qualitative and visual assessment can be made of the nature and respective contributions of different uncertainty types, together with an evaluation of the quality of the information on which the judgement of evidence has been made. Figure 8 gives an immediate visual indication of the strengths and weaknesses of the information in terms of uncertainty, type of uncertainty and quality.

Uncertainty classification

One feature of the evaluation is the ability to discriminate between different sources and types of uncertainty in the information used. Using this approach, uncertainty in the input evidence can be categorized as:

- conceptual uncertainty due to incomplete system understanding
- uncertainty due to lack of information
- uncertainty due to conflict
- uncertainty in the quality of evidence, both for and against the proposition.

Once the process model representing the system has been constructed and parameters for the inference engine have been elicited, the completeness of the process model can be examined by setting the evidence for all lowest level processes to a value of 1, i.e. an assumption of complete support for all the lower level elements of the process tree. Quite commonly, when such sensitivity analysis is carried out, it is found that the top-level outcome does not reach 1 and there remains an element of irreducible uncertainty. This suggests that either the process model is incomplete (i.e. the system is not fully

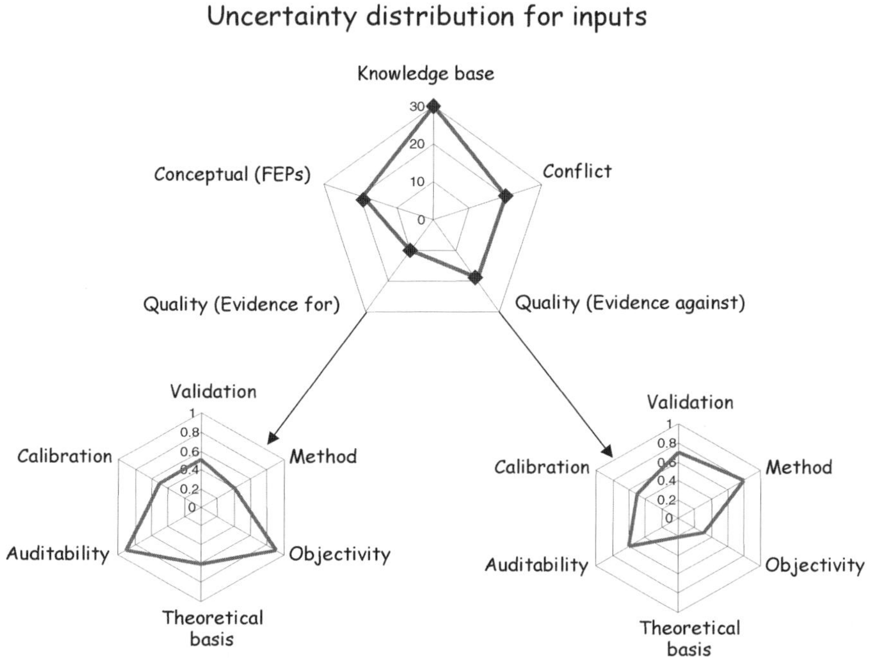

Fig. 8. Radar diagrams showing the distribution of uncertainty and quality scores for the information underpinning the judgements of evidence.

defined), and the remaining uncertainty can be attributed to 'conceptual uncertainty', or that the values assigned to the inference parameters have been misjudged. In some instances, it may be desirable to examine the impact of conceptual uncertainty on the modelled system, in which case unknown processes could be specifically added to the model and sensitivity analysis carried out by systematically varying the influence of the logical operators applied to them.

Uncertainty due to lack of information

Lack of information is a major factor fuelling uncertainty, particularly in the early stages of an investigation. Estimation of the contribution that lack of information makes to overall uncertainty for each process is calculated as the difference between the expert's estimate of the actual information available and the information that the expert would ideally wish to have in order to be fully confident in exercising judgement of that evidence.

Uncertainty due to conflict

For any hypothesis or model there is likely to be some evidence in support and some evidence against. This may arise from conflicting data or from conflicting interpretation of the meaning of the data but, however it arises, it gives rise to uncertainty. In reality the hypothesis is either correct or incorrect, and since both suppositions cannot be correct we can denote the minimum value of 'evidence for' and 'evidence against' as uncertainty due to conflict. For a highly complex model with a large number of contributing FEPs the accumulation of evidence for and against may exceed unity, with the apparent total elimination of 'remaining uncertainty' i.e. $b(\mathrm{A}) + b(\mathrm{Not\ A}) > 1$. What may happen in this case is that 'uncertainty due to lack of knowledge' is reduced at the expense of 'uncertainty due to conflict', i.e. there is an apparent over-commitment of belief resulting from conflict in the evidence. From a decision-maker's point of view, conflict resolution rather than data acquisition becomes the main issue.

Uncertainty due to quality considerations

This uncertainty type can be directly evaluated from the 'quality score' and is a measure of confidence in the quality of the information incorporated into the model, including judgements of the quality of data acquisition, data processing, data interpretation and data integration. As the evidence in support of a proposition may be independent of the evidence against it, separate quality scores are required that give rise to differences in the uncertainty values associated with the evidence for and against the proposition.

Interpreting the outcome

Arguably, the greatest value of analysing uncertainty through evidence-based process modelling lies not in reaching an exact understanding of the expert knowledge inputs, nor in an absolute value of confidence in a particular geological model, but in identifying the major uncertainties in the model, particularly those that provide the greatest impact on confidence in the model. Where there is too much uncertainty for a model to be accepted, the evidence-based evaluation allows identification of the source(s) of uncertainty and hence improved targeting of scarce resources, for example, into further data acquisition, quality improvement or conflict resolution. Sensitivity analysis helps to identify where future investigation is most likely to have the greatest impact in reducing uncertainty due to lack of knowledge. This may be particularly valuable, for instance, when limited resources are available and choices need to be made on which areas to target further data acquisition.

It should be clear by now that, in an uncertain world, we are not dealing with absolute truths or in mathematical terms of accuracy and precision. What we are searching for is some measure of our belief in the dependability of our models, or, for example, in the case of a choice between alternative models, some measure of our relatively greater confidence in one model compared to the alternatives. The top-level result in numerical form is a measure of overall confidence in the geological model. However, we need to be careful about how we interpret this output, avoiding the GIGO garbage in/garbage – or gospel – out epithet given to apparently numerical precise computer output from fuzzy inputs. The initial inputs to this process are linguistic or verbal and hence essentially 'soft' or 'fuzzy' and, ideally, the output should be translated back into verbal qualifiers (in which case FIFO (fuzzy in/fuzzy out) might be a more appropriate acronym for the whole process).

The evidence ratio plot

However expressed, the outcome of the assessment can be used to evaluate the merits or relative merits of models through examination of the three-way relationship between 'evidence for', 'evidence against' and 'remaining uncertainty'.

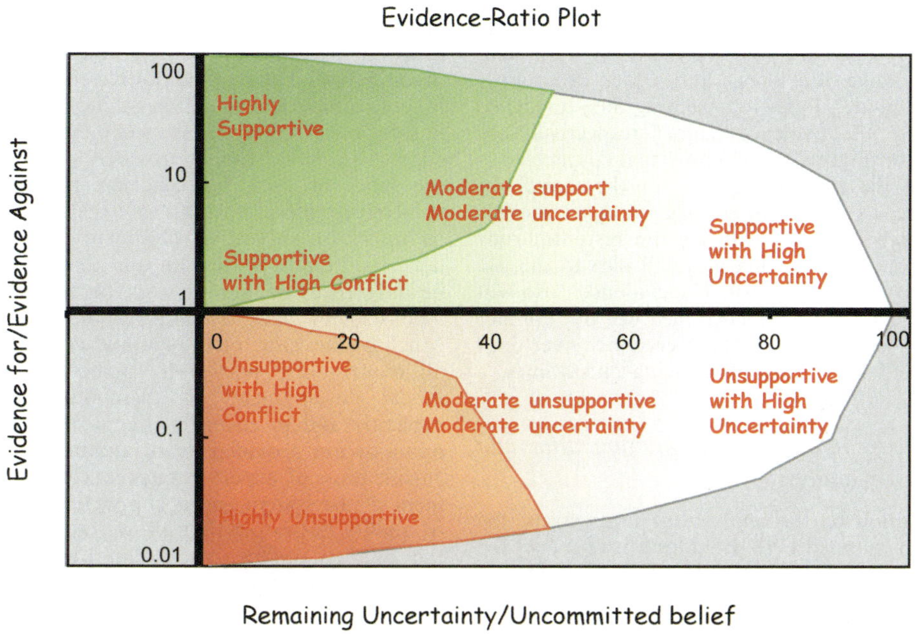

Fig. 9. By plotting the evidence ratio v. remaining uncertainty the evidence ratio plot allows all the contributing processes to be displayed at the same time on the same computer screen or A4-size plot. Different areas of the plot represent different levels of support, enabling unsupportive and highly uncertain processes to be readily identified visually. Sensitivity analysis identifies high-impact processes to be targeted for future research/data acquisition.

The evidence ratio plot introduced here is a useful, novel method of displaying the process model that allows both the top-level outcome and the input evidence to be examined simultaneously on a computer screen or an A4 page. The plot overcomes the requirement for cumbersomely large plot outputs of complex, hierarchical tree structures and allows the user to access and interrogate any subsection of the model. Figure 9 shows the main elements of the ratio plot. The ratio of 'evidence for' to 'evidence against' is plotted on a log scale on the vertical axis; 'remaining uncertainty' is plotted on the horizontal axis. An outcome of [0,1] represents 50% 'evidence for' and 50% 'evidence against' with no remaining uncertainty and plots at the intersection of the graph axes. The value of no 'evidence for' or 'evidence against' with 100% uncertainty plots at the right of the horizontal axis. The line joining the two describes a 'line of conflict' such that the 'evidence for' is always equal to the 'evidence against' with differing values of remaining uncertainty. Values plotting above this line represent a favourable balance of

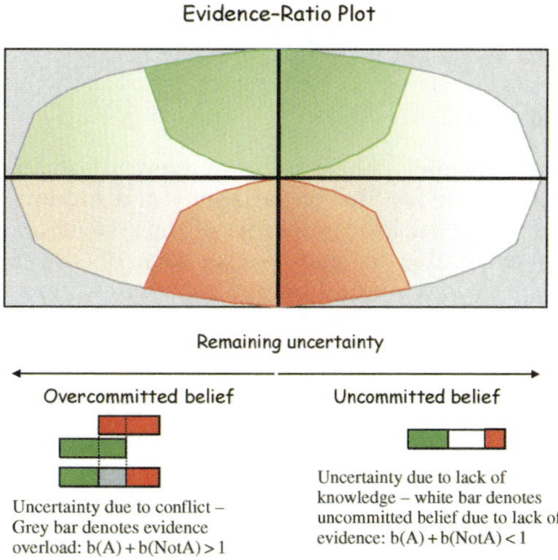

Fig. 10. For large, complex models with high levels of conflicting evidence, the accumulation of evidence may exceed unity. In such cases, processes may plot to the left of the ordinate with conflict displacing lack of knowledge as the dominant cause of uncertainty.

evidence, indicating support for the model; those plotting below the line represent an unfavourable balance of evidence and a lack of support for the model. Lines representing 50% 'evidence for' and 50% 'evidence against' respectively are drawn on Figure 9, enabling visual characterization of the relative support for the model. In practice, as noted earlier, for complex models with high levels of evidence the resultant outcome may exceed unity and will plot to the left of the ordinate (Fig. 10). The evidence ratio will still be a major confidence measure but conflict resolution is likely to take precedence over data gathering as a means of reducing uncertainty.

Knowledge management, communication and sensitivity analysis

In addition to the analysis of uncertainty the method outlined here provides a useful tool for analysis and interrogation of the geological model. The user is able to answer 'Show me the data' or 'Show me the justification' queries by drilling down through the model, exposing different layers of the tree structure and accessing the evidential judgements and associated justification. Used in this way it is a knowledge organization and communication tool. Using sensitivity analysis, the critical path and those FEPs having the greatest relative impact on the top-level uncertainty can be identified. In addition, 'value of information studies' may be carried out to examine the potential impact of alternative data acquisition or conflict resolution strategies on the top-level issue.

Summary

Fundamentally, there are two types of uncertainty: aleatory, due to variability and randomness of the intrinsic properties of the system; and epistemic, due to ignorance or lack of knowledge. Each requires different treatment in terms of evaluation. In general terms, aleatory uncertainty is susceptible to management through 'closed world', traditional, quantitative, probabilistic methods while epistemic uncertainties are best dealt with using 'open-world', qualitative, evidence-based methods.

Evidence-based uncertainty analysis coupled with a formal assessment of the 'quality of knowledge' is a way of demonstrating confidence in a hypothesis or model that is underpinned by uncertain information. The three-value logic provided by using interval analysis provides greater transparency in exposing the underlying causes of uncertainty and allows greater flexibility for sensitivity analysis to identify high-impact uncertainties for future investigation. The method addresses uncertainty in both numerical and non-numerical data, in non-technical as well as technical issues. It enables evidence to be provided not only on the basis of data but also on the different subjective interpretations of the data. The inclusion of subjective judgment in the evidence-gathering process means that there will be differences of opinion and the possibility of inconsistency and conflict in interpretation. Much of the value of the method is in enabling all contributing parties to participate in the model development, in identifying conflicts explicitly and in ensuring that justification for opinions are provided in an auditable fashion. Justification of expert judgements is a requirement of the system and is supported by the use of quality indicators that allow commentary on the quality of the whole interpretation and modelling process.

The approach to building confidence in geological models outlined in this article is largely drawn from two sources. Thanks are due to the researchers at Bristol University for development of the Interval Probability Theory of evidential reasoning and for providing geoscientists with a means of saying 'I do not know'. I am also grateful to J. Ravetz for introducing me to his ideas on post-normal science and the analysis of Pedigree as a means of formalizing 'expert judgement' of the quality of knowledge. I apologize to both sources for any misrepresentation of their work and ideas as presented in this article.

References

BADDELEY, M. C., CURTIS, A. & WOOD, R. 2004. An introduction to prior information derived from probabilistic judgements: elicitation of knowledge, cognitive bias and herding. *In:* CURTIS, A. & WOOD, R. (eds) *Geological Prior Information: Informing Science and Engineering*. Geological Society, London, Special Publications, **239**, 15–27.

BLOCKLEY, D. & GODFREY, P. 2000. *Doing It Differently: Systems for Rethinking Construction.* Thomas Telford, London.

BONHAM-CARTER, G. F. 1994. *Geographic information systems for geoscientists: modelling with GIS.* Pergamon, Ontario, 295–302.

CAPEN, E. C. 1976. The difficulty of assessing uncertainty. *Journal of Petroleum Technology*, **28**, 843–850.

CUI, W. & BLOCKLEY, D. I. 1990. Interval probability theory for evidential support. *International Journal of Intelligent Systems*, **5**, Part 2, 183–192.

CURTIS, A. & WOOD, R. 2004. Optimal elicitation of probabilistic information from experts. *In:* CURTIS, A. & WOOD, R. (eds) *Geological Prior Information: Informing Science and Engineering.*

Geological Society, London, Special Publications, **239**, 127–145.

DAVIS, J. P. & HALL J. W. 2003. A software-supported process for assembling evidence and handling uncertainty in decision-making. *Decision Support Systems*, **35**, 415–433.

FERSON, S. & GINZBURG, L. R. 1996. Different methods are needed to propagate ignorance and variability. *Reliability Engineering and System Safety*, **54**, 133–144.

FOLEY, L., BALL, L., DAVIS, J. P. & BLOCKLEY, D. I. 1997. Fuzziness, incompleteness and randomness: classification of uncertainty in reservoir appraisal. *Petroleum Geoscience*, **3**, 203–209.

FUNTOWICZ, S. O. & RAVETZ, J. R. 1990. *Uncertainty and Quality in Science for Policy*. Kluwer, Dordrecht.

HALL, J. W., BLOCKLEY, D. I. & DAVIS, J. P. 1998. Uncertain inference using interval probability theory. *International Journal of Approximate Reasoning*, **19**, 247–264.

HALPERN, D. H. 1996. *Thought & Knowledge: An Introduction to Critical Thinking*. Lawrence Erlbaum Associates, Mahwah, New Jersey.

HOFFMAN, F. O. & HAMMONDS, J. S. 1994. Propagation of uncertainty in risk assessments. The need to distinguish between uncertainty due to lack of knowledge and uncertainty due to variability. *Risk Analysis*, **14**, 707–712.

HOFFMAN-REIM, H. & WYNN, B. 2002. In risk assessment, one has to admit ignorance. *Nature*, **416**, 123.

JNC TN1410 2000-002. H12: Project to establish the scientific and technical basis for HLW disposal in Japan. Supporting Report 1. *Geological Environment in Japan*. Japan Nuclear Cycle Development Institute, April 2000.

KAHNEMAN, D., SLOVIC, P. & TVERSKY, A. (eds) 1982. *Judgement under Uncertainty: Heuristics and Biases*. Cambridge University Press, Cambridge.

KITAYAMA, K., UMEKI, H. & MASUDA, S. 2001. Japanese HLW disposal program – where we stand. Proceedings of WM'01 Conference, Tucson, Arizona. 25 February–1 March 2001.

NEA. 1999. Nirex methodology for scenario and conceptual model development. *An International Peer Review*. Nuclear Energy Agency. OECD Publications, Paris.

NODA, T. 1987. Anion index as an indicator of geothermal activity. *Journal of the Geothermal Research Society of Japan*, **9**, 131–141. [In Japanese with English abstract]

NUMO. 2002*a*. Siting Factors for the Selection of Preliminary Investigation Areas. *Nuclear Waste Management Organization of Japan*, Publications World Wide Web Address: http://www.numo.or.jp.

NUMO. 2002*b*. Instructions for Application. *Nuclear Waste Management Organization of Japan, Publications* World Wide Web Address: http://www.numo.or.jp.

SHAFER, G. 1976. *A Mathematical Theory of Evidence*. Princeton University Press, Princeton.

SCHRADER-FRECHETTE, K. S. 1993. *Burying Uncertainty: Risk and the Case Against Geological Disposal of Nuclear Waste*. University of California Press, Berkley.

THURY, M., GAUTSCHI, A., MAZUREK, M., MÜLLER, W. H., NAEF, H., PEARSON, F. J., VOMVORIS, S. & WILSON, W. 1994. *Geology and Hydrogeology of the Crystalline Basement of northern Switzerland*. Nagra (National Cooperative for the Disposal of Radioactive Waste), Switzerland, Technical Report **NTB-93-01**.

VAN DER SLUIJS, J. P., POTTING, J., RISBEY, J., VAN VUREN, D., DE VRIES, B., BEUSEN, A., HEUBERGER, P. *ET AL*. 2002. Uncertainty assessment of the IMAGE/TIMER B1 CO2 emissions scenario, using the NUSAP method. *Dutch National Research Programme on Global Air Pollution and Climate Change, Report No.* **410 200**, 104.

WALTZ, E. & LLINAS, J. 1990. *Multisensor Data Fusion*. Artech House, Boston.

Lithospheric structure of the Canadian Shield inferred from inversion of surface-wave dispersion with thermodynamic *a priori* constraints

N. M. SHAPIRO[1], M. H. RITZWOLLER[1], J. C. MARESCHAL[2] & C. JAUPART[3]

[1]*Center for Imaging the Earth's Interior, Department of Physics, University of Colorado, Boulder, CO 80309-0390, USA (e-mail: nshapiro@ciei.colorado.edu)*

[2]*GEOTOP-UQAM-McGill, Centre de Recherche en Géochimie et en Géodynamique, Université du Québec à Montréal, Montréal, Canada*

[3]*Institut de Physique du Globe de Paris, Paris, France*

Abstract: We argue for and present a reformulation of the seismic surface-wave inverse problem in terms of a thermal model of the upper mantle and apply the method to estimate lithospheric structure across much of the Canadian Shield. The reformulation is based on a steady-state temperature model, which we show to be justified for the studied region. The inverse problem is cast in terms of three thermal parameters: temperature in the uppermost mantle directly beneath Moho, mantle temperature gradient, and the potential temperature of the sublithospheric convecting mantle. In addition to the steady-state constraint, prior physical information on these model parameters is based on surface heat flow and heat production measurements, the condition that melting temperatures were not reached in the crust in Proterozoic times and other theoretical considerations. We present the results of a Monte Carlo inversion of surface-wave data with this 'thermal parameterization' subject to the physical constraints for upper mantle shear velocity and temperature, from which we also estimate lithospheric thickness and mantle heat flux. The Monte Carlo inversion gives an ensemble of models that fit the data, providing estimates of uncertainties in model parameters. We also estimate the effect of uncertainties in the interconversion between temperature and seismic velocity. Variations in lithospheric temperature and shear velocity are not well correlated with geological province or surface tectonic history. Mantle heat flow and lithospheric thickness are anti-correlated and vary across the studied region, from 11 mW/m^2 and nearly 400 km in the northwest to about 24 mW/m^2 and less than 150 km in the southeast. The relation between lithospheric thickness and mantle heat flow is consistent with a power law relation similar to that proposed by Jaupart *et al.* (1998), who argued that the lithosphere and asthenosphere beneath the Canadian Shield are in thermal equilibrium and heat flux into the deep lithosphere is governed by small-scale sublithospheric convection.

Studies of the structure, composition and evolution of Precambrian continental lithosphere are the foundation of the current understanding of the processes that have shaped the growth and long-term stability of the continents. The thermal structure of Precambrian continental lithosphere has been studied principally with three methods: inversion of surface heat flow measurements (e.g. Nyblade & Pollack 1993; Pollack *et al.* 1993; Jaupart *et al.* 1998; Jaupart & Mareschal 1999; Nyblade 1999; Artemieva & Mooney 2001), geothermobarometry of mantle xenoliths (e.g. a recent review by Smith 1999), and seismic tomography (e.g. Furlong *et al.* 1995; Goes *et al.* 2000; Röhm *et al.* 2000). Each of these methods has distinct strengths and limitations. Thermobarometry of mantle xenoliths is probably most directly related to deep thermal structure, but high-quality xenoliths are rare and those that are available have recorded the temperature of their formation which may not be representative of the current thermal regime of the lithosphere. In contrast, heat flow measurements directly reflect the recent lithospheric thermal regime, but their ability to resolve the deep structure of the continental lithosphere is limited. Inverting surface heat flow for the mantle geotherm requires strong *a priori* assumptions about the thermal state of the lithosphere and the distribution of heat sources. Seismic data

From: CURTIS, A. & WOOD, R. (eds) 2004. *Geological Prior Information: Informing Science and Engineering*. Geological Society, London, Special Publications, **239**, 175–194. 186239-171-8/04/$15.00

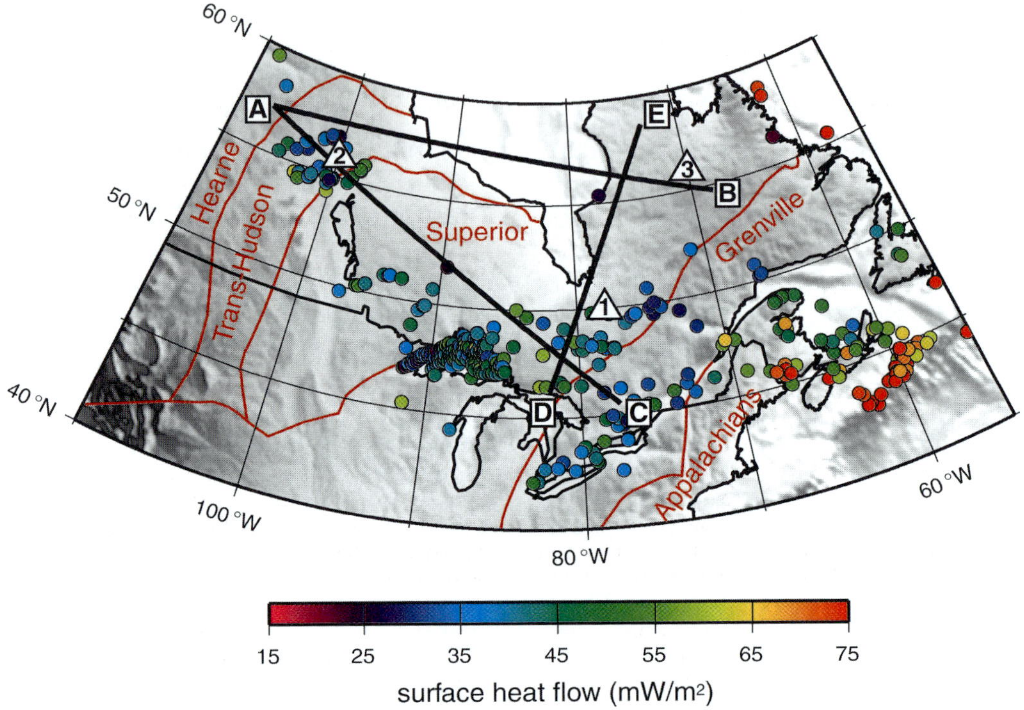

Fig. 1. Reference map of eastern Canada showing the heat flow measurements used in this study as well as the locations of the 1-D (Spatial Points 1, 2, 3) and 2-D profiles (A–B, A–C, D–E) referred to in the study. Red lines, boundaries of principal Precambrian provinces. Adapted from Hoffman (1989).

are directly sensitive to the present deep structure of the lithosphere, but vertical resolution is limited and substantial uncertainties remain in the conversion from seismic velocity to temperature, resulting particularly from ignorance of mantle composition and anelasticity.

The limitations of each of these methods individually lead naturally to exploiting them in combination. For example, Rudnick & Nyblade (1999) describe constraints on the Archean lithosphere that derive from applying xenolith thermobarometry and surface heat flow measurements simultaneously. Shapiro & Ritzwoller (2004) discuss the use of surface heat flux as an *a priori* contraint on inversions of seismic surface-wave dispersion data. The heat flow measurements are used to establish upper and lower temperature bounds in the uppermost mantle directly beneath the Moho discontinuity. The temperature bounds are then converted to bounds on seismic velocity using the method of Goes *et al.* (2000). This approach can improve seismic models beneath continents, particularly beneath cratons and continental platforms, and tighten constraints on mantle temperatures. Shapiro & Ritzwoller also describe an additional thermodynamic constraint that involves replacing *ad hoc* seismic basis functions with a physical model of the thermal state of the upper mantle which is intrinsically a function of temperature. They applied this procedure to the oceanic upper mantle where the thermodynamic model consisted of a shallow conductive layer underlain by a convective mantle. They argued that the constraint produces more plausible models of the oceanic lithosphere and asthenosphere and reduces the uncertainty of the seismic model while negligibly degrading the fit to the seismic data.

This study is an extension of the results of Shapiro & Ritzwoller (2004) in two principal respects. First, Shapiro & Ritzwoller applied heat flow measurements as a prior constraint on seismic inversions only at a few isolated points to test the concept. In the present study, we apply the joint inversion over a wide region of North America, principally in the Canadian Shield, where high-quality heat flow measurements are available and the lithosphere is likely to be in thermal equilibrium. The locations of heat flow measurements used in this study are shown in Figure 1. We refer to these results as deriving from the 'seismic parameterization', because models are constructed in seismic velocity model space (Shapiro & Ritzwoller 2002). The heat flow constraints are converted to seismic velocities from temperature model space.

Heat flow measurements, however, do not cover the entire Canadian Shield, but are clustered mostly in southern Canada. To apply the heat flow constraint broadly across the Canadian Shield, therefore, would require us either to extrapolate existing measurements to other regions or to apply physical constraints derived from the regions where heat flow measurements exist. Here, we use the latter approach as the second extension of the results of Shapiro & Ritzwoller. Based on inversions in regions where heat flow measurements exist, we argue that the uppermost mantle beneath much of the studied region is likely to be in thermal steady state; i.e. the lithosphere is neither heating nor cooling and the surface heat flow is the sum of the heat entering the base of the lithosphere and the heat production in the crust. We reformulate the inverse problem in terms of a physical model of the thermal state of the upper mantle, in which a lithosphere in thermal steady state overlies a convecting mantle. Models are constructed first in temperature model space and are tested to ensure that they satisfy the steady-state constraint, surface heat flow data (within uncertainties), and bounds on the mantle component of heat flow (discussed later). Temperature profiles that satisfy these constraints are converted back to seismic model space where a seismic crust is introduced and the resulting model is tested to see if it fits the seismic data acceptably. We refer to these results as deriving from the 'thermal parameterization'. Figure 2 presents a schematic outline of the method based on the thermal parameterization. We apply this method to estimate the seismic and temperature structure of much of the Canadian Shield, including the mantle component of heat flux and lithospheric thickness.

We begin by discussing the temperature bounds applied on the models based both on the seismic and thermal parameterizations. We also discuss uncertainties in the interconversion between temperature and seismic shear velocity. The joint inversion of surface-wave dispersion and heat flow with the seismic parameterization is then introduced and described followed by the inversion based on the thermal parameterization with the steady-state heat flow constraint. Finally, we discuss the results of the inversion with the thermal parameterization, including estimates of lithospheric thickness and the mantle component of heat flow.

Bounds on temperature and seismic velocities at the top of the mantle

Temperature bounds from heat flow

The assimilation of heat flow data in the seismic inverse problem is accomplished by constraining the uppermost mantle temperatures, T_m estimated from surface heat flow. The Canadian Shield is an ideal location for the first extended application of this method for two reasons. Firstly, the Canadian heat flow data (Fig. 1) are of exceptional quality because, for example, heat flow has been measured using several deep neighbouring boreholes in many cases (e.g. Jessop *et al.* 1984; Drury 1985; Drury & Taylor 1987; Drury *et al.* 1987; Mareschal *et al.* 1989, 1999, 2000; Pinet *et al.* 1991; Guillou *et al.* 1994; Hart *et al.* 1994; Guillou-Frottier *et al.* 1995, 1996; Rolandone *et al.* 2002) and there is an extensive dataset of crustal heat production measurements (see Jaupart & Mareschal 2004). Secondly, as discussed in detail by Shapiro & Ritzwoller (2004), the joint inversion of seismic data and heat flow is most straightforward for cold lithosphere such as that found beneath Precambrian regimes. This is because uncertainties in the anelastic correction are smallest, which is part of the interconversion between temperature and seismic shear velocity. In addition, the volatile content, which can also affect the conversion to temperature, is believed to be small beneath ancient cratons due to the efficiency of partial melting upon their formation (e.g. Pollack 1986).

Even in the best of cases, however, estimating temperatures at the top of the mantle requires determining the crustal geotherm, which depends on thermal conductivity and on the distribution of crustal heat production. Several methods have been used to determine radioactive heat production in the Canadian crust (e.g. Jaupart *et al.* 1998; Jaupart & Mareschal 1999) and they show that a simple linear relation between surface heat flow and crustal heat production is invalid because, in many terranes, such as the greenstone belts, heat production is lower in the upper crust than in the mid-crust. Lower crustal heat production is believed to be relatively homogeneous ($\sim 0.4\,\mu W/m^3$). Based on a simultaneous Monte Carlo inversion of heat flow and gravity data across the Abitibi Belt, Guillou *et al.* (1994) proposed that the mantle component of heat flow lies between 7 and $15\,mW/m^2$. It has been suggested (e.g. Jaupart *et al.* 1998; Jaupart & Mareschal 1999) that such low values for the heat flow from the mantle are characteristic of most of the

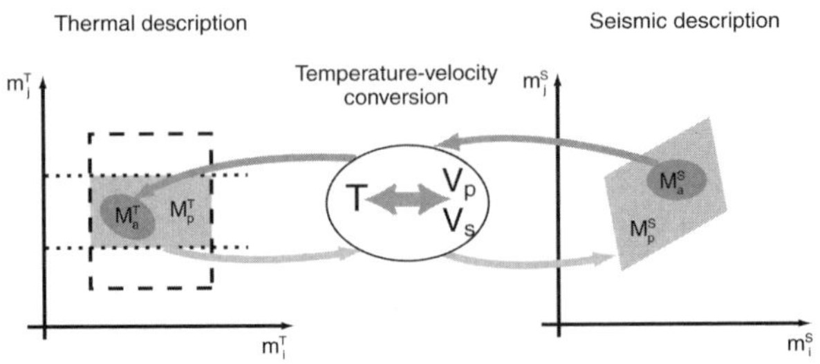

Fig. 2. Schematic representation of the Monte Carlo seismic inversion based on a thermal parameterization with *a priori* constraints. The thermal parameterization (left panel) is constrained by the heat flow data (horizontal dotted lines) where they exist, the steady-state constraint on the thermal structure of the mantle (dashed rectangle), and a lower bound on mantle heat flow (not shown). These constraints delimit the range of physically plausible thermal models M_p^T (light shaded area on left panel). Using a temperature–seismic velocity conversion, this range is converted into a range of physically plausible seismic models M_p^S (light shaded area on right panel) to which a range of crustal seismic models and radial anisotropy are added. Random sampling within M_p^S identifies the ensemble of acceptable seismic models M_a^S (dark shaded area on right panel). Finally, the seismic crust is stripped off and this ensemble is converted back into the ensemble of acceptable temperature models M_a^T (dark shaded area on left panel).

Canadian Shield. The range was further narrowed down by Rolandone *et al.* (2002) who argued that mantle heat flow cannot be lower than 11 mW/m^2 for the crust to have stabilized. We will use the assumption that heat flow from the mantle is relatively low and homogeneous across the region of study to compute crustal geotherms by solving the steady-state heat equation with different models of the distribution of heat production in the crust. The major cause for the uncertainties on the estimated temperature in the upper mantle is the limited knowledge of crustal heat production.

To bound temperatures in the uppermost mantle we consider two end-member models of crustal heat production. The lower bound is set by a two-layer crust that is consistent with a 'cold' uppermost mantle with temperature T_{cold}. Heat production of 0.4 μW/m^3 is used in the 20 km-thick lower crust, and upper crustal heat production is adjusted to match the measured surface heat flow with a constant mantle heat flow of 15 mW/m^2. We fix crustal thickness for the Canadian Shield at 40 km here (e.g. Perry *et al.* 2002) and crustal thermal conductivity at 3.0 W/m/K. A second, single-layer crustal model that produces a 'hot' uppermost mantle with temperature T_{hot} was constructed by assuming that heat production is uniform throughout the crust with a value adjusted to the same constant mantle heat flow of 15 mW/m^2. The crustal thermal conductivity of 2.5 W/m/K is used for the 'hot' model. The resulting two crustal geotherms for a surface heat flux of 45 mW/m^2 are shown in Figure 3.

In regions where the density of heat flow measurements is high, we use these bounds on uppermost mantle temperatures: $T_{\mathrm{min}} = T_{\mathrm{cold}}$ and $T_{\mathrm{max}} = T_{\mathrm{hot}}$. In regions away from heat flow measurements, however, we widen the temperature range by increasing the upper bound on temperature to $T_{\mathrm{max}} = T_{\mathrm{hot}} + (T_{\mathrm{hot}} - T_{\mathrm{cold}})$, but retain $T_{\mathrm{min}} = T_{\mathrm{cold}}$ because this lower bound is already very low. These bounds are varied spatially in a smooth way. Figure 4 displays the spatial variation of these temperature bounds in the uppermost mantle. The temperature limits T_{min} and T_{max} are sufficiently different to account for uncertainties in the crustal thermal parameters but still provide useful constraints on the seismic inver-

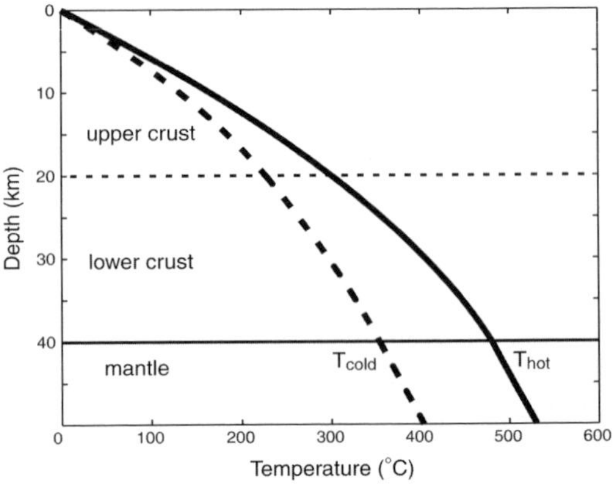

Fig. 3. End-member crustal models that define T_{cold} and T_{hot} at the top of the mantle for a surface heat flux of 45 mW/m^2. For both models, the same values are assumed for the mantle heat flow and thermal conductivity.

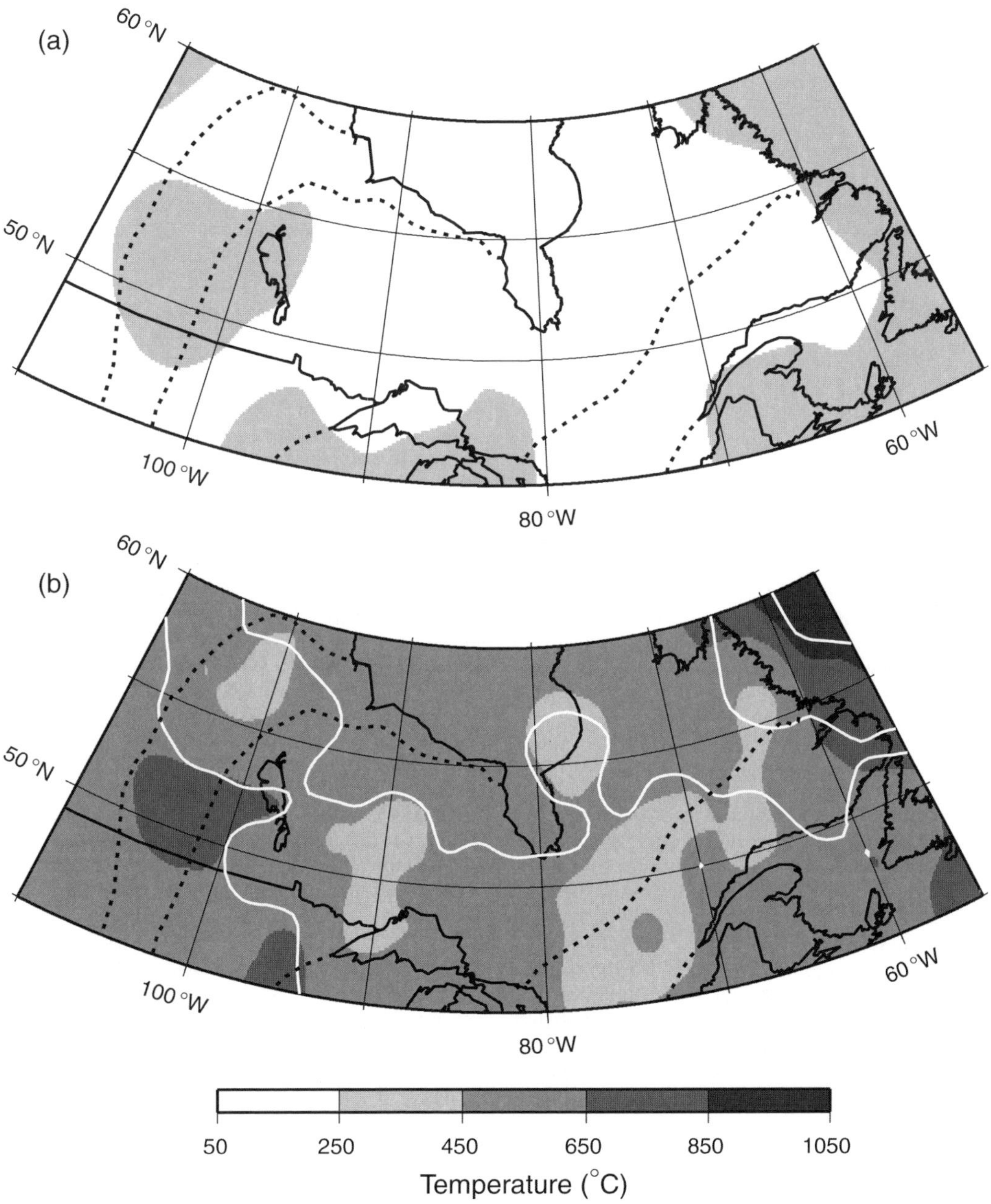

Fig. 4. Uppermost mantle (sub-Moho) temperature bounds inferred from the heat flow data. (**a**) Lower bound T_{min}. (**b**) Upper bound T_{max}. The white contour in (b) encircles the points within 200 km of a heat flow measurement. Temperature bounds are tightest near heat flow measurements. Dashed lines, boundaries of principal Precambrian provinces. Adapted from Hoffman (1989).

sion, as demonstrated by the results below. As a final step, we interpolate the temperature bounds, T_{min} and T_{max} onto the same $2^\circ \times 2^\circ$ geographical grid on which the surface-wave dispersion maps are defined.

Interconversion between temperature and seismic velocity

We convert temperature to shear velocity using the method of Goes *et al.* (2000). This conversion is based on laboratory-measured thermoelastic properties of mantle minerals which are represented as partial derivatives of the elastic moduli with respect to temperature, pressure and composition. The compositional model is the model of the old cratonic mantle proposed by McDonough & Rudnick (1998). This composition includes 83% olivine, 15% orthopyroxene, and 2% garnet with an iron context $X_{Fe} = 0.086$. For the anelastic correction, we follow Sobolev *et al.* (1996) and Goes *et al.* (2000):

$$Q_\mu(P, T, \omega) = A\omega^a \exp[a(H^* + PV^*)/RT] \quad (1)$$

$$v_{\mathrm{anel}}(P, T, \omega) = v(P, T)\left[1 - \frac{2Q_\mu^{-1}(P, T, \omega)}{\tan(\pi a/2)}\right] \quad (2)$$

and set the exponent $a = 0.15$, activation energy $H^* = 500\,\mathrm{kJ/mol}$, and activation volume $V^* = 2.0 \times 10^{-5}\,\mathrm{m^3/mol}$ but, as described in the appendix to Shapiro & Ritzwoller (2004), we set the amplitude $A = 0.049$ in contrast with their value of 0.148.

Uncertainties in the interconversion

Uncertainties in the seismic velocity–temperature relationship result from a number of sources, including uncertainties in mantle composition, in the thermoelastic properties of individual minerals and in the anelastic correction which extrapolates anharmonic mineral properties measured in the laboratory to seismic frequencies. The physical properties of mantle minerals are measured in laboratories with high precision and are, therefore, not major contributors to errors in the velocity–temperature conversion. The most important uncertainties relate to mantle mineralogical composition and the anelastic correction. The presence of substantial quantities of melt and/or water in the mantle would also affect seismic velocities. These effects are expected to be negligible beneath old continental lithosphere, which is believed to have been largely dessicated during multiple episodes of melting during cratonic formation and is too cold for substantial quantities of melt currently to reside in the uppermost mantle.

To bound the effect of uncertainties in mantle mineralogical composition, we consider a pair of mantle compositional models proposed by McDonough & Rudnick (1998), one for on-cratonic (see previous section) and the other for off-cratonic (68% olivine, 18% orthopyroxene, 11% clinopyroxene, and 3% garnet with an iron content $X_{Fe} = 0.1$) mantle. An assessment of the uncertainty in shear velocity converted from temperature is shown in Figures 5a–c. We start with a cratonic temperature model that is composed of a conductive steady-state linear geotherm with a sub-Moho temperature of $T_m = 437\,^\circ\mathrm{C}$ and mantle heat flow $Q_M = 15\,\mathrm{mW/m^2}$ overlaying a 1300 °C adiabat (Fig. 5a). This temperature model is converted to a shear velocity model using the method of Goes *et al.* (2000) applied to the on-cratonic and off-cratonic compositions, as shown in Figure 5b. The difference in the resulting shear velocity curves provides a conservative estimate of the uncertainty in the temperature–velocity conversion within a single tectonic province.

We also consider two different models of the anelastic correction. Model Q_1 is the model used by Shapiro & Ritzwoller (2004) and is described in the section above. For a contrasting model, we define Model Q_2 which is based on values taken from Berckhemer *et al.* (1982) from experiments on forsterite ($a = 0.25$, $A = 2.0 \times 10^{-4}$, $H^* = 584\,\mathrm{kJ/mol}$, $V^* = 2.1 \times 10^{-5}\,\mathrm{m^3/mol}$). These Q models are shown in Figure 6a computed from the temperature model in Figure 5a. Model Q_2 has weaker attenuation and, therefore, will have a smaller anelastic correction, as shown in Figure 6b, where the strength of the anelastic correction is $2Q^{-1}/\tan(\pi a/2)$, by Equation 2. These models represent fairly extreme values for the anelastic correction, so the difference in shear velocities obtained from these models provides a conservative estimate of uncertainties in the temperature–velocity conversion caused by our ignorance of Q.

Figure 5c shows that, near to the surface, where the temperatures are relatively low and Q is high, the uncertainty in the anelastic correction is small and compositional uncertainty dominates. Deeper in the mantle, temperature increases, Q reduces, the anelastic correction strengthens and uncertainties in the anelastic correction become appreciably more important. Overall, the estimated uncertainty grows from about $\pm 0.5\%$ of the seismic velocity in the uppermost mantle near the Moho to about $\pm 1\%$ in the asthenosphere. A similar assessment of the uncertainty in temperature coverted from shear velocity is shown in Figures 5d–f. Again, uncertainty in composition dominates at shallow depths and uncertainty in Q is more important deeper in the upper mantle. The resulting uncertainty in temperature converted

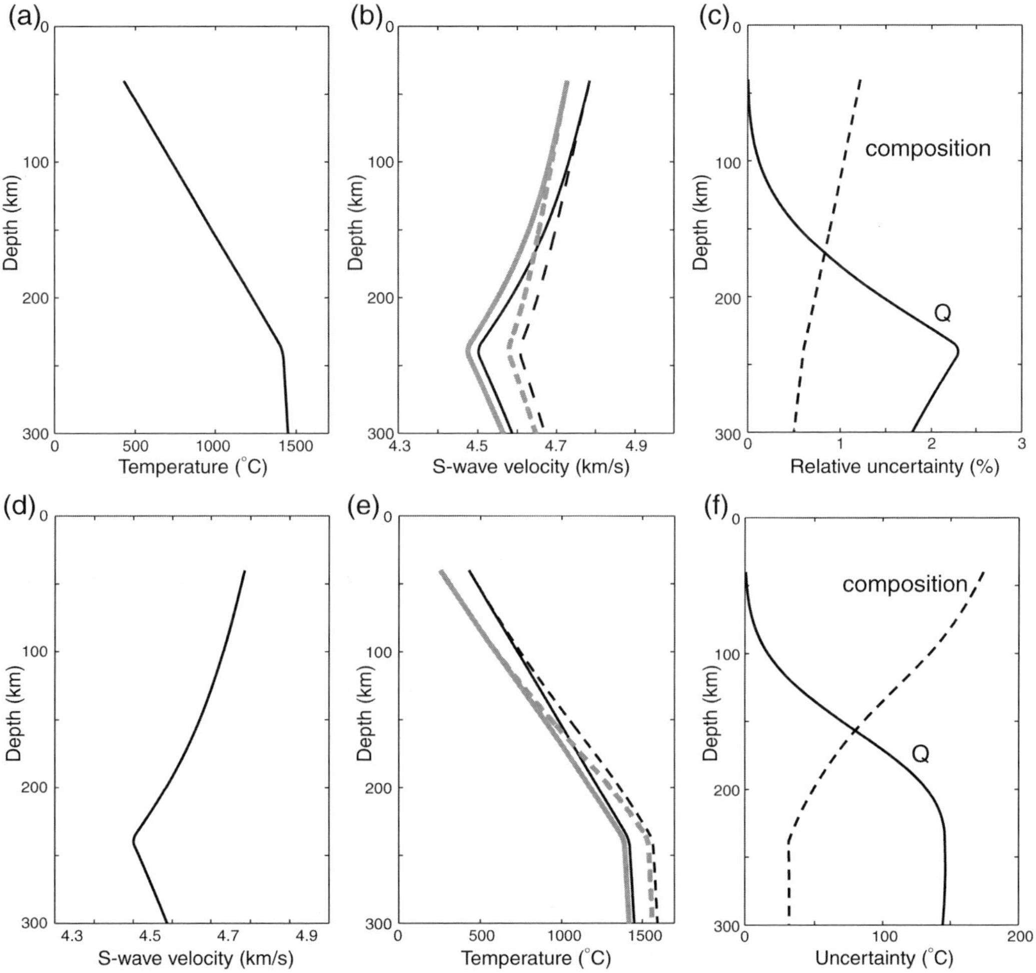

Fig. 5. Assessment of uncertainties in the interconversion between temperature and shear velocity. (**a**) Input mantle temperature profile. (**b**) Shear velocity profiles converted from temperature with variable compositional and Q models. Blackline, on-cratonic composition; grey line, off-cratonic composition; solid line, mantle model Q_1; dashed line, mantle model Q_2. (**c**) Relative uncertainties in the shear velocity produced by uncertainties in composition (dashed line) and Q (solid line). (**d**) Input mantle shear velocity profile. (**e**) Temperature profiles converted from shear velocity with variable compositional and Q models, as in (b). (**f**) Relative uncertainties in temperature produced by uncertainties in composition (dashed line) and Q (solid line).

from shear velocity is about $\pm 100\,^{\circ}\mathrm{C}$ at all depths.

It is important to account for these uncertainties in the inversions both with the seismic parameterization and the thermal parameterization presented below, because both involve interconversion between temperature and shear velocity. With the seismic parameterization, we convert the temperature bounds to bounds in seismic velocities explicitly. To account for the uncertainty of this conversion, we increase the range of seismic velocities by $\pm 0.5\%$ of the seismic velocity. The width of the bounds on seismic velocity, therefore, approximately doubles. With the thermal parameterization, trial models are constructed in temperature space so we introduce these uncertainties in the conversion to shear velocity by increasing the bounds on the temperature model parameters. We describe how we introduce these uncertainties into the range of allowed thermal parameters in following sections.

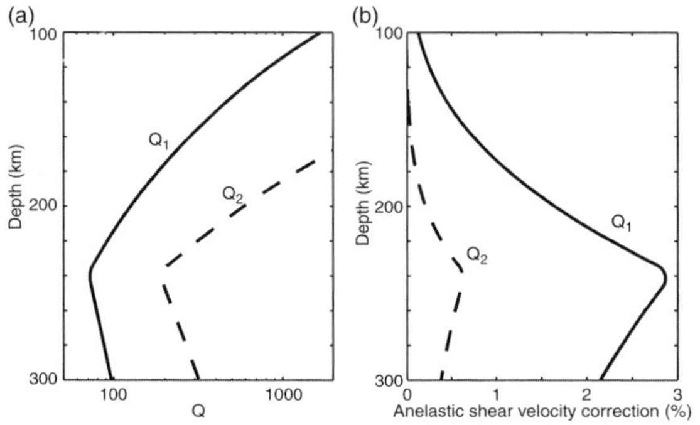

Fig. 6. (**a**) Two alternative Q models, Q_1 and Q_2, described in the text, plotted here for the temperature profile shown in Figure 5a. (**b**) The strength of the anelastic correction from models Q_1 and Q_2 presented as the percent perturbation to the anharmonic shear velocity. Solid line, Model Q_1; dashed line, Model Q_2.

Joint inversion: seismic parameterization

Inversion procedure

The seismic dataset is composed of fundamental mode surface-wave phase (Trampert & Woodhouse 1996; Ekström *et al.* 1997) and group velocity (e.g. Ritzwoller & Levshin 1998; Levshin *et al.* 2001) measurements that are used to produce surface-wave dispersion maps on a $2^\circ \times 2^\circ$ geographical grid using 'diffraction tomography' (Ritzwoller *et al.* 2002), a method that is based on a physical model of lateral surface-wave sensitivity kernels. As described by Shapiro & Ritzwoller (2002), at each node of the grid, the Monte Carlo seismic inversion produces an ensemble of acceptable shear velocity models that satisfy the local surface-wave dispersion information, as illustrated in Figure 7 for Spatial Point 1 (whose location is indicated on Fig. 1). The model is radially anisotropic in the mantle ($V_{sv} \neq V_{sh}$) to a depth of about 200 km, on average. The model is constructed to a depth of 400 km. We summarize the ensemble of acceptable seismic models with the 'Median Model', which is the centre of the corridor defined by the ensemble, and assign an uncertainty at each depth equal to the half-width of the corridor. When converting to temperature, we need the effective isotropic velocity in the upper mantle, which we define as $V_s = (V_{sv} + V_{sh})/2$.

Figure 8 illustrates how assimilating heat flow information into the surface-wave inversion affects the inversion at Point 1, which is located south of Hudson Bay within the Superior Province (see Fig. 1). The heat flow constraint is shown in Figure 8a and b as a small box through which all models that satisfy the heat flow constraint must pass. In temperature space, this box has a width equal to the temperature extremes shown in Figure 4. In seismic velocity space, these extremes have been augmented by 0.5% in accordance with the estimate of the uncertainty in the conversion between temperature and seismic velocity described above. The small box in Figure 8a–d shows this increase.

The models that fit the heat flow constraint are shown in Figure 8c and d and those that do not satisfy the constraint are shown in Figure 8e and f. Both the seismic and thermal models that fit the heat flow constraint are less oscillatory than those that do not satisfy the constraint. In the absence of the heat flow constraint, shear velocities display a minimum directly beneath the Moho and the geotherm exhibits a physically implausible minimum at about 100 km depth. The introduction of the heat flow constraint eliminates most of the models with this non-physical behaviour, systematically favouring models with very high seismic velocities in the uppermost mantle directly beneath the Moho. This is consistent with the results of LITHOPROBE refraction studies which have shown very fast P_n (≈ 8.2 km/s) velocity beneath most of the Canadian Shield (e.g. Perry *et al.* 2002; Viejo & Clowes 2003). In converting from seismic velocity to temperature in constructing Figure 8b, d and f we have not included the uncertainty in the conversion from shear velocity to temperature. This uncertainty would widen the ensemble of temperatures somewhat.

Similar behaviour can been seen at Points 2 and 3, located in northern Manitoba within the Trans-Hudson Orogen and in the Ungava Peninsula within the Superior Province, respectively (Fig. 1). The results for these two sites are presented in Figures 9 and 10. In the Ungava

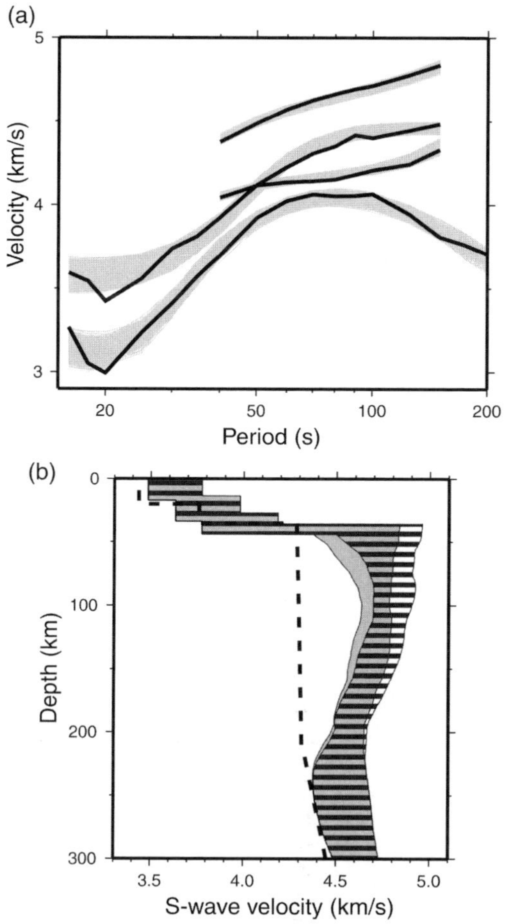

Fig. 7. Monte Carlo inversion with the seismic parameterization, but without thermodynamic constraints. (**a**) Surface-wave dispersion curves (black lines) at Spatial Point 1 (whose location is indicated on Fig. 1) and the range of curves (grey lines) predicted from the constitutive seismic models in (b). (**b**) The ensemble of radially anisotropic models that acceptably fit the dispersion curves found in (a). Solid grey corridor, V_{sv}; cross-hatched corridor, V_{sh}; dotted line, V_s from the 1-D model ak135 (Kennett *et al.* 1995).

Peninsula (Point 3), which is remote from heat flow measurements, the heat flow bounds on the seismic model are weaker and the seismic and temperature profiles displayed in Figure 10a and b are substantially more oscillatory and, hence, more physically questionable than those shown in Figures 8 and 9.

Characteristics of the upper mantle geotherm

As shown in Figure 8d, the temperature profiles that satisfy the heat flow constraint near heat flow measurements are consistent with a nearly linear shallow mantle geotherm. At Point 1, the temperature gradient $dT/dz \approx 5.5\,\mathrm{K/km}$, which translates to a mantle heat flux $Q_M = kdT/dz \approx 16.5\,\mathrm{mW/m^2}$ for thermal conductivity $k = 3\,\mathrm{W/m/K}$. The linearity of the mantle geotherm is consistent with a 'steady-state' thermal regime with no mantle heat sources. The shallow geotherm in Figure 8d displays a knee at about 200 km, below which the geotherm has a different nearly linear temperature gradient. This gradient, $\approx 0.5\,\mathrm{K/km}$, is similar to the mantle adiabatic gradient. The shallower temperature gradient defines a thermal boundary layer whose thickness we will refer to as the lithospheric thickness. The definition of 'lithospheric thickness' is somewhat arbitrary but, for simplicity, we will define it as the depth where the shallow linear gradient intersects the mantle adiabat. For Point 1 (Fig. 8d), therefore, lithospheric thickness is estimated to be about 200 km. The main sources of errors in this estimate are due to uncertainties in the shallow geothermal gradient and in the potential temperature of the convecting mantle (i.e. the horizontal position of the mantle adiabat).

Figures 9 and 10 present two wrinkles in characterizing the mantle geotherm. In Figure 9, which shows the results for the Trans Hudson Orogen (Point 2), the lithosphere is so thick that the transition to the mantle adiabat is not observable. Thus, lithospheric thickness cannot be directly constrained more than by a lower bound of about 300 km. Mantle heat flux, however, is fairly well constrained to about $11\,\mathrm{mW/m^2}$ and the temperature gradient is consistent with thermal steady state. At Point 3, in the Ungava Peninsula away from heat flow measurements (Fig. 10), the vertical oscillations in the temperature profile make it difficult to estimate either the mantle heat flux or the lithospheric thickness, or to test the steady-state hypothesis.

The problems exemplified by Figures 9 and 10 motivate us to apply further constraints in the inversion that are based on tightening physically reasonable bounds on allowed temperature models. These constraints are designed to allow us to obtain better estimates of mantle heat flux and lithospheric thickness.

Joint inversion: thermal parameterization

Inversion procedure

Figures 9 and 10 motivate us to introduce a parameterization based on a physical model in thermodynamic steady state. The linearity of the

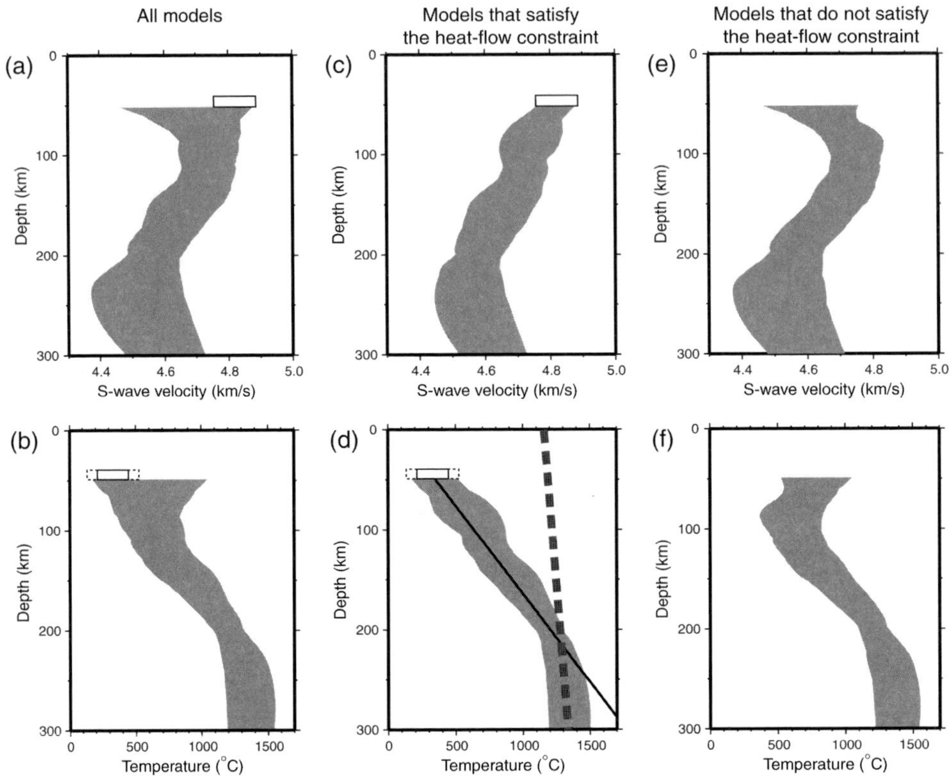

Fig. 8. Results of the Monte Carlo inversion with the seismic parameterization for the Superior Province south of Hudson Bay (Point 1 on Fig. 1), illustrating the effect of the application of the heat flow constraint. (**a**) and (**b**) The ensemble of acceptable seismic ($V_s = (V_{sv} - V_{sh})/2$) and temperature models that fit the seismic dispersion curves acceptably. The small box at the top shows the bounds derived from heat flow. (**c**) and (**d**) The models that fit both the local dispersion curves and the heat flow constraint. (**e**) and (**f**) The models that fit the local dispersion curves but not the heat flow constraint. In (d), the best-fitting linear geotherm (solid line, $Q_M = 16.5\,\mathrm{mW/m^2}$) is shown as the solid line. The thick dashed line indicates the adiabat, whose horizontal offset is determined from the deep part of the temperature profile and whose vertical gradient is 0.5 °C/km.

shallow mantle geotherm, which is consistent with the steady-state hypothesis, is a general feature of the seismic model near heat flow measurements. The physical model we have adopted is schematized in Figure 11. The thermal parameterization consists of a linear gradient in the shallow mantle over a deeper adiabatic gradient set equal to 0.5 K/km. These two gradients meet in a narrow transition region to eliminate a kink in the temperature profile. The Monte Carlo procedure randomly generates three numbers: the mantle temperature directly beneath Moho (T_m), the shallow mantle temperature gradient (dT/dz), and the potential temperature (T_p). Uppermost mantle temperature and the shallow gradient define the lithospheric geotherm. The potential temperature (i.e. the upward continuation of the adiabatic temperature profile to the surface) sets temperatures in the asthenosphere. Lithospheric thickness, L, is defined by the intersection between the lithospheric geotherm and the adiabat. Other parameters could also be varied within some bounds, for example, the adiabatic gradient, but doing so does not significantly increase the range of temperature profiles retained. As with the seismic parameterization, the inversion is performed at each point on a $2^\circ \times 2^\circ$ grid across the region of study.

Each of the three parameters is subjected to a constraints. Firstly, the uppermost mantle temperature is within the same temperature bounds as for the seismic inversion, $T_\mathrm{max} \leqslant T_\mathrm{m} \leqslant T_\mathrm{min}$, where T_max and T_min are shown in Figure 4. This constraint remains tightest near the heat flow measurements. In accordance with Figure 5, to

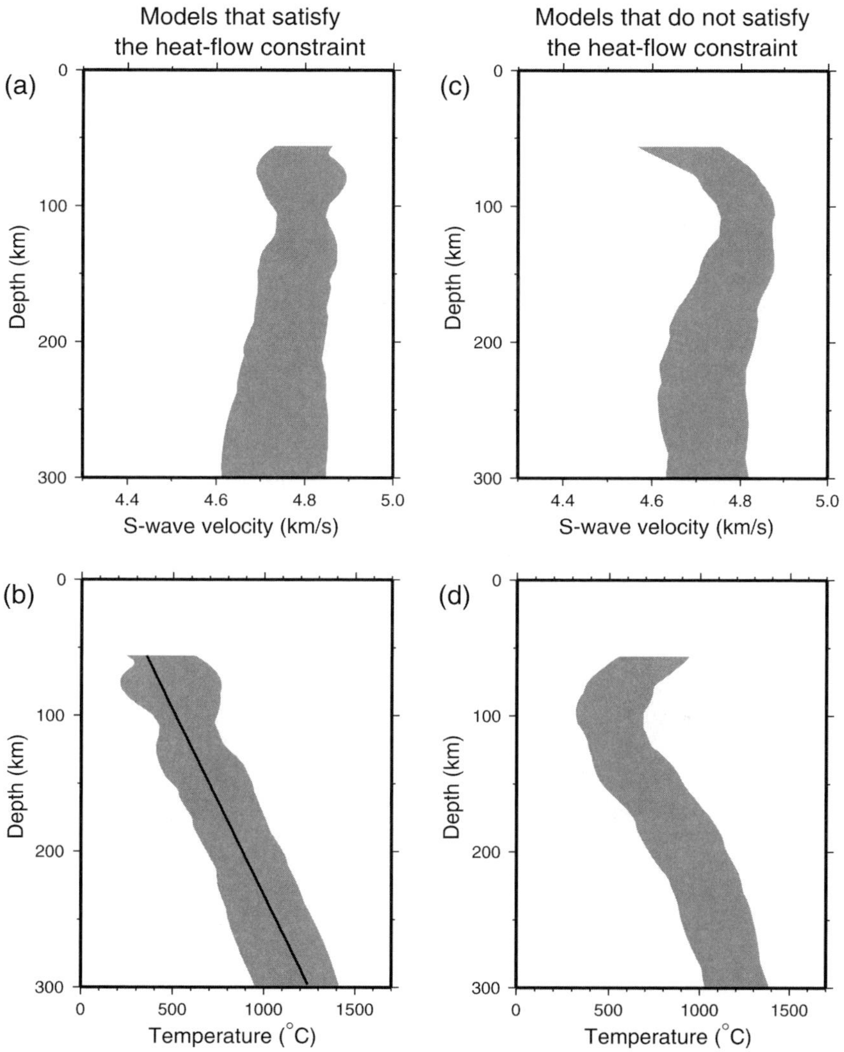

Fig. 9. Results of the joint inversion with the seismic parameterization at the point in the Trans-Hudson Orogen (Point 2 in Fig. 1). Here (**a**) and (**b**) are the models that fit both the local dispersion curves and the heat flow constraint and (**c**) and (**d**) are the models that do not fit the heat flow constraint. The mantle adiabat cannot be discerned, as the knee in the temperature profile appears to be deeper than the extent of the model (i.e. 300 km). As in Figure 8, in (b) the best-fitting linear geotherm ($Q_M = 11\,\mathrm{mW/m^2}$) is shown.

account for uncertainty in the conversion to shear velocity, we increase these bounds by $\pm 0.5\%$ in the seismic velocity, as was also done in Figure 9a–d. Secondly, following Rolandone *et al.* (2002), lithospheric heat flux $Q_M = kdT/dz$ is constrained to be larger than $11\,\mathrm{mW/m^2}$. It is also constrained to be less than surface heat flux, Q_s. Thus, $3.67\,\mathrm{K/km} \leqslant T/dz \leqslant Q_s/k$, where thermal conductivity $k = 3.0\,\mathrm{W/m/K}$. Finally, the range of allowed potential temperatures is somewhat difficult to quantify. McKenzie & Bickle (1988) have proposed a mean upper mantle potential temperature of 1280 °C and Jaupart *et al.* (1998) argue that uncertainties in this value are at least ± 50 °C. The mean value beneath continents may be somewhat lower than the value proposed by McKenzie and Bickle. To be conservative, we extend the range somewhat and apply the following intrinsic bounds on potential tempera-

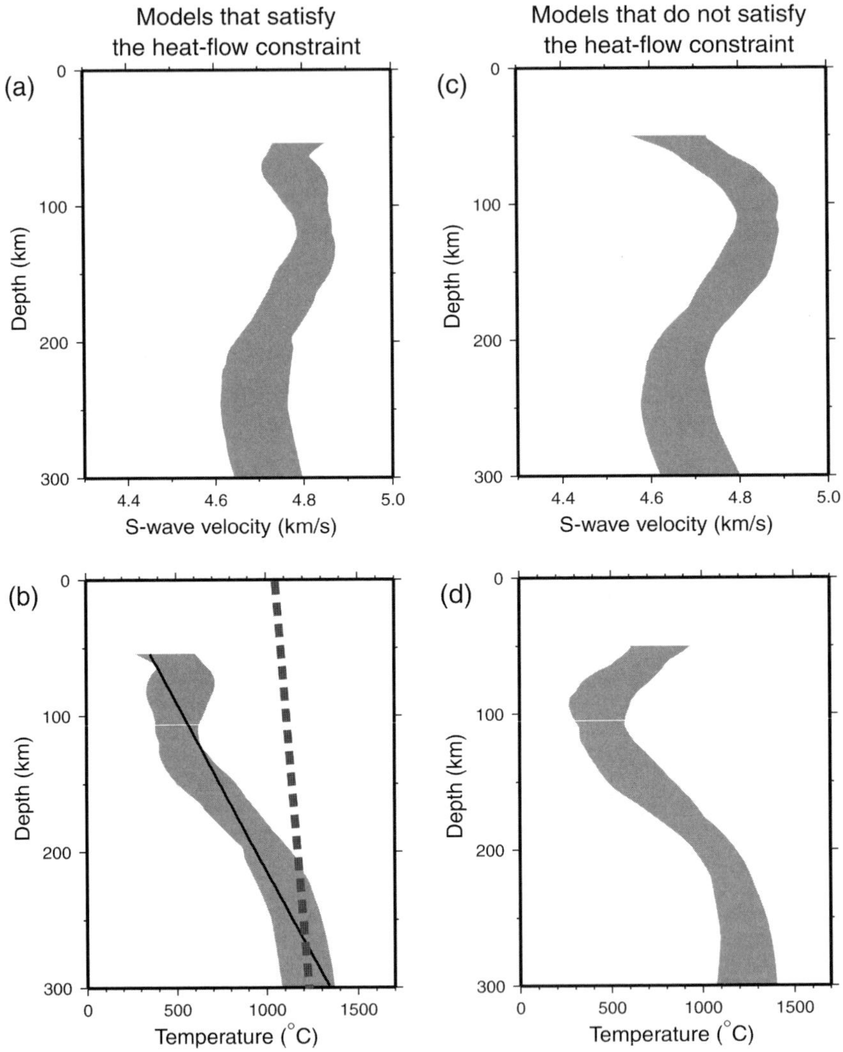

Fig. 10. Results of the joint inversion with the seismic parameterization at the point in the Superior Province in the Ungava Peninsula of New Quebec (Point 3 in Fig. 1). In contrast to Spatial Points 1 and 2, this point is remote from heat flow measurements, which results in a weak heat flow constraint so that the oscillations in the seismic and temperature profiles have not been eliminated. As in Figures 8 and 9, in (b) the best-fitting linear geotherm (solid line, $Q_M = 12\,\mathrm{mW/m^2}$) is shown with the mantle adiabat, but there are large uncertainties due to irregularities in the temperature profile with depth.

ture: $1100\,°\mathrm{C} \leqslant T_p \leqslant 1300\,°\mathrm{C}$. To account for uncertainty in the temperature to shear velocity conversion, we expand these bounds by $\pm 100\,°\mathrm{C}$ to $1000\,°\mathrm{C} \leqslant T_p \leqslant 1400\,°\mathrm{C}$.

One of the principal advantages of the thermal parameterization is the possibility to apply physically based constraints on the model parameters. Although the bounds on potential temperature are poorly known, the lower bound of $11\,\mathrm{mW/m^2}$ on mantle heat flow strongly constrains the seismic model.

After a trial model is constructed in temperature space, it is converted to shear velocity using the method of Goes *et al.* (2000). Trial seismic crustal structures are introduced as well as mantle radial anisotropy similar to the generation of these features of the model in the seismic parameterization. At each grid node, some

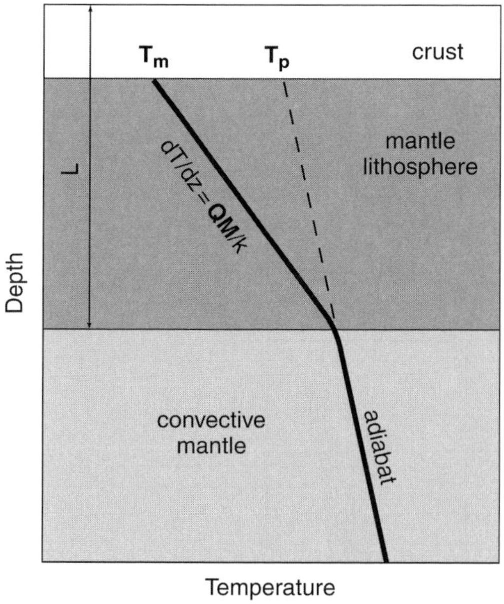

Fig. 11. Schematic representation of the thermal model which is defines by three parameters. The steady-state lithospheric geotherm is defined by temperature directly beneath the Moho (T_m) and the linear gradient (dT/dz) which is simply related to mantle heat flow (Q_M). Below the lithosphere the temperature gradient in the convecting mantle is adiabatic with potential temperature T_p and an adiabatic gradient of 0.5 K/km. These two gradients join smoothly through a narrow transition region in order to eliminate a non-physical kink in the temperature profile.

temperature profiles will be rejected entirely, but some will be found to fit the seismic data acceptably for an appropriate subset of seismic crustal models and models of radial anisotropy. These profiles define an ensemble of acceptable profiles in temperature space. They are also combined with the crustal and radial anisotropic models to define an ensemble of acceptable models in seismic space.

Results for Points 1 and 3 (Fig. 1) are shown in Figure 12. At Point 1, the results are very similar to those obtained with the seismic parameterization (Fig. 8b, d). The estimated average mantle component of heat flow is 15 mW/m^2 and its standard deviation is 2.0 mW/m^2. Average lithospheric thickness is 246 km with a standard deviation of 33 km. At Point 3, the thermal parameterization yields an ensemble of models that fit the surface-wave data but do not display the physically questionable vertical oscillations that appear in Figure 10. The estimated average mantle heat flow and average lithospheric thickness are 12.2 mW/m^2 and 294 km with corresponding standard deviations of 1.4 mW/m^2 and 46 km.

Seismic and thermal models

The middle of the ensemble of acceptable models in the seismic and temperature model spaces are the Median Models. Slices of the Median Models of shear velocity and temperature are shown in Figures 13 and 14. Although shear velocity and temperature are fairly homogeneous at 80 km depth, at greater depths the variability is greater, reflecting variations in lithospheric thickness across the study area, as discussed below. The lithosphere is warmer and thinner to the south-east, in the Appalachians, and becomes thicker toward the north and, especially, towards the northwest, as can be seen in Figure 14b and f. Vertical oscillations that plague the seismic parameterization are absent from the model, as the cross-sections in Figure 14 attest.

Mantle heat flux and lithospheric thickness

At each point, we construct an ensemble of mantle heat flow (Q_M) and lithospheric thickness (L) estimates derived from the ensemble of acceptable temperature profiles ($Q_M = kdT/dz$, L = depth where the lithospheric geotherm intersects the mantle adiabat). The average of the ensemble of acceptable mantle heat flow and lithospheric thickness estimates is shown in Figure 15.

We assign uncertainties to Q_M and L equal to the standard deviation of the ensemble of acceptable values. Figure 16a shows that the uncertainty in mantle heat flow is greatest when mantle heat flow is high. This is because, for a 1 mW/m^2 change, there is a bigger change in the temperature profile when heat flow is low (nearly vertical temperature profile) than when it is high (steep temperature profile). It appears that the uncertainty saturates at 2.5 mW/m^2. Figure 16b shows that the uncertainty in lithospheric thickness is also greatest for the thickest lithosphere. This is because the lithospheric temperature gradient and the slope of the mantle adiabat are almost equal. The uncertainty increases almost linearly with lithospheric thickness, but becomes much more scattered for thick lithosphere.

Discussion

The seismic velocities and temperatures displayed in Figures 13 and 14 demonstrate

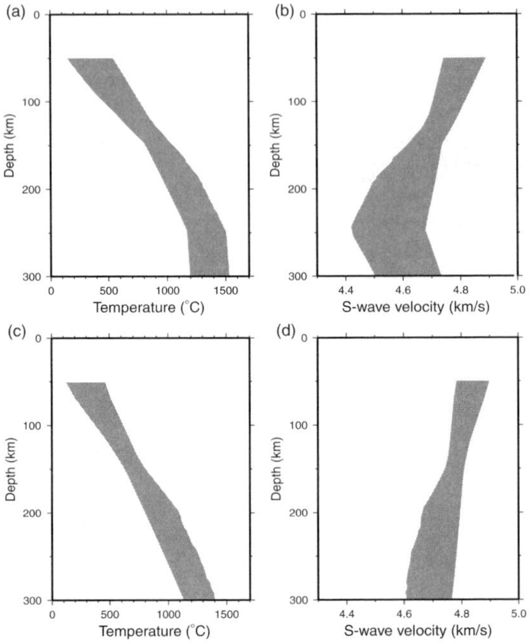

Fig. 12. Example results of the Monte Carlo inversion with the thermal parameterization at Spatial Points 1 and 3. (**a**) Ensemble of acceptable temperature models at Point 1. (**b**) Ensemble of acceptable seismic models at Point 1. (**c**) Ensemble of acceptable temperature models at Point 3. (**d**) Ensemble of acceptable seismic models at Point 3.

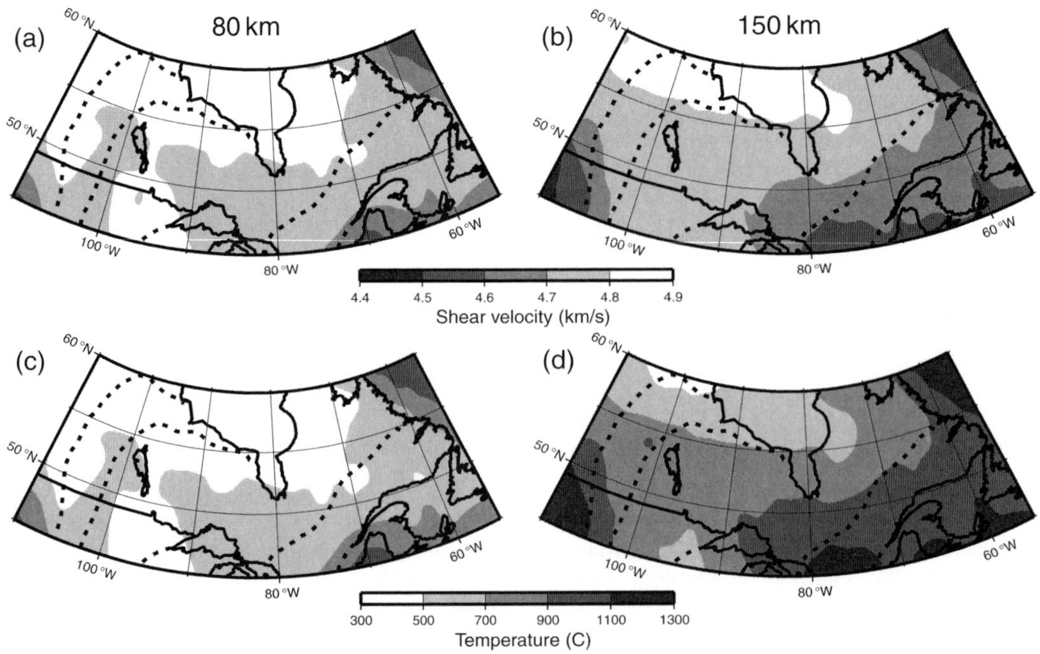

Fig. 13. (**a**) and (**b**) Horizontal slices of the seismic $V_s = (V_{sh} + V_{sv})/2$ model and (**c**) and (**d**) the temperature model at depths of (a, c) 80 km and (b, d) 150 km. Dashed lines, boundaries of principal Precambrian provinces. Adapted from Hoffman (1989).

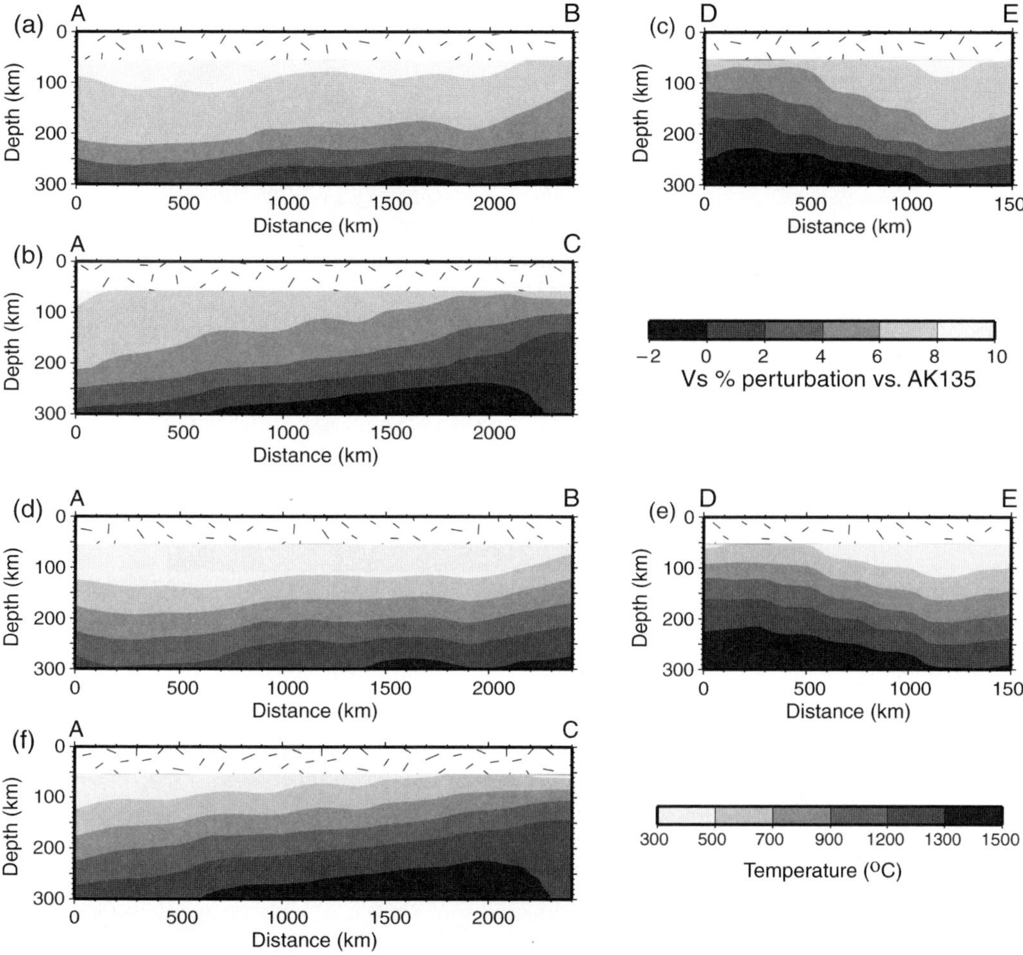

Fig. 14. Vertical slices of the mantle seismic and temperature models along the three profiles shown in Figure 1. **(a)–(c)** Isotropic shear velocity ($V_s = (V_{sh} + V_{sv})/2$) presented as percent perturbation relative to the global 1-D model ak135. **(d)–(f)** Temperature in °C.

considerable variability across the study area. This information is probably best summarized by mantle heat flow and by the lithospheric thickness (Figure 15). The mantle heat flow appears to increase and lithospheric thickness to decrease beneath the Appalachians to the southeast. Within the Canadian Shield, mantle heat flow from seismic inversions ranges between 11 and 18 mW/m^2, with apparently higher values in the Grenville Province. Within most of the Superior Province, mantle heat flow ranges between 11 and 15 mW/m^2, with small amplitude, short wavelength (< 1000 km) spatial variations. Because of horizontal diffusion of heat, such variations in mantle heat flow are damped at the surface and cannot be resolved by the heat flow data. Pinet *et al.* (1991) had concluded from their analysis of heat flow and heat production data that the mantle heat flow is the same beneath the Grenville Province and the Superior Province. Within the Canadian Shield, the variations are not always well correlated with geological provinces. The lowest mantle heat flow values are found beneath the Archean Rae Province and the Palaeo–Proterozoic Trans Hudson Orogen, in northern Manitoba and Saskatchewan, where

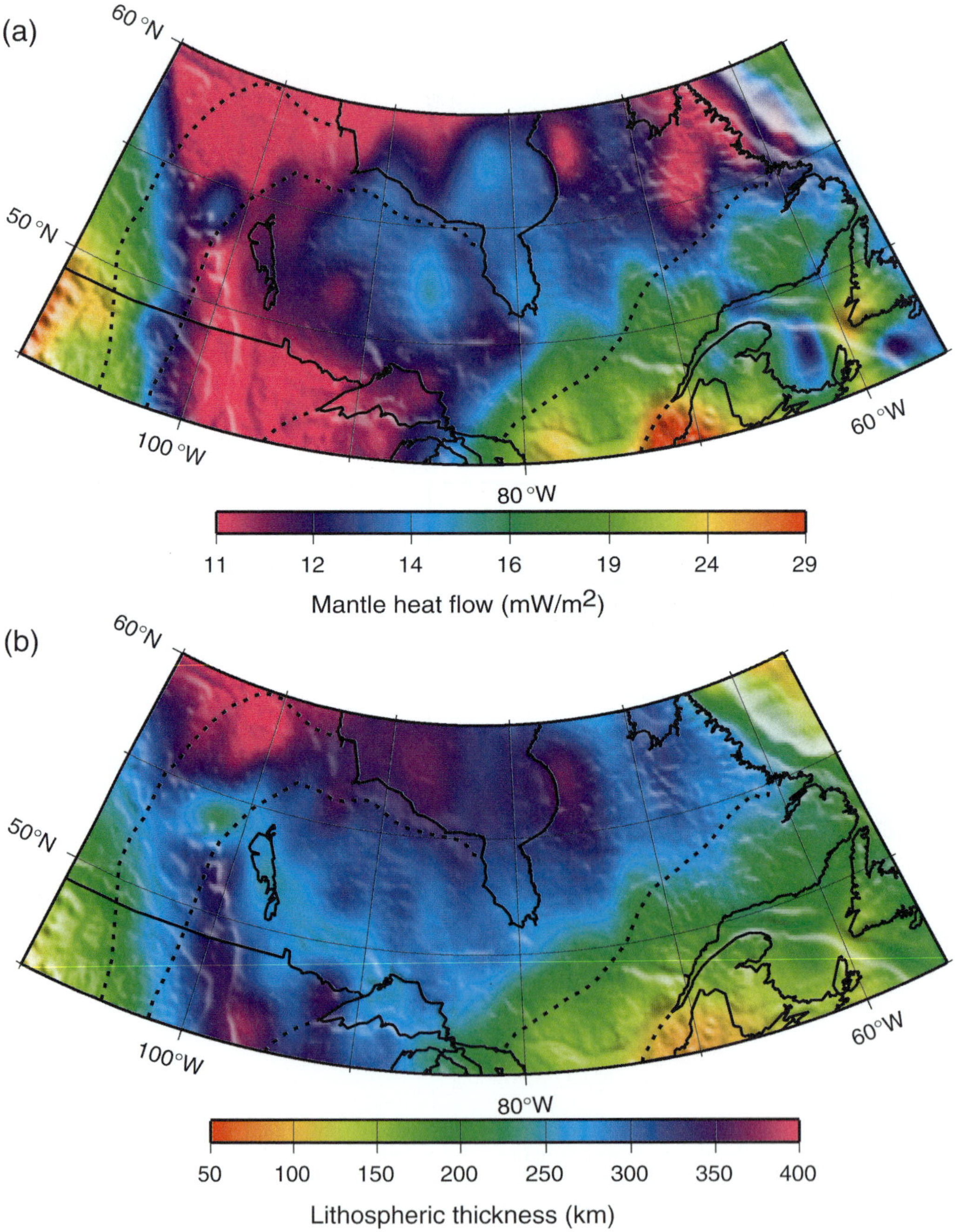

Fig. 15. (**a**) The estimated mantle component of heat flow, Q_M. (**b**) The estimated lithospheric thickness, L. Dashed lines, boundaries of principal Precambrian provinces. Adapted from Hoffman (1989).

juvenile crust is thrust over the Archean Sask craton. Although the thermal regime of the Canadian Shield does not simply reflect its surface geology and is not simply related to the last tectonomagmatic event, it reveals the deeper structure of the lithosphere.

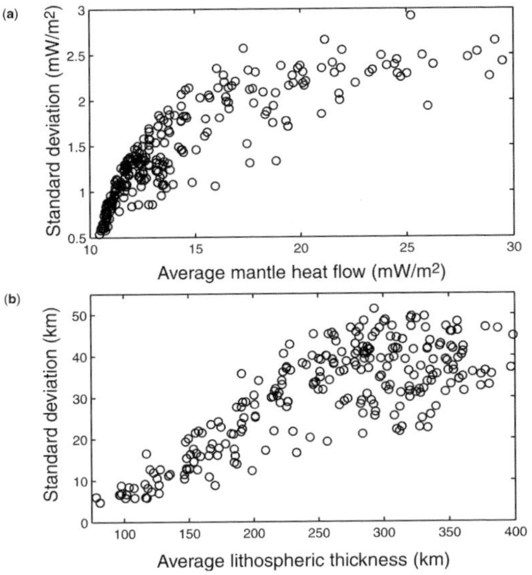

Fig. 16. (**a**) Estimated standard deviation of mantle heat flow plotted v. mantle heat flow. (**b**) Estimated standard deviation of lithospheric thickness plotted v. lithospheric thickness. Values are taken at the model nodes on a 2° × 2° grid across the region of study.

Patterns of lithospheric variability do emerge, however. As Figure 17 shows, lithospheric thickness and mantle heat flow are anti-correlated. A scaling law between mantle temperature and heat flow from the convecting mantle was used by Jaupart *et al.* (1998) to determine the lithospheric thickness. The analysis assumes that the heat flow at the base of the lithosphere is supplied by small-scale convection (e.g. Davaille & Jaupart 1993), and results in an approximately power law relation between mantle heat flow and lithospheric thickness. Results of our seismic inversion shown in Figure 17 are also well approximated by a power law curve (Fig. 17) that is relatively close to the shape of the Q_M to L relationship given by Jaupart *et al.* (1998). However, Jaupart *et al.* pointed out that changes in lithospheric thickness do not require changes in mantle heat flow and that Moho temperatures (determined also by crustal heat production) also control lithospheric thickness. This is consistent with Figure 17 showing much more variability in lithospheric thickness than in mantle heat flow.

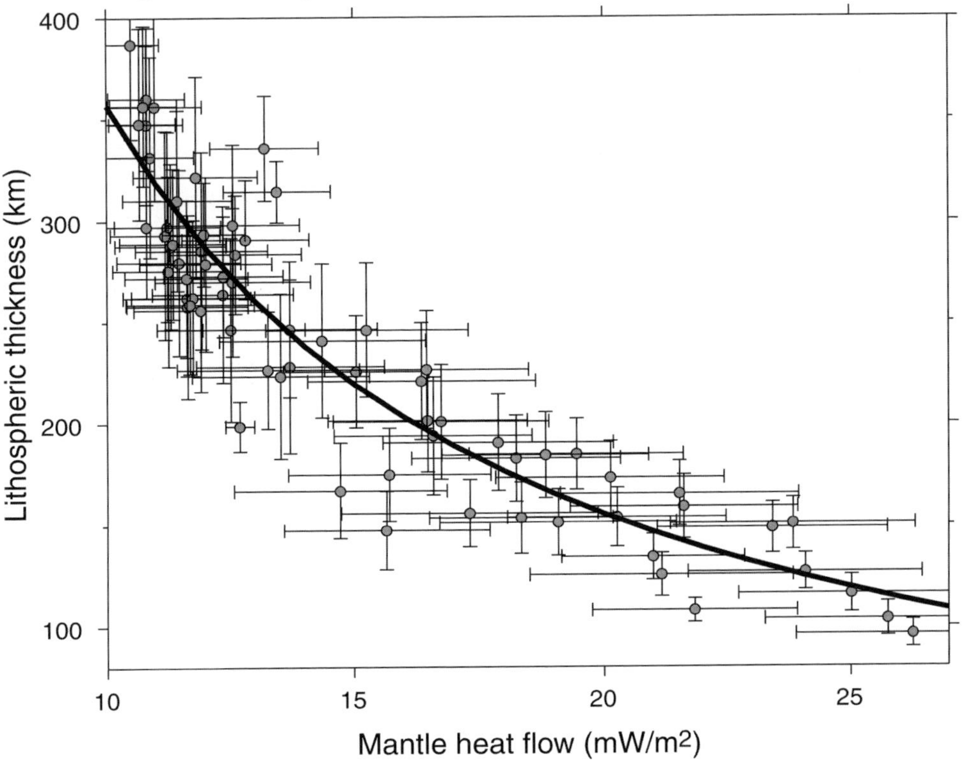

Fig. 17. Lithospheric thickness (L) v. mantle heat flow (Q_M) taken from model nodes near the heat flow measurements shown in Figure 1. One standard deviation error bars in both L and Q_M are shown. The solid line is a power law curve that fits the data well ($L = 5660Q_M^{-1.2}$).

Conclusions

The primary conclusion of this work is that seismic surface-wave data and surface heat flow observations can be reconciled over broad continental areas; i.e. both types of observations can be fit simultaneously with a simple steady-state thermal model of the upper mantle. This has motivated the reformulation of the seismic surface-wave inverse problem in terms of a thermal model described by three parameters: temperature in the uppermost mantle directly beneath Moho, the mantle temperature gradient, and the potential temperature of the sublithospheric convecting mantle from which we also estimate lithospheric thickness and mantle heat flux. In addition to the steady-state thermal constraint, prior physical information based on surface heat flow measurements is applied.

The results of a Monte Carlo inversion of the surface-wave data with this 'thermal parameterization' across the Canadian Shield demonstrate that lithospheric temperature and shear velocity are not well correlated with surface tectonic history, which implies that the tectonic regime of the crust is not simply related to the thermal regime of the deeper mantle lithosphere. At the same time, however, the inferred relation between lithospheric thickness and mantle heat flow is consistent with a hypothesis of Jaupart *et al.* (1998), who argued that the lithosphere and asthenosphere beneath the Canadian Shield are in thermal equilibrium and heat flow into the deep lithosphere is governed by small-scale sublithospheric convection.

We would like to thank J. Trampert and I. Main for helpful referee reports. The phase velocity measurements used in the inversion were generously donated by J. Trampert of Utrecht University and M. Antolik, A. Dziewonski and G. Ekström at Harvard University. All maps were generated with the Generic Mapping Tools (GMT) data-processing and display package (Wessel & Smith 1991, 1995). This work was supported in part by a grant from the US National Science Foundation, NSF-OPP-0136103. JCM is grateful for the continuous support of the Natural Sciences and Engineering Research Council (Canada) through a discovery grant.

References

ARTEMIEVA, I. M. & MOONEY, W. D. 2001. Thermal thickness and evolution of Precambrian lithosphere: a global study. *Journal of Geophysical Research*, **106**, 16 387–16 414.

BERCKHEMER, H., KAMPFMAN, W., AULBACH, E. & SCHMELING, H. 1982. Shear modulus and Q of forsterite and dunite near partial melting from forced oscillation: experiments. *Physics of the Earth and Planetary Interior*, **29**, 30–41.

DAVAILLE, A. & JAUPART, C. 1993. Transient high-Rayleigh number thermal convection with large viscosity variations. *Journal of Fluid Mechanics*, **253**, 141–166.

DOIN, M. P. & FLEITOUT, L. 1996. Thermal evolution of the oceanic lithosphere: an alternative view. *Earth and Planetary Science Letters*, **142**, 121–136.

DRURY, M. J. 1985. Heat flow and heat generation in Churchill Province of the Canadian Shield and their paleotectonic significance. *Tectonophysics*, **115**, 25–44.

DRURY, M. J. & TAYLOR, A. E. 1987. Some new measurements of heat flow in the Superior Province of the Canadian Shield. *Canadian Journal of Earth Sciences*, **24**, 1486–1489.

DRURY, M. J., JESSOP, A. M. & LEWIS, T. J. 1987. The thermal nature of the Canadian Appalachians. *Tectonophysics*, **11**, 1–14.

EKSTRÖM, G., TROMP, J. & LARSON, E. W. F. 1997. Measurements and global models of surface waves propagation. *Journal of Geophysical Research*, **102**, 8137–8157.

FURLONG, K. P. SPAKMAN, W. & WORTEL, R. 1995. Thermal structure of the continental lithosphere: constraints from seismic tomography. *Tectonophysics*, **244**, 107–117.

GOES, S., GOVERS, R. & VACHER, R. 2000. Shallow mantle temperatures under Europe from P and S wave tomography. *Journal of Geophysical Research*, **105**, 11 153–11 169.

GUILLOU, L., MARESCHAL, J. C., JAUPART, C., GARIÉPY, C., BIENFAIT, G. & LAPOINTE, R. 1994. Heat flow gravity and structure of the Abitibi Belt, Superior Province, Canada: implications for mantle heat flow. *Earth and Planetary Science Letters*, **122**, 103–123.

GUILLOU-FROTTIER, L., MARESCHAL, J. C., JAUPART, C., GARIÉPY, C., LAPOINTE, R. & BIENFAIT, G. 1995. Heat flow variations in the Grenville Province, Canada. *Earth and Planetary Science Letters*, **136**, 447–460.

GUILLOU-FROTTIER, L., MARESCHAL, J. C., JAUPART, C., GARIÉPY, C., BIENFAIT, G., CHENG, L. Z. & LAPOINTE, R. 1996. High heat flow in the Thompson Belt of the Trans-Hudson Orogen, Canadian Shield. *Geophysical Research Letters*, **23**, 3027–3030.

HART, S. R., STEINHART, J. S. & SMITH, T. J. 1994. Terrestrial heat flow in Lake Superior. *Canadian Journal of Earth Sciences*, **31**, 698–708.

HOFFMAN, P. F. 1989. Precambrian geology and tectonic history of North America. *In:* BALLY, A. W. & PALMER, E. R. (eds) *The Geology of North America: An Overview*. Geological Society of America, Boulder, 447–512.

JAUPART, C. & MARESCHAL, J. C. 1999. The thermal structure and thickness of continental roots. *Lithos*, **48**, 93–114.

JAUPART, C. & MARESCHAL, J. C. (2004). Crustal heat production. *In:* RUDNICK, R. (ed.) *Textbook of Geochemistry*, Vol. 3. *Composition of the Continental Crust*, Elsevier Publishing Company, Amsterdam, 65–84.

JAUPART, C., MARESCHAL, J. C., GUILLOU-FROTTIER, L. & DAVAILLE, A. 1998. Heat flow and thickness of the lithosphere in the Canadian Shield. *Journal of Geophysical Research*, **103**, 15 269–15 286.

JESSOP, A. M., LEWIS, T. J., JUDGE, A. S., TAYLOR, A. E. & DRURY, M. J. 1984. Terrestrial heat flow in Canada. *Tectonophysics*, **103**, 239–261.

KENNETT, B. L. N., ENGDAHL, E. R. & BULAND, R. 1995. Constraints on seismic velocities in the Earth from travel times. *Geophysical Journal International*, **122**, 403–416.

LEVSHIN, A. L., RITZWOLLER, M. H., BARMIN, M. P., VILLASENOR, A. & PADGETT, C. A. 2001. New constraints on the Arctic crust and uppermost mantle: surface wave group velocities, Pn and Sn. *Physics of the Earth and Planetary Interior*, **23**, 185–204.

MARESCHAL, J. C., PINET, C., GARIÉPY, C., JAUPART, C., BIENFAIT, C., DALLA-COLETTA, G., JOLIVET, J. & LAPOINTE, R. 1989. New heat flow density and radiogenic heat production data in the Canadian Shield and the Quebec Appalachians. *Canadian Journal of Earth Sciences*, **26**, 845–853.

MARESCHAL, J. C., JAUPART, C. CHENG, L. Z. COLANDONE, F. RARIÉPY, C. GIENFAIT, C. BUILLOU-FROTTIER L. G & LAPOINTE, R. 1999. Heat flow in the Trans-Hudson Orogen of the Canadian Shield: implications for Proterozoic continental growth. *Journal of Geophysical Research*, **104**, 29 007–29 024.

MARESCHAL, J. C., POIRIER, A., ROLANDONE, F., BIENFAIT, G., GARIÉPY, C., LAPOINTE, R. & JAUPART, C. 2000. Low mantle heat flow at the edge of the North American continent, Voisey Bay, Labrador. *Geophysical Research Letters*, **27**, 823–826.

MCDONOUGH, W. F. & RUDNICK, R. L. 1998. Mineralogy and composition of the upper mantle. *In:* HEMLEY, R. J. (ed.) *Ultrahigh-Pressure Mineralogy: Physics and Chemistry of the Earth's Deep Interior*. Mineralogical Society of America, Washington, DC, 139–164.

MCKENZIE, D. P. & BICKLE, M. J. 1988. Volume and composition of melt generated by extension of the lithosphere. *Journal of Petrology*, **29**, 625–679.

NYBLADE, A. A. 1999. Heat flow and the structure of the Precambrian lithosphere. *Lithos*, **48**, 81–91.

NYBLADE, A. A. & POLLACK, H. N. 1993. A global analysis of heat flow from Precambrian terrains: implications for the thermal structure of Archean and Proterozoic lithosphere. *Journal of Geophysical Research*, **98**, 12 207–12 218.

PERRY, H. K. C., EATON, D. W. S. & FORTE, A. M. 2002. LITH5.0: a revised crustal model for Canada based on LITHOPROBE results. *Geophysical Journal International*, **150**, 285–294.

PINET, C., JAUPART, C., MARESCHAL, J. C., GARIÉPY, C., BIENFAIT, G. & LAPOINTE, R. 1991. Heat flow and structure of the lithosphere in the eastern Canadian Shield. *Journal of Geophysical Research*, **96**, 19 941–19 963.

POLLACK, H. N. 1986. Cratonization and thermal evolution of the mantle. *Earth and Planetary Science Letters*, **80**, 175–182.

POLLACK, H. N., HURTER, S. J. & JOHNSON, J. R. 1993. Heat flow from the Earth's interior: analysis of the global data set. *Reviews of Geophysics*, **31**, 267–280.

RITZWOLLER, M. H. & LEVSHIN, A. L. 1998. Eurasian surface wave tomography: group velocities. *Journal of Geophysical Research*, **103**, 4839–4878.

RITZWOLLER, M. H., SHAPIRO, N. M., BARMIN, M. P. & LEVSHIN, A. L. 2002. Global surface wave diffraction tomography. *Journal of Geophysical Research*, **107**, 2335. [DOI:10.1029/2002JB001777]

RÖHM, A. H. E., SNIEDER, R., GOES, S. & TRAMPERT, J. 2000. Thermal structure of continental upper mantle inferred from *S*-wave velocity and surface heat flow. *Earth and Planetary Science Letters*, **181**, 395–407.

ROLANDONE, F., JAUPART, C., MARESCHAL, J. C., GARIÉPY, C., BIENFAIT, G., CARBONNE, C. & LAPOINTE, R. 2002. Surface heat flow, crustal temperatures and mantle heat flow in the Proterozoic Trans-Hudson Orogen, Canadian Shield. *Journal of Geophysical Research*, **107**, 2341. [DOI:10.1029/2001JB000698]

RUDNICK, R. L. & NYBLADE, A. A. 1999. The thickness and heat production of Archean lithosphere: constraints from xenoliths thermobarometry and surface heat flow. *In:* FEI, Y., BERTKA, C. M. & MYSEN, B. O. (eds) *Mantle Petrology: Field Observations and High-Pressure Experimentation: A Tribute to Francis R. (Joe) Boyd.* Geochemical Society, St Louis, Special Publications, **6**, 3–12.

RUDNICK, R. L., MCDONOUGH, W. F. & O'CONNELL, R. J. 1998. Thermal structure, thickness and composition of continental lithosphere. *Chemical Geology*, **145**, 395–411.

RUSSELL, J. K., DIPPLE, G. M. & KOPYLOVA, M. G. 2001. Heat production and heat flow in the mantle lithosphere, Slave Craton, Canada. *Physics of the Earth and Planetary Interior*, **123**, 27–44.

SHAPIRO, N. M. & RITZWOLLER, M. H. 2002. Monte-Carlo inversion for a global shear velocity model of the crust and upper mantle, *Geophysical Journal International*, **151**, 88–105.

SHAPIRO, N. M. & RITZWOLLER, M. H. (2004). Thermodynamic constraints on seismic inversions. *Geophysical Journal International*, **157**, 1175–1188, [DOI: 10.1111/j1365–246X.2004.02254.X].

SMITH, D. 1999. Temperatures and pressures of mineral equilibration in peridotite xenoliths: review, discussion, and implications. *In:* FEI, Y., BERTKA, C. M. & MYSEN, B. O. (eds) *Mantle Petrology: Field Observations and High-Pressure Experimentation: A Tribute to Francis R. (Joe) Boyd.* Geochemical Society, St Louis, Special Publications, **6**, 171–188.

SOBOLEV, S. V., ZEYEN, H., STOLL, G., WERLING, F., ALTHERR, R. & FUCHS, K. 1996. Upper mantle temperatures from teleseismic tomography of

French Massif Central including effects of composition, mineral reactions, anharmonicity, anelasticity and partial melt. *Earth and Planetary Science Letters*, **157**, 193–207.

TRAMPERT, J. & WOODHOUSE, J. H. 1995. Global phase velocity maps of Love and Rayleigh waves between 40 and 150 s period. *Geophysical Journal International*, **122**, 675–690.

VIEJO, G. F. & CLOWES, R. M. 2003. Lithospheric structure beneath the Archean Slave Province and Proterozoic Wopmay Orogen, northwestern Canada, from LITHOPROBE refraction/wide-angle reflection survey. *Geophysical Journal International*, **153**, 1–19.

WESSEL, P. & SMITH, W. H. F. 1991. Free software helps map and display data. *Eos, Transactions of the American Geophysical Union*, **72**, 441.

WESSEL, P. & SMITH, W. H. F. 1995. New version of the generic mapping tools released. *Eos, Transactions of the American Geophysical Union*, **76**, 329.

Beyond kriging: dealing with discontinuous spatial data fields using adaptive prior information and Bayesian partition modelling

JOHN STEPHENSON[1], K. GALLAGHER[1] & C. C. HOLMES[2]

[1]*Department of Earth Science and Engineering, Imperial College London, South Kensington, London SW7 2AS, UK (e-mail: john.stephenson@imperial.ac.uk)*

[2]*Department of Mathematics, Imperial College London, South Kensington, London SW7 2AS, UK*

Abstract: The technique of kriging is widely known to be limited by its assumption of stationarity, and performs poorly when the data involve localized effects such as discontinuities or nonlinear trends. A Bayesian partition model (BPM) is compared with results from ordinary kriging for various synthetic discontinuous 1-D functions, as well as for 1986 precipitation data from Switzerland. This latter dataset has been analysed during a comparison of spatial interpolation techniques, and has been interpreted as a stationary distribution and one thus suited to kriging. The results demonstrate that the BPM outperformed kriging in all of the datasets compared (when tested for prediction accuracy at a number of validation points), with improvements by a factor of up to 6 for the synthetic functions.

The technique of kriging is used extensively in spatial interpolation problems and has found many applications in geology as well as ecology and environmental sciences. For example, predicting ore grades in mining, building heterogeneous reservoir models from well and seismic data, and inferring pollution levels from sparsely distributed data points.

The critical drawbacks to kriging lie in the assumptions that the spatial process results from a deterministic mean, either constant in the case of ordinary kriging (OK) or a linear combination of parameters as in universal kriging (UK), combined with a zero mean, spatially correlated stationary process. The principal work in kriging is defining this correlation process, which is usually achieved via modelling the semi-variogram. This framework does not allow for discontinuities in the spatial process, or other localized effects. The current Bayesian applications to this problem have revolved around optimal estimation of the covariance matrix (e.g. Handcock & Stein 1993).

A Bayesian partition model (BPM) (Denison *et al.* 2002*a*) avoids the kriging limitations. The BPM defines a set of disjoint regions, whose position and shape is defined using a Voronoi tessellation, such that the positions of the Voronoi centres form the parameters of the model. The choice of the Voronoi parameterization is not essential to the mechanics of the BPM, though it does provide an extremely simple implementation that is applicable in any number of spatial dimensions. The data is then assumed to be stationary only within an individual partition, thus avoiding the key pitfall of kriging, and can thereby deal gracefully with discontinuities. This parameterization allows for arbitrarily complex surfaces (in 2-D or 3-D, or indeed higher dimensions if required).

The model space is explored using Markov chain Monte Carlo (MCMC) and a Bayesian framework, so that each iteration will either move, remove or create one partition. The prior information is fairly loosely defined, through hyperpriors on the mean level of the variables and regression variance within a partition, and the hyperpriors are adapted as the algorithm proceeds. The model is naturally biased towards simpler models (fewer regions), and there is no need to specify how many regions the model will have(thus providing models with varying dimension). Once convergence is achieved, the desired point predictions are then made by sampling from the Markov chain, and taking the average over all the samples. Using a Bayesian framework leads directly to the inclusion of prior information into the modelling framework, and provides immediate access to the uncertainties of the predictions.

In this contribution, we illustrate the application of Bayesian partition modelling to spatial

From: CURTIS, A. & WOOD, R. (eds) 2004. *Geological Prior Information: Informing Science and Engineering*. Geological Society, London, Special Publications, **239**, 195–209. 186239-171-8/04/$15.00

data that would create problems using a conventional kriging approach, as well as to stationary data where we would expect kriging to be well adapted.

Geostatistics and kriging

The name synonymous with geostatistics is George Matheron, who, building on the foundational work of Daniel Krige in statistically estimating average ore grades, developed the now fundamental process of kriging (Matheron 1963). The technique has found many diverse applications such as time-lapse seismic noise characterization (Coleou *et al.* 2002) and image processing (Decenciere *et al.* 1998). We shall concentrate here simply on the spatial interpolation properties.

This introduction to kriging is not meant to be exhaustive, rather, we want to provide enough information to convey the key limitations of the process. For full coverage of geostatistics and kriging, we direct the reader either to Cressie (1993) for a theoretical approach, or to Isaaks & Srivastava (1989) for more accessible and practical methods.

Spatial interpolation and terms

The most obvious use of geostatistics involves using a dataset to generate a map of values across a surface. Throughout this paper, we shall refer to the dataset D, as comprising n data values $y_1 \ldots y_n$, having been sampled at the corresponding irregularly spaced locations $\boldsymbol{x}_1 \ldots \boldsymbol{x}_n$ from a subset $\mathcal{D}$ of space $\mathbb{R}^d$ with dimension d. In the case of spatial statistics d is typically 2, for example, northings and eastings, though the methods described are not limited by this. Traditionally, we then seek to estimate the new values on a regularly spaced grid across $\mathcal{D}$, so as to allow contour maps or surfaces to be generated.

To illustrate these approaches, we present a simple example using the heavily analysed Walker Lake sample dataset from (Isaaks & Srivastava 1989), comprising values at 470 2-D spatial locations, drawn from an exhaustive dataset of 78 000. The data values we use are referred to as set V in the text, and are derived from a digital elevation model of the Walker Lake region in California. Their positions are shown in Figure 1. For full details of these datasets, see Isaaks & Srivastava (1989).

The task now is to use the data D to make a prediction of a new data value y_{n+1} at position $\boldsymbol{x}_{n+1}$. One approach is to estimate y_{n+1}, using a weighted sum of n available data, scaled by the weights $w_1 \ldots w_n$.

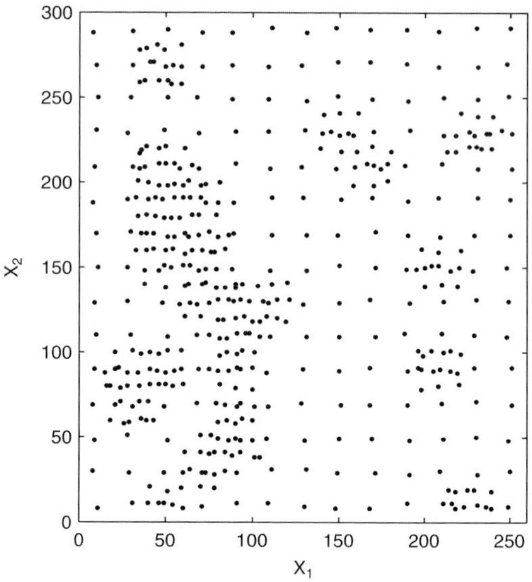

Fig. 1. Spatial locations of the Walker Lake sample dataset.

$$y_{n+1} = \sum_{i=1}^{n} w_i y_i \tag{1}$$

Common sense dictates that the best approximations to y_{n+1} will probably come from those points closest to it, and thus we expect to assign them higher weights. One way of achieving this deterministically would be via the inverse distance weighting scheme (IDW), i.e.

$$y_{n+1} = \frac{\sum_{i=1}^{n} \frac{1}{d_i} y_i}{\sum_{i=1}^{n} \frac{1}{d_i}} \tag{2}$$

where the distances $d_1 \ldots d_n$, are given by $d_i = \|\boldsymbol{x}_i - \boldsymbol{x}_{n+1}\|$. It is common practice to place even greater emphasis on closer points by squaring these distances. Other simple deterministic weighting approaches include polygonal and triangulation methods. All of these methods suffer when the data is clustered (thus not accounting for redundant information when samples are close together) and make no allowance for noise in the data (they are exact interpolators) (Isaaks & Srivastava 1989).

Kriging approach

Kriging takes a more stochastic approach and, using the same predictor as in Equation 1, seeks to choose optimal weights so as to minimize the variance of the estimation error. Before this is possible, many simplifications and assumptions must be made.

The starting point for Matheron (1963) was to assume that any observed data can be seen as the realization of a random variable, or more formally

$$\{Y(\boldsymbol{x}) : \boldsymbol{x} \in \mathscr{D}\}. \tag{3}$$

so that y will represent a realization from this random function, and $\hat{y}$ will be our estimate of y, which can never truly be known.

From this formulation, we are able to write the estimation error as another random variable R:

$$R(\boldsymbol{x}_{n+1}) = \hat{Y}(\boldsymbol{x}_{n+1}) - Y(\boldsymbol{x}_{n+1}) \tag{4}$$

$$= \sum_{i=1}^{n} w_i Y(\boldsymbol{x}_i) - Y(\boldsymbol{x}_{n+1}) \tag{5}$$

It is usual at this stage to split Equation 3 further into two separate parts so that

$$Y(\boldsymbol{x}) = \mu(\boldsymbol{x}) + \delta(\boldsymbol{x}) \tag{6}$$

where μ is a deterministic trend while δ is a zero mean stochastic process with an unknown covariance matrix. The various forms of kriging differ only in how they treat the trend. For example, OK assumes a constant mean μ, while UK seeks a linear combination of known functions. For the purposes of this article, we shall only consider the mathematically simpler technique of OK.

Stationarity

Before we can continue, we must make further assumptions about the stochastic process $\delta(\boldsymbol{x})$. To make the mathematics tractable, and in order to estimate a covariance function from the data D, we must assume that δ is second-order stationary, such that its covariance matrix can be written as

$$Cov(\delta(\boldsymbol{x}_i), \delta(\boldsymbol{x}_j)) = C(\boldsymbol{x}_i - \boldsymbol{x}_j) \tag{7}$$

$$= C(\boldsymbol{h}_{ij}) \tag{8}$$

Thus the covariance of two random variables is a function only of the vector between them. Furthermore, if we assume isotropy, C becomes a function of $h = \|\boldsymbol{h}\|$, i.e. the distance between them. We are now in a position to use the data to calculate this covariance function $C(h)$.

Covariance function

For computational reasons, the covariance function is usually calculated by using the semi-variance estimation, which can be approximated using

$$\hat{\gamma}(h) = \frac{1}{2N_h} \sum_{(ij)|h_{ij} \approx h} (y_i - y_j)^2 \tag{9}$$

where N_h represents the discrete number of point combinations that roughly correspond to the distance h, $\pm$ a specified tolerance. For example, in Figure 2a, the $470^2/2$ data pairs from the Walker Lake dataset were grouped by distance into bins with interval 5 and tolerance $\pm$ 2.5. The semi-variance is related to the covariance by $C(h) = \gamma(\infty) - \gamma(h)$, and can be seen graphically in Figure 2b.

Equation 9 approximates the semi-variance by grouping together points a similar discrete distance h apart, and then finding their mean square difference. When the data are irregularly spaced (as is usually the case), we are forced to model the semi-variance to provide the required values of $\gamma(h)$ not covered by the approximation in Equation 9. The model must be chosen so as to provide a positive definite covariance matrix. This is achieved by plotting a semi-variogram (h against $\hat{\gamma}(h)$) and then fitting one of several well known semi-variogram models which can be combined linearly to give nested approximations to $\gamma(h)$.

Some common semi-variogram models include:

- **Gaussian :**

$$\gamma(h) = 1 - exp\left(-\frac{h^2}{a^2}\right) \tag{10}$$

- **Exponential :**

$$\gamma(h) = 1 - exp\left(-\frac{h}{a}\right) \tag{11}$$

- **Spherical :**

$$\gamma(h) = \frac{3}{2}\left(\frac{h}{a}\right) - \frac{1}{2}\left(\frac{h}{a}\right)^3 \tag{12}$$

where a is a parameter commonly called the range, which reflects the degree of correlation between the data. It should be noted that many resources, including (Isaaks & Srivastava 1989), include an extra factor 3 on gaussian and

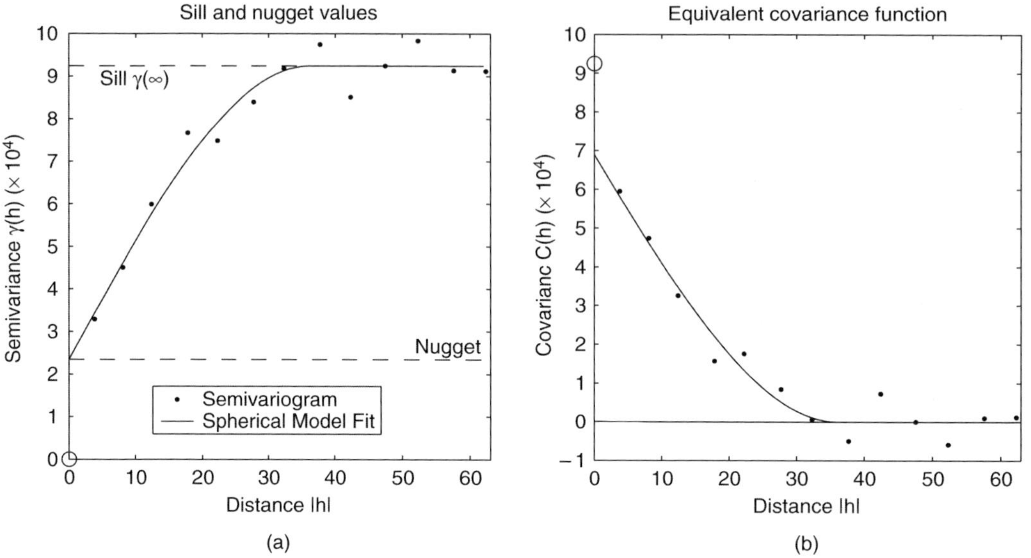

Fig. 2. (**a**) Omni-directional semi-variogram of the Walker Lake sample dataset, with a spherical model fitted, to demonstrate the values of the sill and the nugget. (**b**) The covariance function that corresponds to the model in (a).

exponential variograms so that a reflects the distance at which the data is uncorrelated. Here we have adopted the form consistent with the GSTAT software (Pebesma & Wesseling 1998) to be used later in prediction analysis. These models are then scaled by another value, termed the 'sill', which is effectively the value of the semi-variogram at infinite h (see Fig. 2a). Three different variogram models are shown in Figure 3 using the Walker Lake datasets. Each of these models also include a nugget effect which effectively prevents kriging from being an exact interpolator, thus making allowance for noise in the sample measurements (see Fig. 2a). This corresponds to a non-zero semi-variance at $h = 0$.

Prior information

It cannot be over-stated that the quality of prediction, generated using kriging, depends entirely on how the semi-variogram is modelled. For example, prior information regarding the smoothness of a function would lead to a choice of a Gaussian model, while a noisy dataset will usually require a nugget effect to prevent exact interpolation. While computer fitting methods do exist, such as Cressie's weighted least square's method (Cressie 1993) or restricted maximum likelihood (REML) (Kitanidis 1983), they are rarely relied upon. It is usual that a great deal of user information is built into the choice of models and parameters, and the process has been called more of a black art than a science.

Anisotropy can be included by limiting $\boldsymbol{h}$ to only being in either the major or minor anisotropic direction, modelling these semi-variograms separately, and then transforming the distance vector appropriately.

Using the model structure defined by Equations 3, 6 and 8, combined with the properties of a weighted sum of random variables, allows us to write explicitly the estimation error variance σ_R^2. The weights w_i from Equation 1 are then calculated by finding the partial derivatives of σ_R^2 with respect to each of the unknown weights w_i, and then setting each equation equal to zero. This linear system of equations can then be solved analytically to give the optimum set of weights $\boldsymbol{w}$, to predict y_{n+1} at position $\boldsymbol{x}_{n+1}$, and the minimized error variance σ_{OK}^2. For details of this process, we direct the reader to the standard text (Cressie 1993).

As an example, Figure 4 demonstrates the kriging prediction surface for values of $\boldsymbol{x}_{n+1}$ that cover a grid with density 1, using the Walker Lake data and the spherical covariance function shown in Figure 2. Figure 5 presents the values of the kriging variance, σ_{OK}^2, that have been minimized for each of the desired locations. As we would expect, it is evident that the kriging variance is much lower close to samplelocations, reflecting our increased certainty in these areas (see Fig. 1 for the sample locations).

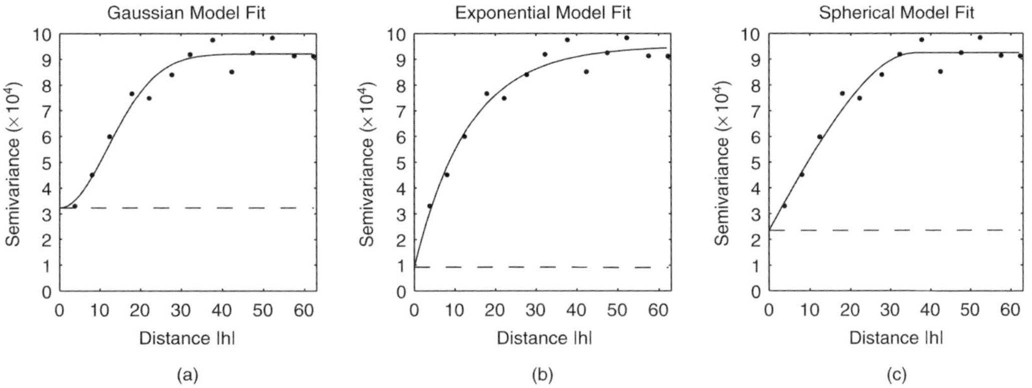

Fig. 3. Three different models fitted to the Walker Lake sample semi-variance. (**a**) Gaussian, (**b**) exponential, and (**c**) spherical. Each model also includes a nugget effect.

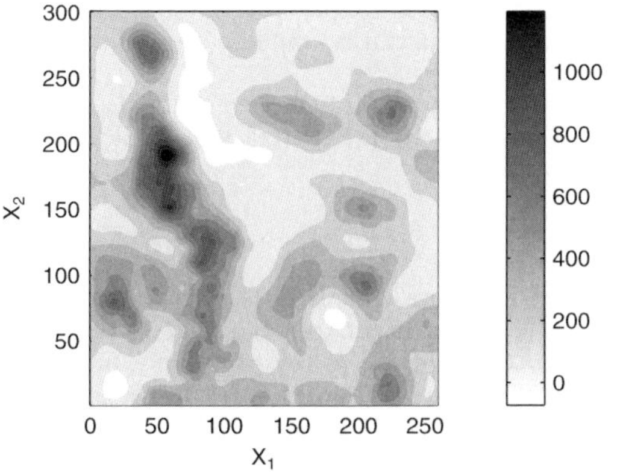

Fig. 4. Contour plot of the ordinary kriging prediction for the Walker Lake dataset using a spherical covariance function. The predictions are made over a grid, with density 1.

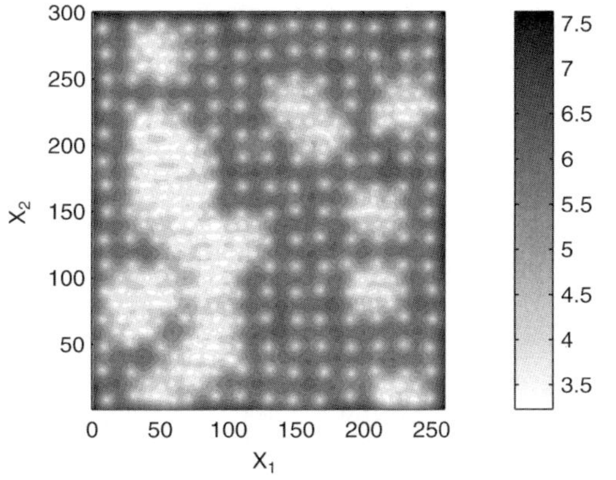

Fig. 5. Contour plot of the ordinary kriging error variance σ^2_{OK} evaluated at each point on a grid with density 1 (units $\times$ 10^4).

Kriging is known as the best linear unbiased estimator (BLUE), where it is linear in the sense that the predictor is a linear combination of the available data, unbiased as it sets the mean error equal to zero, and best in that it minimizes the error variance.

Other kriging methods

It should be noted that numerous methods have been put forward to evade the inherent stationarity problems of kriging. Notable methods include lognormal, disjunctive kriging, IK and kriging of residuals (Cressie 1993), as well as, more recently, model-based kriging (Fuentes 2001). A good empirical assessment of some of these methods can be found in (Moyeed & Papritz 2002), and their method of assessing prediction quality has been adopted in this work. Their assessment of prediction accuracy on the dataset of McBratney *et al.* (1982), concluded that the accuracy of OK was comparable if not better than any of the tested non-linear methods. For this reason and to minimize complexity, we shall only compare the BPM to OK.

Bayesian partition model

The aims of the BPM are similar to those of UK, i.e. to reduce the influence of the deterministic trend μ in Equation 6 by fitting a more complex model to it. UK fails, however, due to its reliance on simple linear polynomial models across the whole surface, whereas the underlying surface may contain various localized effects or clustering not taken into account, for example, discontinuities from a fault in a digital elevation model. By modelling the trend in such a simple way, the assumption of stationarity becomes invalid and will lead to spurious results.

The BPM works around these limitations by fitting a far more flexible non-linear trend to the data and then treating the errors as strictly stationary, rather than second-order stationary (i.e. regression). Again here we present an introduction to the techniques used, and for full details and other examples, we refer the reader to (Denison *et al.* 2002*a, b*).

Model description

The BPM seeks to separate the domain $\mathscr{D}$ into disjunct regions, within which the data $\mathscr{D}$ is modelled using linear regression. This simple framework allows a non-linear trend to be formed in any number of dimensions. For example, in the 1-D case, the trend could be modelled as a series of discontinuous linear regression fits to the data (e.g. Fig. 6).

However, a single model, even after optimization over all the parameters, will struggle to fit the smoothly varying parts of the function in Figure 6, and will introduce extra discontinuities not present in the true function. We adopt a prior distribution on models that reflects our relative beliefs on the accuracy of each. This leads to a model averaging approach to prediction, i.e. we can use many models from the MCMC sampling to produce a final model. By taking enough samples, a smoothly varying final model is possible, while still allowing for sharp discontinuities if required by the data, as these will be present in most, if not all, accepted models over which we average.

Model parameters

The first step in the BPM is to parameterize the partitions in terms of Voronoi tessellations. This requires a set of k centres $\boldsymbol{\theta} = \boldsymbol{t}_1 \ldots \boldsymbol{t}_k$. Every point in the domain $\mathscr{D}$ is then assigned to its closest centre, creating disjoint regions whose boundaries are the perpendicular bisectors with the surrounding centers. Hence the vector $\boldsymbol{y}_j$ would comprise the subset of n_j points from D that are closest to the center $\boldsymbol{t}_j$, and so belong to the region R_j. 1-D and 2-D examples of these are given in Figures 6 and 7 respectively. It is not essential for the Voronoi tessellation to be used; however, it does reduce the number of model parameters and can be applied to systems of any dimensionality.

A linear regression fit is then calculated within each of these regions so that, when $\boldsymbol{x}$ lies within region R_j,

$$y \mid \boldsymbol{x} \in R_j = \boldsymbol{B}(\boldsymbol{x})\boldsymbol{\beta}_j + \delta \tag{13}$$

where, as before, δ represents a random error, $\boldsymbol{\beta}_j$ a $p \times 1$ vector of unknown scalars (regression parameters) specific to the partition j, and $\boldsymbol{B}(\boldsymbol{x})$ the $1 \times p$ vector of values of the known basis function $\boldsymbol{B}$. For example, in 1-D linear regression, $\boldsymbol{\beta}_j$ will contain an intercept and a slope, and $\boldsymbol{B}(\boldsymbol{x}) = [1 \; x]$, thus giving p a value of 2. For ease of notation later, we assign $\boldsymbol{B}_j$ the $n_j \times p$ basis

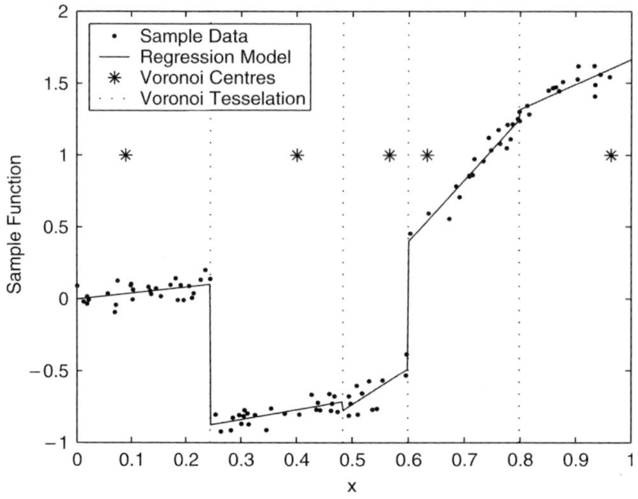

Fig. 6. Construction of a single model via partitioning and regression fits to the data.

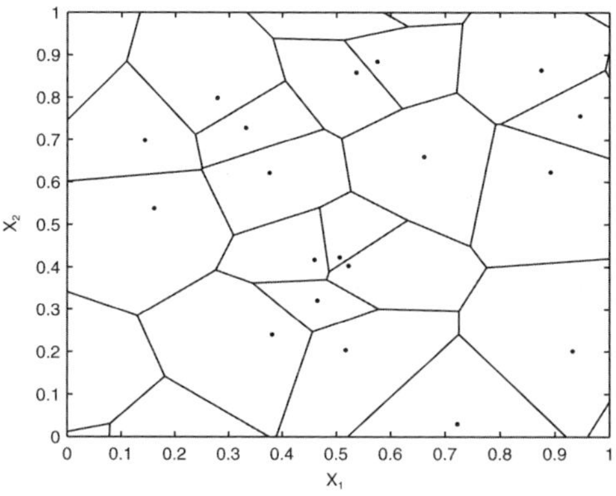

Fig. 7. Example of a 2-D Voronoi tessellation.

matrix comprising the value of $\boldsymbol{B}$ evaluated for the n_j data locations that lie within region j.

In contrast to the approach in UK, where a similar regression is used to approximate the trend, here the error δ (see Equation 6) is assumed to be strictly stationary with a covariance function of $\boldsymbol{I}\sigma^2$. σ^2 represents the regression variance and is assumed to be the same within all partitions, and $\boldsymbol{I}$ is the identity matrix. By making these assumptions, we only ever have to invert $p \times p$ matrices.

This fills the full complement of unknown parameters, which we will group as $\boldsymbol{\varphi} = [\boldsymbol{\theta}, \boldsymbol{\beta}, \sigma^2]$. It should be noted that there is no fixed number of partitions, and the value of k will vary from model to model, thus allowing the dimensionality of the model space to vary accordingly.

Model selection

We are now in a position to make inferences of the value y_{n+1} at a new position $\boldsymbol{x}_{n+1}$ given our data samples D. From the general representation theorem, we represent the density of this new value as the integral

$$p(y_{n+1} \mid D) = \int_{\Phi} p(y_{n+1} \mid \boldsymbol{\varphi}) p(\boldsymbol{\varphi} \mid D) d\boldsymbol{\varphi}, \qquad (14)$$

over the entire parameter space Φ. This represents model averaging over every possible set of model parameters. The first part of the integrand in Equation 14 is the likelihood of the new data value, and is easily defined once $\boldsymbol{\varphi}$ has been chosen. The second part represents the posterior distribution of model parameters and is analytically intractable.

It is at this point we turn to Bayesian statistics and MCMC techniques to approximate Equation 14. We write Equation 14 as the Monte Carlo integral

$$p(y_{n+1} \mid D) \approx \frac{1}{M} \sum_{i=1}^{M} p(y_{n+1} \mid \boldsymbol{\varphi}_i) \qquad (15)$$

so that as the number of samples, M, increases, the summation will converge to the integral in Equation 2, providing that we choose a sampling method that draws the parameter samples $\boldsymbol{\varphi}_i$ from their posterior distribution $p(\boldsymbol{\varphi} \mid D)$. This distribution is ensured by creating a Markov chain whose stationary distribution is the desired posterior density.

The posterior distribution via Bayes' theorem

To begin this process, we must define the posterior distribution in terms of Bayes' theorem so that

$$p(\boldsymbol{\varphi} \mid D) \propto p(D \mid \boldsymbol{\varphi}) p(\boldsymbol{\varphi}) \qquad (16)$$

or, using common terminology,

Posterior $\propto$ *Likelihood* $\times$ *Prior*.

We will now deal with each of the likelihood and prior definitions in turn

Likelihood: The likelihood is a density that is returned by considering the errors of our model δ as coming from a normal distribution. In principle, we can relax this assumption, but it does provide practical advantages in terms of evaluating the integrals (as we demonstrate below). Hence following from Equation 13, the likelihood probability density function of the data D_j within the partition R_j is given by

$$p(D_j \mid \boldsymbol{\beta}_j, \boldsymbol{\theta}_j, \sigma^2) = N(\boldsymbol{B}_j \boldsymbol{\beta}_j, \sigma^2) \qquad (17)$$

so that, as each partition is independent, we can write the overall likelihood of the data, for all k partitions, as

$$p(D \mid \boldsymbol{\beta}, \boldsymbol{\theta}, \sigma^2) = \prod_{j=1}^{k} N(\boldsymbol{B}_j \boldsymbol{\beta}_j, \boldsymbol{I}\sigma^2). \qquad (18)$$

This form for the likelihood leads directly to our choice of function for the prior distribution over our parameters $\boldsymbol{\varphi}$. We are able to facilitate the mathematics of the problem greatly by choosing a prior that is conjugate to the normal distribution. These conjugate priors are of such a form that the resulting posterior distribution will be from the same distribution family as the prior (Bernardo & Smith 1994).

Priors: We must place a joint prior on all of the unknown parameters in $\boldsymbol{\varphi}$, so here we seek $p(\boldsymbol{\beta}, \boldsymbol{\theta}, \sigma^2)$. The prior placed on $\boldsymbol{\theta}$ is chosen to be flat, so that all possible positions and numbers of partitions are equally probable, and hence allows the dimension of $\boldsymbol{\theta}$ to be variable, up to some set maximum number of partitions K. This allows us to express the joint prior as

$$p(\boldsymbol{\beta}, \boldsymbol{\theta}, \sigma^2) = p(\sigma^2)p(\boldsymbol{\beta} \mid \boldsymbol{\theta}, \sigma^2) \tag{19}$$

By analogy with the Bayesian linear model (Bernardo & Smith 1994), the conjugate prior for the normal likelihood is the normal inverse gamma distribution, so that for each partition we get

$$p(\boldsymbol{\beta}_j \mid \boldsymbol{\theta}, \sigma^2) = N(\mathbf{0}, \sigma^2 \boldsymbol{V}) \tag{20}$$

with $\boldsymbol{V} = v\boldsymbol{I}$ where v is a scalar value, giving the prior variance over the regression parameters. As the dimensions of $\boldsymbol{\beta}$ are fixed, $\boldsymbol{V}$ will be the same for each partition. Taking into account an inverse gamma prior over the common regression variance $p(\sigma^2) = IG(a, b)$ where a and b are the shape parameters of the distribution, as well as the independence of partitions, we get a full prior via

$$p(\boldsymbol{\beta}, \boldsymbol{\theta}, \sigma^2) = p(\sigma^2)p(\beta \mid \theta, \sigma^2) \tag{21}$$

$$= IG(a, b) \prod_{i=1}^{k} N(\mathbf{0}, \sigma^2 \boldsymbol{V}) \tag{22}$$

so that, in effect, we are left with three prior parameters a, b and v, which must be given values at the start. The parameter with the greatest influence is v, so this too is given a set of priors (or hyperpriors), and is updated throughout the MCMC run.

The posterior: The power of using conjugate priors can be seen in evaluating the posterior density, so that by using the result from the standard Bayesian linear model, and taking a model with k partitions, we can give the posterior distribution as

$$p(\boldsymbol{\beta}, \boldsymbol{\theta}, \sigma^2 \mid D) = IG(a^*, b^*) \prod_{j=1}^{k} N(\boldsymbol{m}_j^*, \sigma^2 \boldsymbol{V}_j^*) \tag{23}$$

where the prior values of a, b, m and V have been updated by information in the likelihood to give

$$\boldsymbol{m}_j^* = (\boldsymbol{V}^{-1} + \boldsymbol{B}_j'\boldsymbol{B}_j)^{-1}(\boldsymbol{B}_j'\boldsymbol{y}_j) \tag{24}$$

$$\boldsymbol{V}_j^* = (\boldsymbol{V}^{-1} + \boldsymbol{B}_j'\boldsymbol{B}_j)^{-1} \tag{25}$$

$$a^* = a + n/2 \tag{26}$$

$$b^* = b + \left\{ \boldsymbol{y}'\boldsymbol{y} - \sum_{j=1}^{k} (\boldsymbol{m}_j^*)'(\boldsymbol{V}_j^*)^{-1}\boldsymbol{m}_j^* \right\} \tag{27}$$

This updating property is the primary reason for the choice of the conjugate prior, as we have evaded the need to evaluate the integral

$$p(D) = \int_{\Phi} p(D \mid \boldsymbol{\varphi})p(\boldsymbol{\varphi})d\boldsymbol{\varphi}, \tag{28}$$

which is the required constant of proportionality in Equation 16, and provided a posterior distribution that we can easily integrate using standard integrals when we later consider the marginal likelihood.

Reversible jump Markov chain Monte Carlo

The remarkable feature of MCMC, is its ability to converge to a desired stationary distribution, provided the probability of transition to a new state in the chain is correctly defined. The technique used here is the Metropolis–Hastings reversible jump of Green (1995), which is capable of handling high and variable dimensional models in a very efficient manner.

Starting from a model with one randomly placed partition, the chain moves forward, around the model space by one of either of three steps with equal probability:

- **Birth:** A new Voronoi centre is created at a random position within the domain $\mathscr{D}$, thus increasing the dimensionality of the model by one.
- **Death:** A randomly selected centre is removed from $\boldsymbol{\theta}$, thus reducing the dimensionality of the model by one.
- **Move:** A randomly selected centre within $\boldsymbol{\theta}$ is given a new random position within $\mathscr{D}$.

The acceptance probability for our model under reversible jump algorithm is given by

$$\alpha = min\left\{1, \frac{p(D \mid \boldsymbol{\beta}_{m+1}, \boldsymbol{\theta}_{m+1}, \sigma^2)}{p(D \mid \boldsymbol{\beta}_m, \boldsymbol{\theta}_m, \sigma^2)}\right\}, \tag{29}$$

which is a ratio of the likelihoods of the data given two different model choices, and in turn are called 'marginal likelihoods'. The models compared are the current member of the chain m

and the proposed sample $m+1$. Note that this ratio is the same as the Bayes' factor, when we assume that the prior probabilities of each set of model parameters are the same (See Bernardo & Smith 1994).

Figure 8 demonstrates this process of model selection and model averaging (see Equation 15), with 10 different samples taken from the posterior distribution via reversible jump MCMC. This figure also demonstrates the varying dimensionality of the model from sample to sample, as seen by the differing numbers of partitions used.

Marginal likelihood

The marginal likelihood of a particular model $\boldsymbol{\varphi}_i$, required in Equation 29, can easily be defined by rearranging Bayes' theorem (Denison *et al.* 2002*a*) to give

$$p(D \mid \boldsymbol{\varphi}_i) = \frac{p(D \mid \boldsymbol{\beta}_i, \boldsymbol{\theta}_i, \sigma^2) p(\boldsymbol{\beta}_i, \sigma^2)}{p(\boldsymbol{\beta}_i, \boldsymbol{\theta}_i, \sigma^2 \mid D)} \tag{30}$$

Because of the nature of the conjugate priors we have chosen previously, it becomes possible to integrate out the unknown model parameters $\boldsymbol{\beta}$ and σ^2, which, for the BPM as described before, gives the following

$$p(D \mid \boldsymbol{\varphi}_i) = \frac{(b)^a \Gamma(a^*)}{(b^*)^{a^*} \Gamma(a)} \pi^{-n/2} \prod_{j=1}^{k} \frac{|\boldsymbol{V}_j^*|^{1/2}}{|\boldsymbol{V}_j|^{1/2}} \tag{31}$$

where b, a, V and their starred updates are given in Equations 24–27, and are specific to this ith model. This means that the marginal likelihood is dependent only on the position of the Voronoi centres, defined for model i by $\boldsymbol{\theta}_i$, and the predefined parameters a, b and v.

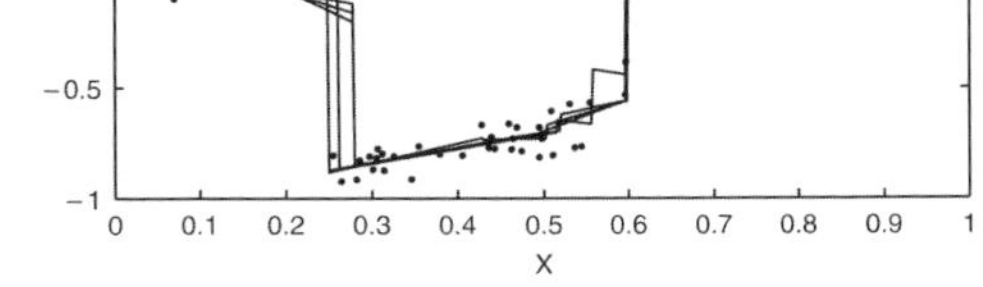

Fig. 8. Ten different samples taken from the posterior distribution.

Adaptive priors

In order to ensure greater independence from the prior regression parameter variance v, the reversible jump method allows the data to determine suitable values. This comes about by assuming that v itself has a distribution of values, and as such can be assigned a prior (the hyperprior) and consequently a posterior distribution from which suitable values can be drawn. Full discussion can be found in Denison *et al.* (2002b), but we shall give the result here:

$$p(v^{-1}) = Ga(\alpha_1, \alpha_2) \tag{32}$$

$$(v^{-1} \mid D, \boldsymbol{\beta}, \sigma^2, \boldsymbol{\theta}) = Ga\left(\alpha_1 + \frac{p \times k}{2}, \alpha_2 + \frac{1}{2\sigma^2} \sum_{j=1}^{k} \boldsymbol{\beta}'\boldsymbol{\beta}\right). \tag{33}$$

Here, Ga represents the usual gamma distribution, and α_1 and α_2 are two constants that define the distribution of the hyperprior $p(v^{-1})$ (in this case a noninformative hyperprior was used, by giving α_1 and α_2 a value of 0.1). First however, as we have integrated out σ^2 and β, we need to obtain up-to-date values for these parameters for use in Equation 33, so they are drawn from their updated distributions: $IG(a^*, b^*)$ and $N(\boldsymbol{m}^*, \sigma^2 \boldsymbol{V}^*)$ respectively.

Computational advantages

The key computational advantage of partition models over any of the forms of kriging is that, for each sample of parameters $\boldsymbol{\varphi}_i$, to calculate the marginal likelihood it is only necessary to invert k $(p \times p)$ matrices where, as before, k is the number of partitions in the model, and p the order of $\boldsymbol{B}$ in Equation 13. This value will depend on the spatial dimension we are working in and the order of the regression being fitted. For example, in a 2-D spatial system with coordinates x_1 and x_2, and fitting linear planes inside each partition, $\boldsymbol{B}$ will take values of $\boldsymbol{B} = [\mathbf{1}, \boldsymbol{x}_1, \boldsymbol{x}_2,]$, and thus give p a value of 3.

This is in stark contrast to kriging, where we must invert one large $n \times n$ matrix. This is another of the highlighted drawbacks to kriging, and typically leads only to kriging subsets of the data for each prediction point. The drawback of using the BPM, is that we must iterate the process many times to provide a large enough number of samples in the Monte Carlo integral (Equation 15), though these calculations take little time due to the size of the necessary inversions. It should also be mentioned that many applications in the oil industries now require that kriging itself be iterated many times, for example, sequential Gaussian simulation in reservoir characterization (Hohn 1999).

Prediction uncertainties

Another positive feature of the BPM is the ease with which prediction uncertainties can be assigned. These are generated by considering all of the predictions at each validation point generated over the course of the MCMC sampling procedure and then returning the 95% credible intervals of their distributions.

With kriging in the ordinary kriging form we have presented here, uncertainty can be adjudged by use of the minimized kriging variance σ^2_{OK}, as displayed in Figure 5. By assuming multi-variate normal distributions, 95% confidence intervals can be assigned in the usual way so that we get the range $\hat{y} \pm 2\sigma_{OK}$ (Isaaks & Srivastava 1989). The results from such a treatment, however, are largely unsatisfactory and usually represent only a measure of the local sample distribution (Journel 1986). In addition, kriging uncertainties are always underestimated, as the parameters of the covariance function are assumed to be known, despite them having been estimated when modelling the semi-variogram. More realistic assessments of uncertainties are reviewed in Goovaerts (2001) and necessitate either the more complex approaches of IK or disjunctive kriging or, finally, an iterative stochastic simulation approach.

The data sets

In order to demonstrate the main benefits of the BPM over OK, we have compared the two methods for various synthetic examples, as well as for a real dataset. The former were chosen as examples best suited to the BPM (i.e. where the assumptions of stationarity are challenged), while the latter represents a stationary dataset and is thus suited to a kriging approach.

Synthetic examples

Two synthetic 1-D functions, A and B, were chosen both with large and significant discontinuities in their functions. For the following we define the boxcar step function $Box(c_1, c_2)$ as

$$Box(c_1, c_2) = \begin{Bmatrix} 1 & \text{if} & c_1 < x < c_2 \\ 0 & & otherwise \end{Bmatrix} \quad (34)$$

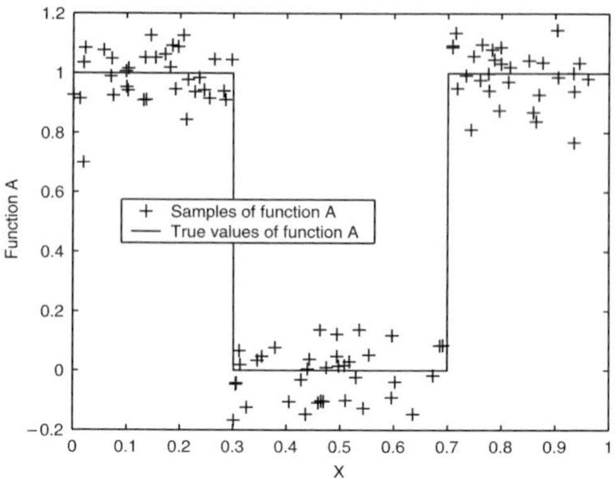

Fig. 9. Sample and validation datasets for function A. Sample set comprises 100 random locations, with added Gaussian noise N(0, 0.1). Validation set comprises 1000 regularly spaced locations.

and the precise definitions of the two functions are given below.

- **Function A:** An upside down boxcar function:

$$y(x) = 1 - Box(0.3, 0.7). \quad (35)$$

- **Function B:** An addition of a plane, Gaussian and another upside down boxcar. Explicitly, this gives:

$$y(x) = 0.6x + exp\left(-\frac{(x-0.9)^2}{(0.2)^2}\right) - Box(0.25, 0.6) \quad (36)$$

The sample datasets were formed by taking 100 random x positions within the range $0 < x < 1$, and then adding Gaussian noise with standard deviations of 0.1 and 0.08 respectively to the y values of functions A and B. The validation datasets comprise 1000 and 500 data points for functions A and B respectively, along an equally spaced grid. The BPM and OK algorithms were then used to give predictions at the validation dataset positions and compared. The data sets, together with the noise free functions, are shown graphically in Figures 9 and 10.

Rainfall data from Switzerland

The rainfall dataset comprises 100 measurements of daily rainfall in Switzerland on 8 May 1986, taken randomly from a full dataset of 467 points. This dataset is freely available, and formed the basis of the 1997 Spatial Interpola-

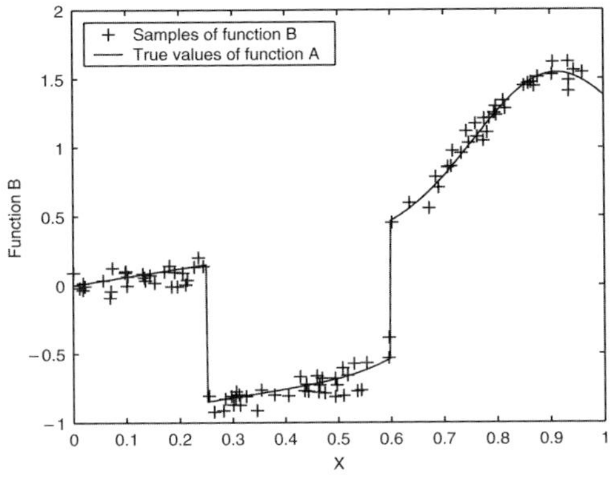

Fig. 10. Sample and validation datasets for function B. Sample set comprises 100 random locations, with added Gaussian noise N(0, 0.08). Validation set comprises 500 regularly spaced locations.

tion Comparison (SIC97), a comparison of numerous geostatistical techniques. Indeed, to ensure the validity of the BPM, we compare it to the published results using OK and IK (Atkinson & Lloyd 1998).

The values are in units of 0.1 mm. The importance of such a dataset derives from rainfall being used as the prime measurement of radioactivity deposition, after the Chernobyl disaster in 1986.

Further information regarding elevation was also provided separately, but this information was not incorporated in either model. Figure 11 shows the relative densities and positions of sample and validation datasets.

Prediction quality

We first define the criteria used in comparing OK to the BPM,

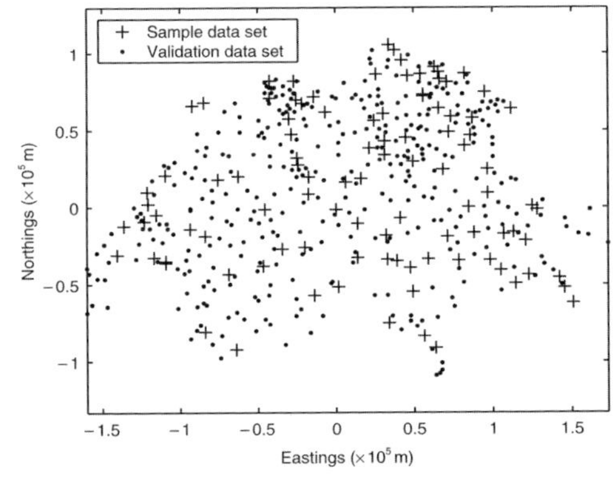

Fig. 11. Map showing positions of the sample and validation data sets for the rainwater datasets.

- **Root mean square error.**

$$RMSE = \left\{ \frac{1}{N_v} \sum_{i=1}^{N_v} (\hat{y}_i - y_i)^2 \right\}^{\frac{1}{2}} \tag{37}$$

- **Mean absolute error:**

$$MAE = \frac{1}{N_v} \sum_{i=1}^{N_v} |\hat{y}_i - y_i| \tag{38}$$

- **Relative bias:**

$$\tau BIAS = \frac{\sum_{i=1}^{N_v} (\hat{y}_i - y_i)}{N_v \bar{y}_p} \tag{39}$$

- **Relative mean separation:**

$$\tau MSEP = \frac{\sum_{i=1}^{N_v} (\hat{y}_i - y_i)^2}{\sum_{i=1}^{N_v} (\bar{y}_p - y_i)^2} \tag{40}$$

where $\hat{y}_i$ represents the prediction of the true function value y_i over the N_v validation data points and $\bar{y}_p$ is the arithmetical mean of the prediction dataset. These tests are a combination of those used by Atkinson & Lloyd (1998) and Moyeed & Papritz (2002). In all cases, a better prediction model is one that produces a lower value of these measures.

Synthetic function: model specification and Results

For each of the test functions, A and B, an OK and a BPM prediction were made. In the latter case, it was enough to give the data to the program and let it run, while, for OK, numerous decisions had to be made to ensure a viable fit.

Ordinary kriging

All of the kriging results were produced using GSTAT, a highly flexible program and a standard geostatistical package (Pebesma & Wesseling 1998). The first step for any kriging analysis is the modelling of the variogram. The assumptions and models made during this process influence the results greatly. The two semi-variograms are given in Figures 12 and 13 for functions A and B respectively, along with their corresponding models. It should be noted that the parameters for these models assume there is not a factor 3 in the exponential see covariance function (p. 197). The models were originally fitted with the weighted least squares

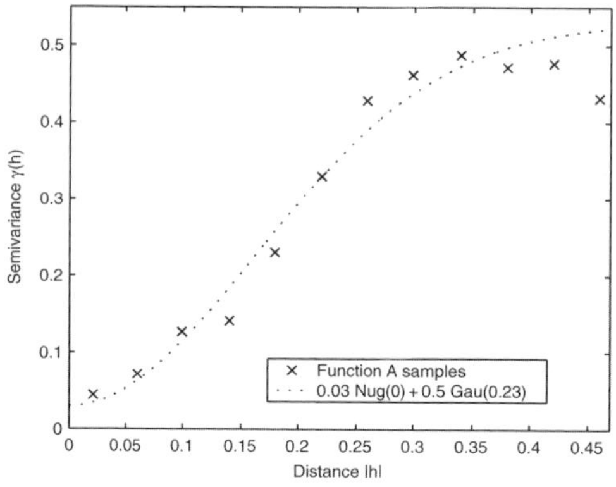

Fig. 12. Omni-directional variogram and fitted Gaussian/nugget model for function A (steplength 0.04).

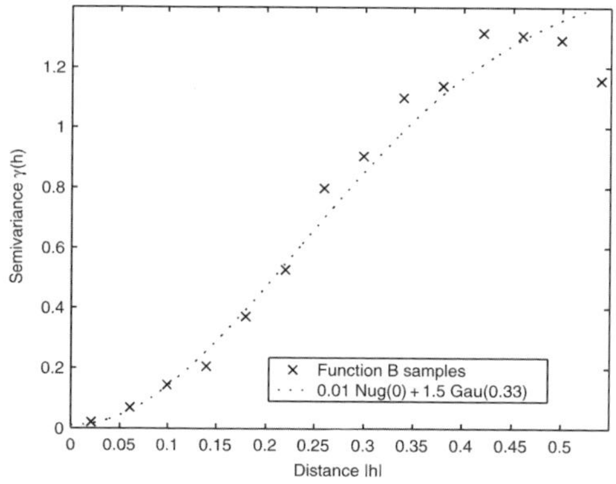

Fig. 13. Omni-directional variogram and fitted Gaussian/nugget model for function B (steplength 0.04).

method of Cressie (1993), and then adjusted manually to provide a more consistent fit.

Although numerous more complicated nested models were considered, it was found that the best results were produced using simple nugget and Gaussian model combinations. This is consistent with the known smoothness and noise characteristics of the two functions we were considering.

Predictions were then made for the validation datasets, again using GSTAT, with a maximum search radius of 0.25. This search radius was used to take into account our knowledge of the evident localized clustering in the sample data.

Partition model

The only parameters required for the BPM were the number of samples to be taken and the order of the regression fit within the partitions. For function A, 20 000 samples were taken, with regressions of order 0, while for function B, 50 000 samples were taken with regressions of order 1. These choices reflect the differing complexities of the two functions.

The number of partitions k for function A ranged between 3 and 5, whilst for function B, values ranged from 3 to 9, with 5 being by far the most frequent. These numbers illustrate the model's desire to reduce complexity, as well as the fact that the dimension of the model space is allowed to vary from sample to sample. This property is built into any Bayesian inference process, and is commonly called *Ockham's razor*. For further details of this effect with regard to the BPM, we refer the reader to Denison *et al.* (2002*b*).

Results

We are now able to analyse the resulting predictions using both visual and the statistical values described previously in (see prediction quality, p. 205). Comparative plots of the two techniques are given in Figures 14 and 15 for functions A and B respectively.

It is evident that, in terms of overall prediction, the BPM has outperformed OK for the given synthetic examples. Table 1 gives the prediction measures outlined and the improved performance of BPM is clear.

In both functions, the prediction from BPM is between 3 and 6 times better, and the bias is reduced. However, these simple measures do not get across the principal advantage of the BPM: that of dealing gracefully with discontinuities. This is best evidenced in Figures 14 and 15,

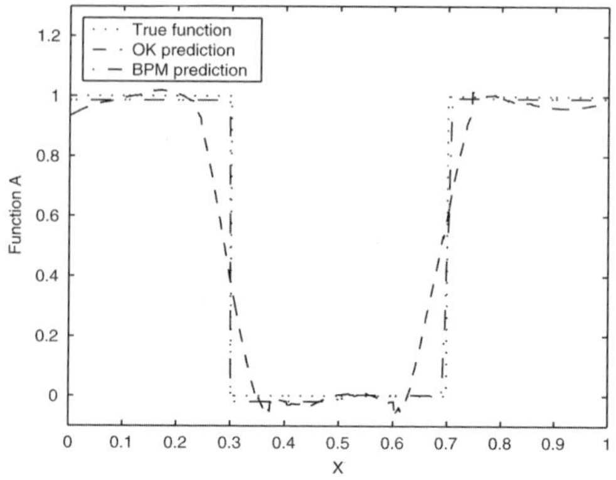

Fig. 14. Predictions of function A test data via OK and BPM (20 000 samples).

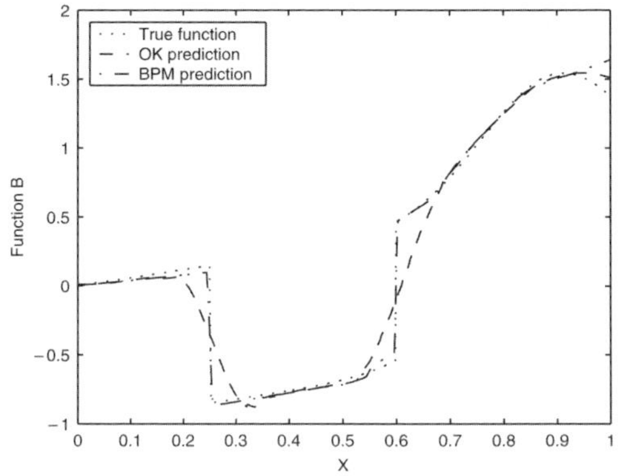

Fig. 15. Predictions of function B test data via OK and BPM (50 000 samples).

Table 1. *Prediction errors for functions A and B*

	Function A		Function B	
Test	OK	BPM	OK	BPM
RMSE	0.1455	0.0507	−0.1062	−0.0254
MAE	0.0736	0.0165	0.0284	0.0048
rBIAS	−0.0208	−0.0184	0.1428	0.0585
rMSEP	0.0880	0.0107	0.0801	0.0336

MAE, mean absolute error; rBIAS, relative bias; RMSE, root mean square error; rMSEP, relative mean separation.

where, for the BPM, the edge is sharply defined, while for OK, the values are smeared across the discontinuity. This is easily explained when considering a point on the discontinuity. The points either side of the discontinuity at similar distances will be treated with equal weight, thus producing the smoothing effect. This effect is impossible to remove with OK, unless you implicitly tell it there are discontinuities.

Rainwater results

Ordinary kriging implementation

The variogram models used for analysis of the rainwater dataset are taken directly from Atkinson & Lloyd (1998). These models take into account the evident anisotropies at 45°, fitting two models to each of the major and minor axes of anisotropy. The models fitted are a nested combination of Gaussian and spherical models (Equations 10 and 12 respectively). The prediction values for their models were re-created using GSTAT and verified against the values given in the work of Atkinson & Lloyd to ensure our correct implementation. Values are also given for the analysis of IK for two of the statistics we are concerned with (RMSE and MAE), and are included for comparison.

Partition model implementation

Again, implementing the BPM was merely a matter of giving data to the program and ensuring a high enough number of samples were taken. The regression order was taken to be 1, and thus linear planes were fitted within each partition.

Result comparison

Four different models were compared for how well they predicted on the 367 members of the validation set. These were OK, IK, IDW (see Equation 2) and the BPM.

The IDW estimate is displayed for a benchmark, and the entire sample dataset was used in the calculation of each prediction point. This comparison is given in Table 2.

It is clear again from the results that, in terms of prediction, the BPM has provided the best estimates, although the performance relative to OK is not as dramatic, as a consequence of the dataset being more suited to kriging than the synthetic examples considered earlier. In addition, when compared to OK, the distribution statistics of the BPM were found to be closer to the true values of the prediction dataset. The true mean of the data is given as 185.36, with OK predicting a value of 181.87 and BPM a value of 186.53. This is clear further when comparing the rBIAS values and demonstrate that OK has under-estimated values in this case.

Summary

The fundamental limitation of kriging is the assumption of stationarity. This is coupled with difficulties in choosing appropriate covariance

Table 2. *Comparison of prediction errors for four different methods*

Test	OK	IK	BPM	IDW
rBIAS	−0.0194	—	0.0065	0.0001
rMSEP	0.2887	—	0.2625	0.3824
RMSE*	59.71	60.0	56.94	68.72
MAE*	41.10	42.6	41.10	50.82

MAE, mean absolute error (in units of 0.1 mm); rBIAS, relative bias; RMSE, root mean square error (in units of 0.1 mm); rMSEP, relative mean separation.

function models as well as a common need to analyse only subsets of the data due to computational considerations. The synthetic 1-D results demonstrate effectively that OK performs poorly when faced with discontinuities which are common in geological data. The model is not flexible enough to incorporate the smooth underlying function, as well as the sharp cut-off points of the discontinuities, resulting in a smearing over the edges. This was despite good data support close to the boundaries of the discontinuities.

For all of the synthetic functions, the BPM outperformed OK by several orders of magnitude and identified the boundaries of each of the discontinuities to an accuracy of the scale of the validation grid. These datasets were well adapted to the properties of the BPM and the predictions produced left very little room for improvement. The most evident example of this is found in function B, where the model was able to regenerate both the smooth Gaussian and plane combination towards the right of the function, as well as the sharp boundaries of the discontinuity.

Furthermore the BPM was able to model the stationary rainwater data. Despite the data lacking an underlying trend, the BPM was able to effectively model the validation set, as well as give good estimates of its distributions. This demonstrates that BPM performs well in situations where OK performs well, but requires less model specification.

In summary, some of the advantages of BPM over OK are:

- It is easy to access prediction uncertainties (see Prediction uncertainties, p. 204, for a discussion regarding uncertainties with OK).
- There is no need to invert $n \times n$ matrices (for 2-D linear plane fitting in BPM, only ever need to invert 3×3 matrices). This is in stark contrast with the model-based kriging method of Fuentes (2001), where large covariance matrices must be inverted at each step of the MCMC chain.
- Only limited knowledge of the dataset is required in order to produce high-quality predictions. This is possible due to the automatic updating of the priors via the hyperprior structure.
- Very flexible models can be produced.
- The model can be defined over the entire space. It is an important but subtle statistical point that, by being forced for computational reasons to krige using only a small number of closely surrounding points (a maximum of 16 in the case of the rainwater data), it means that the model is not defined over the entire region.

Some further improvements we are looking at implementing include using alternatives to the standard Voronoi tessellation for partitioning. This method, although extremely easy to implement, is found to be restrictive in that it precludes certain data configurations. For example, data which evidently belong to the same partition are sometimes forced to be separated by the position of surrounding Voronoi centres. This results in regions being split into more partitions than necessary and reduces the efficacy of the regression calculation.

Another of the current disadvantages of the model is the inability to model microscale trends easily, and a relevant example may be boulders in a digital elevation model. These will probably be treated as noise in the regression fit and not modelled correctly. This can be taken into account by either letting the Markov chain run for longer periods of time, or perhaps starting to use some of the stationary properties of kriging within partitions.

The authors would like to thank the reviewers, A. Curtis and A. Malinverno, for their helpful and insightful comments on all aspects of this paper, as well as P. Williamson of Total for his guidance and ongoing support. This research has been supported by the combined National Environment Research Council and Total CASE award NER/S/C/2002/10625.

References

ATKINSON, M. & LLOYD, D. 1998. Mapping precipitation in Switzerland with ordinary and indicator kriging. *Journal of Geographic Information and Decision Analysis*, **2**, 65–76.

BERNARDO, J. M. & SMITH, A. F. M. 1994. *Bayesian Theory*. Wiley, Chichester and New York Probability and Mathematical Statistics Series.

COLEOU, T., HOEBER, H. & LECERF, D. 2002. Multivariate geostatistical filtering of time-lapse seismic data for an improved 4d signature. *Society of Exploration Geophysicists, International Exposition and 72nd Annual Meeting; Technical Program, Expanded Abstracts with Authors' Biographies*. Society of Exploration Geophysicists, Tulsa, 1662–1665.

CRESSIE, N. A. C. 1993. *Statistics for Spatial Data*. Revised edition. J. WILEY, New York, Probability and Mathematical Statistics Series: Applied Probability and Statistics.

DECENCIERE, E., DE-FOUQUET, C. & MEYER, F. 1998. Applications of kriging to image sequence coding. *Signal Processing: Image Communication*, **13**, 227–49.

DENISON, D. G. T., ADAMS, N. M., HOLMES, C. C. & HAND, D. J. 2002*a*. Bayesian partition modelling.

Computational Statistics and Data Analysis, **38**, 475–485.

DENISON, D. G. T., HOLMES, C. C., MALLICK, B. & SMITH, A. F. M. 2002*b*. *Bayesian methods for nonlinear classification and regression*. Wiley, Chichester and New York. Probability and Statistics Series.

FUENTES, M. 2001. A high frequency kriging approach for non-stationary environmental processes. *Environmetrics*, **12**, 469–483.

GOOVAERTS, P. 2001. Geostatistical modelling of uncertainty in soil science. *Geoderma*, **103**, 3–26.

GREEN, P. J. 1995. Reversible jump Markov chain Monte Carlo computation and Bayesian model determination. *Biometrika*, **82**, 711–732.

HANDCOCK, M. S. & STEIN, M. L. 1993. A Bayesian-analysis of kriging. *Technometrics*, **35**, 403–410.

HOHN, M. E. 1999. *Geostatistics and Petroleum Geology*. 2nd edition. Kluwer, Dordrecht.

ISAAKS, E. H. & SRIVASTAVA, R. M. 1989. *Applied Geostatistics*. Oxford University Press, New York.

JOURNEL, A. G. 1986. Geostatistics: models and tools for the earth-sciences. *Mathematical Geology*, **18**, 119–140.

KITANIDIS, P. K. 1983. Statistical estimation of polynomial generalized covariance functions and hydrologic applications. *Water Resources Research*, **19**, 909–921.

MATHERON, G. 1963. Principles of geostatistics. *Economic Geology and the Bulletin of the Society of Economic Geologists*, **58**, 1246–1266.

MCBRATNEY, A. B., WEBSTER, R., MCLAREN, R. G. & SPIERS, R. B. 1982. Regional variation of extractable copper and cobalt in the topsoil of southeast scotland. *Agronomie*, **2**, 969–982.

MOYEED, R. A. & PAPRITZ, A. 2002. An empirical comparison of kriging methods for nonlinear spatial point prediction. *Mathematical Geology*, **34**, 365–386.

PEBESMA, E. J. & WESSELING, C. G. 1998. GSTAT: a program for geostatistical modelling, prediction and simulation. *Computers & Geosciences*, **24**, 17–31.

Using prior subsidence data to infer basin evolution

NICKY WHITE

Bullard Laboratories, Department of Earth Sciences, University of Cambridge, Madingley Rise, Madingley Road, Cambridge, CB3 0EZ, UK. (e-mail: nwhite@esc.cam.ac.uk)

Abstract: Quantitative models, which predict the structural and thermal evolution of sedimentary basins and margins, can be used to extracted information from subsidence data derived from discrete and noisy stratigraphic records. Although many basin-modelling algorithms exist, most of them solve the forward problem and many assume that rifting is instantaneous. A 2-D optimization strategy, which calculates spatial and temporal variation of strain rate, is outlined. This general approach should help to elucidate the dynamical evolution of sedimentary basins but it also addresses three issues of interest to the hydrocarbon industry. First, the residual misfit between observed and predicted basin geometries allows competing structural and stratigraphic interpretations to be objectively tested. Second, the animated evolution of sedimentary basin and passive continental margins can be produced using the retrieved strain-rate tensor. Thirdly, spatial and temporal variations of strain rate control basal heat flow, which in turn constrains the temperature and maturation histories of the sedimentary pile. Here, a small selection of 2-D results are presented and the basis of a 3-D formulation is described.

Introduction

One of the most important risks in hydrocarbon exploration is the thermal evolution of sedimentary basins or continental margins (White *et al.* 2003). This risk can be addressed by developing quantitative thermal and structural models which are based upon prior information about subsidence histories. Here, a strategy for analysing the 2-D and 3-D structural and thermal evolution of extensional sedimentary basins and passive continental margins is outlined. The cornerstone of this approach is to use inverse theory to extract strain-rate histories from subsidence observations. These observations are subject to differing degrees of uncertainty and it is important that such uncertainties are incorporated into any modelling strategy. A crucial function of inverse modelling is to investigate how observational errors are mapped into the solution space. The aim is to develop a set of 2-D and 3-D inverse algorithms which determine the strain-rate histories of sedimentary basins and continental margins as a function of time and space. These histories will define both structural and thermal evolution and thus help to refine maturation and fluid flow models of margins. Inverse algorithms can be used to carry out a systematic analysis of conjugate margins systems located worldwide. I anticipate that these algorithms will be useful de-risking tools, especially as hydrocarbon exploration is directed toward deeper, more highly extended, continental margins.

The strategy outlined here has developed out of two independent approaches pioneered by two groups over the last 5 years. The first approach is concerned with a general quantitative description of the way in which continental crust actively deforms. Haines & Holt (1993) showed that strain-rate data based upon earthquake focal mechanisms and global-positioning system (GPS) measurements can be inverted to obtain the complete horizontal motions within zones of active distributed deformation. This approach has also been used to analyse Quaternary fault slip rate data (Holt *et al.* 2000). Application of these inverse algorithms is generating important insights into active shortening (e.g. India, Tibet and southeast Asia: Holt *et al.* 2000), active oblique deformation (e.g. New Zealand, California: Flesch *et al.* 2000; Beavan & Haines 2001) and active extension (e.g. Aegean Region, Basin and Range Province: Holt *et al.* 2000). A growing body of accurate GPS measurements will make the determination of a global strain rate model tractable (Kreemer *et al.* 2000; http://archive.unavco.ucar.edu).

The second approach has grown out of a long-standing interest in the development of sedimen-

From: CURTIS, A. & WOOD, R. (eds) 2004. *Geological Prior Information: Informing Science and Engineering*. Geological Society, London, Special Publications, **239**, 211–224. 186239-171-8/04/$15.00

tary basins (McKenzie 1978). White (1994) has shown how 1-D subsidence data derived from well log information can be successfully inverted by permitting strain rate to vary as a function of time. This algorithm has been applied to ~ 2000 stratigraphic sections derived from field logging, to industry wells and to seismic reflection data by Newman & White (1999). The resultant strain-rate distributions have been tested using independent information about the duration and magnitude of stretching periods. Strain-rate distributions can then be used to constrain the rheological properties of stretching lithosphere. Subsequently, Bellingham (1999) and Bellingham & White (2000) developed a 2-D inverse model, which extracts the spatial and temporal variation of strain rate from individual basin transects. These 1-D and 2-D algorithms make no assumptions about the number, duration, intensity and distribution of rifting episodes within a basin. Instead, strain rate is allowed to vary smoothly through time and space until the summed misfit between observed and predicted stratigraphy is minimized. Potentially important 2-D effects, such as lateral heat flow and varying elastic thickness, are incorporated. Bellingham & White (2002) have applied their 2-D inverse model to a set of well-known extensional sedimentary basins.

These recent and significant advances can be exploited to tackle the challenging problem of 3-D finite deformation as a function of geological time. In essence, the planform analysis of Haines & Holt (1993) can be combined with the cross-sectional analysis of Bellingham & White (2000). In its general form, 3-D inverse modelling will be a significant advance that will extract considerable added value from 3-D seismic reflection volumes. Although Gemmer & Nielsen (2000, 2001) have developed a 3-D inverse model for the thermal evolution of extensional sedimentary basins, their model does not include finite deformation.

Nature of the problem

In the physical and medical sciences, inverse modelling of prior information within the framework of a quantitative model plays a crucial role in extracting information from data. A familiar example is medical tomography, whereby detailed images of a human body are generated by inversion of scanning information. The medical profession does not use forward modelling to generate images of body structure since this approach is both slow and unreliable. Forward-modelled calculations of the thermal and kinematic evolution of passive margins should be treated with similar scepticism unless the model space is systematically explored. Despite the explosive growth of 3-D seismic reflection coverage, no serious attempt has been made to pose and solve the inverse problem of 3-D basin/margin evolution. Instead, a variety of different 2-D forward models have been built, many of which assume instantaneous rifting. This restrictive assumption requires strain rate to be infinite, which is physically unrealistic. Furthermore, it is difficult to assess the quality of solutions and to investigate how observational error affects the results obtained. Here, the inverse problem is solved so that the following questions can be answered:

Does a solution exist?

Is there a unique solution or are there many possible solutions?

If a solution is obtained, how well resolved is it?

Are solutions affected by trade-off between different parameters?

How do stratigraphic and crustal thickness uncertainties map formally into solution space?

To solve for 3-D finite deformation, the detailed geometry of a sedimentary basin must be calculated from any given spatial and temporal distribution of strain rate. This calculation is defined as the forward problem. Strain-rate patterns determine the horizontal and vertical components of the deformation velocity field which, in turn, determine the evolution of the crustal and lithospheric stretching factor, $\beta(x, y, z, t)$, which is defined as the ratio of unthinned to thinned crust or lithosphere. The inverse problem is solved by determining the spatial and temporal variation of strain rate required to fit 3-D subsidence and crustal-thinning observations according to appropriate misfit and smoothing criteria (Bellingham & White 2000). Inversion is only computationally tractable if the forward problem can be calculated rapidly, since the 3-D inverse problem typically has $\sim 10^6$ dimensions (e.g. 20 grid points in each of the x, y, z and t directions).

Why is this problem important to industry? Successful exploration for hydrocarbons relies upon accurate risk analysis. In an extensional system, stratigraphic interpretation, structural evolution and thermal maturation are important sources of risk. An inverse strategy seeks to quantify these three sources of risk at deep-water, highly extended passive margins where drilling costs are considerable.

Continuum deformation in three dimensions

The fundamental quantity that governs the thermal and structural evolution of passive continental margins is the variation of strain rate through time and space (Fig. 1). 3-D subsidence, well log information, normal faulting, crustal thinning and free-air gravity datasets can be inverted by allowing the strain-rate tensor to vary smoothly through time and three spatial dimensions. The importance of strain-rate distributions cannot be under-estimated: these patterns directly determine the manner by which basins and continental margins grow. Strain rate also modulates the temporal and spatial variation of basal heat flux across continental margins together with decompression melting of the asthenosphere when the base of the lithosphere is advected upwards (White & Bellingham 2002). In 3-D, it should be possible to analyse margin development without *a priori* knowledge of extension direction. Results will thus enable animations of the combined thermal, structural and stratigraphic evolution to be made.

Finite deformation, velocity fields and strain rates are a central concern for 3-D basin modelling. A continuum approach is used here to model the average or smooth (over wavelengths of 10–20 km) deformation of the lithosphere as a function of time and space. For now, the normal faulting process will not be included. Accordingly, we must solve

$$\frac{\mathbf{DF}}{\mathrm{D}t} = \mathbf{LF} \tag{1}$$

where **F** is the deformation gradient tensor and L is the velocity gradient tensor whose elements are

$$L_{ij} = \frac{\partial v_i}{\partial x_j} \tag{2}$$

where v_i are the velocities in the x_j directions (Malvern 1969). The expression $\mathrm{D}/\mathrm{D}t$ in Equation 1 is the substantive or Lagrangian derivative, which means that the time derivative is applied to a vector joining one pair of particles. At any time t, the deformation of a short line,

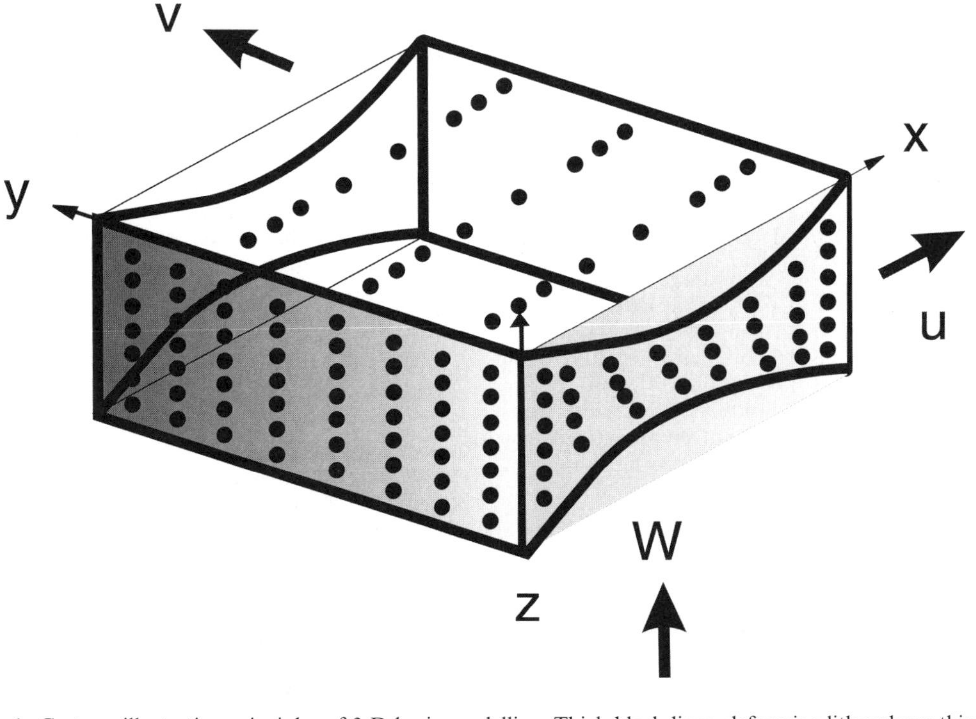

Fig. 1. Cartoon illustrating principles of 3-D basin modelling. Thick black lines, deforming lithosphere; thin black lines, *xyz* axes; *u*, *v* and *w*, particle velocities; *xz* cross-section of lithosphere where solid dots illustrate particles which have moved according to velocity field defined by *u* and *w*; *yz* cross-section of lithosphere where stationary solid dots indicate that $w = v = 0$ in this plane; white horizontal panel, surface of lithosphere where solid dots indicate that displacement is governed by *u* alone.

$\mathbf{p}(t)$, within the continuum is given by

$$\mathbf{p}(t) = \mathbf{F}(t)\mathbf{p}(0), \tag{3}$$

where $\mathbf{p}(0)$ is the vector joining the two particles at $t = 0$. Equation 1 is valid for any temporally and spatially varying velocity field. In Eulerian form, this equation is written as

$$\frac{D\mathbf{F}}{Dt} = \left(\frac{\partial}{\partial t} + \mathbf{v}\cdot\nabla\right)\mathbf{F} = \mathbf{LF} \tag{4}$$

where $\mathbf{v}\cdot\nabla$ are advective terms which take account of the transport of elements within the continuum. Simple analytical solutions to these equations only exist if these advective terms are zero. Because $\mathbf{v}$ cannot be zero, this condition means that the spatial gradients of $\mathbf{F}$ would have to be zero everywhere. Thus $\mathbf{L}$ is constant and it follows that all velocity gradients must be constant. Of course, this simplifying assumption cannot apply to extensional sedimentary basins and continental margins where velocity gradients necessarily vary as a function of time and space. Consequently, Equation 1 must be solved in full.

Equation 4 describes the 3-D finite deformation of the lithosphere. For any velocity field, we must also solve the evolving thermal structure of the lithosphere, $T(x, y, z, t)$. T is calculated by solving the 3-D heat flow equation with appropriate advective terms,

$$\left(\frac{\partial}{\partial t} + \mathbf{v}\cdot\nabla\right)T = \frac{\partial}{\partial x_i}\left(k_{ij}\frac{\partial T}{\partial x_j}\right) \tag{5}$$

Crustal heat production and sediment blanketing, which have a secondary effect on the thermal evolution of basins that extend and fill at typical rates (McKenzie 1981), are ignored. A future implementation could easily include these secondary effects. Equations 4 and 5 are cornerstones of 2-D and 3-D inversion algorithms but neither can be solved analytically for arbitrary velocity fields. Standard finite-element and/or finite-difference models are used (see, e.g., Bellingham & White 2000; Beavan & Haines 2001).

Approximations

A generalized 3-D inversion algorithm does not yet exist because of computational limitations. For a low-density grid with 20 points in the x, y, z, and t directions, the inverse problem has 160 000 dimensions. Because the forward problem can take several seconds of CPU time to solve, a generalized 3-D problem is substantial but tractable (cf. the tomographic problem of global seismology).

Here, two simplified and complementary implementations that tackle different aspects of the simpler and faster 2-D problem are described. The first was developed by Haines & Holt (1993) and is concerned with retrieving velocity fields from earthquake and global-positioning system (GPS) data. The second was developed by Bellingham & White (2000) and obtains the spatial and temporal distribution of strain rate from subsidence and crustal thinning observations. Elements of both approaches will form the basis of a generalized 3-D approach.

Active deformation in planform

Haines & Holt (1993) described a general method for obtaining relative horizontal motions on the surface of a sphere from strain-rate data. Strain-rate data are extracted from earthquake moment tensor analysis, from GPS and very long baseline interferometry (VLBI) data, and from Quaternary rates of deformation on faults. The approach stems from earlier work by Haines (1982), who showed that, if the spatial distribution of strain rates is everywhere defined, then the full velocity gradient tensor is uniquely defined. In other words, if a generally variable strain field in which the rates of horizontal shear strain are everywhere known then the complete horizontal velocity field and inferred rotation can be recovered if one line is known to remain unstrained and unrotated. By using a least-squares inversion, measured strain rates are used to recover the active deformation field. This approach has been used to calculate velocity fields of actively deforming regions on Earth (for review see Holt *et al.* 2000).

This planform method is not concerned with the temporal variation of velocity fields. The obvious reason for this simplification is the nature of the observations. GPS and earthquake moment tensor data are only available for a limited time period of 10–100 years. In the Aegean Sea, independent geological constraints suggest that extension has lasted for $\sim$5 Ma (Jackson *et al.* 1992). Although estimates of slip upon Quaternary faults can lengthen the sampling period, it is unlikely that strain-rate data can be obtained over time scales that are representative of typical rift durations. In any case, most extensional systems are no longer active.

Finite deformation in section

Within inactive basins and continental margins, strain-rate data and velocity fields must be indirectly determined. At present, the most promising approach uses the history of the sedimentary basin fill itself to extract information about the strain-rate tensor. Over the last 10 years, simple 1-D and 2-D methods, which estimate strain rate by inverting subsidence data, have been developed (White 1994; White & Bellingham 2002). Three features characterize the simplest 2-D inverse model (Fig. 2). First, strain rate is allowed to vary through space and time. Secondly, the evolving temperature structure of the lithosphere is solved using the 2-D heat flow equation with appropriate advective terms. Thirdly, loads are imposed on the lithosphere using regional, as opposed to local, isostasy. The philosophy is to keep the 2-D model as simple as possible while incorporating the essential elements; complexity can be added if required by observation. The most important simplifications are that the horizontal velocity is constant as a function of depth and that short-wavelength (< 20 km) normal faulting is not included. Both of these assumptions could be relaxed in a future implementation. The effects of sediment blanketing have not been included in the thermal modelling. The starting point is a lithospheric template whose thermal properties are defined by the parameter values listed in Table 1. As in most models, density is assumed to be a linear function of the thermal expansion coefficient and temperature is assumed to vary linearly with depth.

The algorithm is divided into four parts. First, the variation of strain rate through space and time is defined and used to calculate the velocity field. Secondly, this velocity field determines the evolving thermal structure. Thirdly, the temperature structure constrains the changing density structure, which defines the loading history. Finally, subsidence history is calculated by imposing loads through the flexural equation.

Velocity fields

In two dimensions, we assume that material does not flow out of the plane of section (i.e. $v_1 = u, v_2 = 0, v_3 = v, x_1 = x, x_2 = 0$, and $x_3 = z$; Figs 2 & 3). The problem is further simplified if the horizontal velocity does not vary with depth (i.e. $\partial u/\partial z = 0$). These assumptions mean that Equation 1 becomes

$$\frac{\partial \mathbf{F}}{\partial t} + u\frac{\partial \mathbf{F}}{\partial x} + v\frac{\partial \mathbf{F}}{\partial z} = \mathbf{LF} \qquad (6)$$

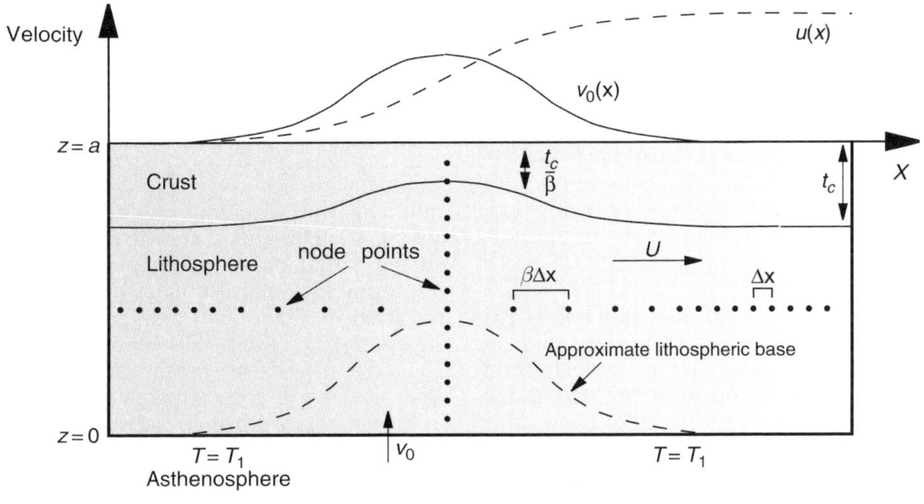

Fig. 2. Cartoon corresponding to xz panel in Figure 1 and illustrating principles which underlie 2-D strain-rate inverse modelling. For a given strain-rate distribution, $G(x, t)$, the vertical and horizontal velocities, $u(x, t)$ and $v(x, z, t)$, are calculated (note that notation has changed slightly from Fig. 1). v decreases from v_o at the base of the lithosphere to zero at the surface. This 2-D velocity field is used to solve the heat flow equation on a finite-difference grid. Boundary conditions for the linear temperature structure are $T = T_1$ at $z = 0$ (base of lithosphere) and $T = 0$ at $z = a$ (top of lithosphere and reference level). Vertical, horizontal and temporal node spacing is governed by Von Neumann stability criteria (Press *et al.* 1992). Horizontal node spacing increases as a function of $\beta(x, t)$. The resultant temperature field varies through space and time. Other parameters are listed in Table 1.

Table 1. *Definitions and values of model parameters*

Symbol	Parameter	Value	Units
a	Lithospheric thickness	120–125	km
t_c	Pre-rift thickness of continental crust		km
G	Lithospheric strain rate		Ga^{-1}
u	Horizontal advective velocity for 2-D modelling		$km\,s^{-1}$
v	Vertical advective velocity for 2-D modelling		$km\,s^{-1}$
β	Stretching factor		none
τ_e	Lithospheric elastic thickness		km
T	Temperature		°C
T_1	Temperature (real) at base of lithosphere	1333	°C
α	Lithospheric thermal expansion coefficient	3.28×10^{-5}	$°C^{-1}$
κ	Thermal diffusivity of the lithosphere	8.04×10^{-7}	$m^2\,s^{-1}$
ρ_a	$= \rho_m(1 - \alpha T_1)$, asthenospheric density	3.20	$g\,cm^{-3}$
ρ_c	Density of continental crustal material at STP	2.78	$g\,cm^{-3}$
ρ_m	Density of mantle material at STP	3.35	$g\,cm^{-3}$
ρ_w	Density of seawater	1.03	$g\,cm^{-3}$
σ	Poisson's ratio	0.25	
E	Young's modulus	70.0	GPa
g	Gravitational acceleration	9.8	$m\,s^{-2}$

where

$$\mathbf{L} = \begin{pmatrix} \frac{\partial u}{\partial x} & 0 & 0 \\ 0 & 0 & 0 \\ \frac{\partial v}{\partial x} & 0 & \frac{\partial v}{\partial z} \end{pmatrix}. \tag{7}$$

Thus any vector $\mathbf{p}(0)$ is deformed to $\mathbf{p}(t)$ where

$$\begin{pmatrix} x_t \\ z_t \end{pmatrix} = \begin{pmatrix} F_{11} & F_{13} \\ F_{31} & F_{33} \end{pmatrix} \begin{pmatrix} x_0 \\ z_0 \end{pmatrix} \tag{8}$$

By definition, the spatial and temporal variation of the stretching factor is $\beta(x,t) = F_{11}$. $\mathbf{F}$ is initially a unit matrix and Equation 8 reduces to

$$\frac{\partial \beta}{\partial t} + u\frac{\partial \beta}{\partial x} = \frac{\partial u}{\partial x}\beta. \tag{9}$$

These simplifications mean that just one of the nine components of the deformation gradient tensor is directly related to the vertical strain rate and to the strain. Equation 9 describes the 2-D finite deformation of the lithosphere. For any velocity field, subsidence of the Earth's surface is calculated by solving this equation in conjunction with the appropriate heat flow equation. Given a horizontal strain-rate distribution, $G(x,t)$, we must calculate the velocity field that governs lithospheric deformation. By definition

$$G(x,t) = \frac{\partial u}{\partial x}. \tag{10}$$

Therefore, the horizontal velocity is given by

$$u(x,t) = \int_0^x G(x,t)dx. \tag{11}$$

Since the amount of material which flows sideways must be balanced by an equal amount of material which flows across $z = 0$, the compatability condition

$$\frac{\partial u}{\partial x} = -\frac{\partial v}{\partial z} \tag{12}$$

applies and so

$$v(x,z,t) = G(x,t)(a - z). \tag{13}$$

This velocity field (u, v) prescribes the spatial and temporal variation of crustal and lithospheric stretching, $\beta(x,t)$, which is obtained by solving Equation 9. $\beta(x,t)$ is solved on a deforming grid and details of the methodology are given in White & Bellingham (2002). Once the strain-rate pattern has been defined, extension across the basin can be calculated at any time. $\beta(x,t)$ will grow through time and space, reflecting the horizontal advection of lithospheric material. The ability to calculate the structural and thermal development of a basin as a function of time and space has considerable commercial potential.

Temperature structure

The second part of the algorithm solves the temperature history as a function of x, z and t. $T(x,z,t)$ is calculated by solving the heat flow

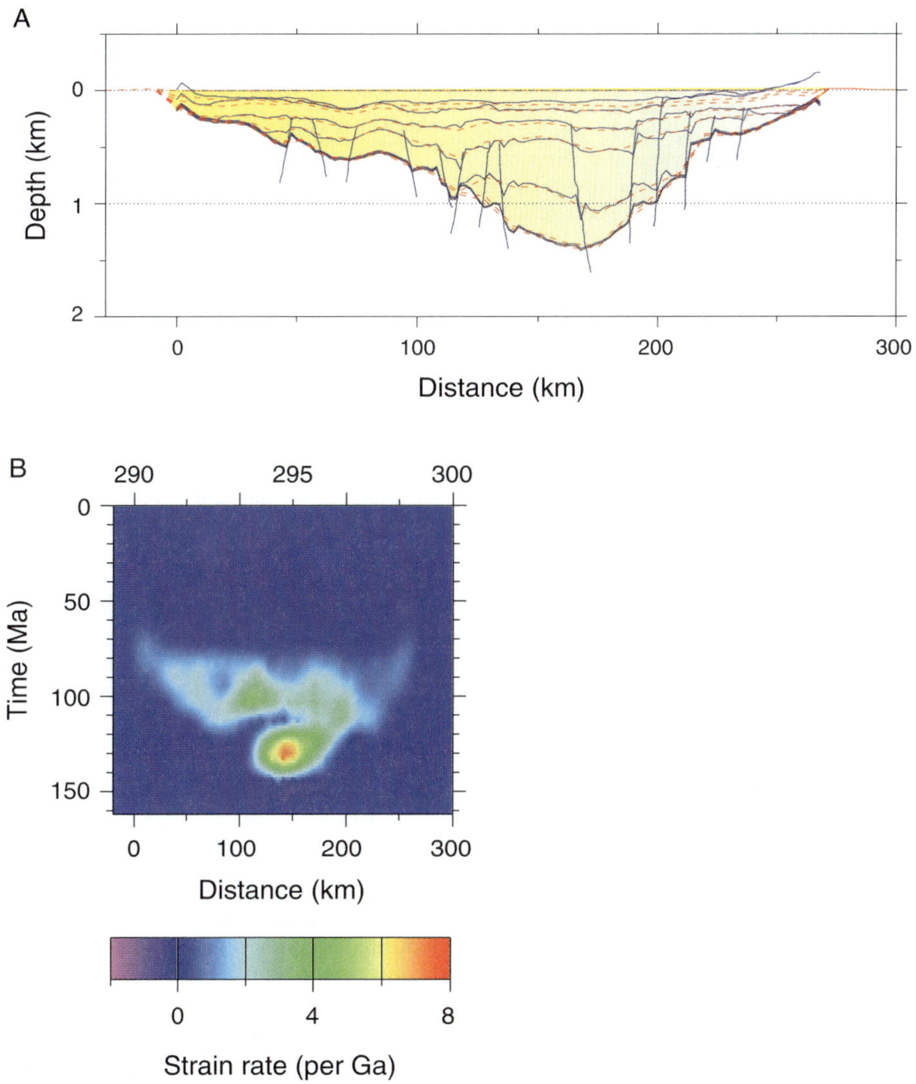

Fig. 3. San Jorgé Basin, offshore Argentina (see Bellingham & White 2002 for further details and location). (**a**) Depth-converted and interpreted cross-section. Yellow zone, modelled basin; thin lines, decompacted and water-loaded stratigraphic horizons (intra-Callovian (162 Ma), Upper Valanginian (132 Ma), Upper Barremian (121 Ma), intra-Cenomanian (96 Ma), Upper Coniacian (86 Ma), intra-Maastrichtian (69 Ma), intra-Paleocene (58 Ma), and sea bed). Dashed lines, best-fitting synthetic horizons generated by inverse modelling. Errors in palaeobathymetry were included in the inversion but they are generally small ($\pm$ 50 m) and have been omitted from the profile for clarity. Thick lines, normal faults. (**b**) Spatial and temporal variation of strain rate for $\tau_e = 0$ km, which yields the synthetic horizons shown in (a). See Bellingham & White (2002) for discussion of elastic thickness. Note localized primary event at $\sim$130 Ma and more diffuse secondary event with much lower strain rate at 100 Ma.

equation

$$\frac{\partial T}{\partial t} + u\frac{\partial T}{\partial x} + v\frac{\partial T}{\partial z} = k\left(\frac{\partial^2 T}{\partial z^2} + \frac{\partial^2 T}{\partial x^2}\right) \qquad (14)$$

The boundary conditions are $T = 0$ at $z = a$, and $T = T_1$ at $z = 0$. Other parameters are listed in Table 1. The initial thermal structure is given by $T = T_1(a - z)$. This second-order partial differential equation has horizontal and vertical advective terms, which vary as a function of space and time. This equation has coupled terms and is not amenable to analytical attack. It is solved on a finite-difference grid using a combination of the forward time-centred space and lax methods (Press *et al.* 1992). Numerical stability is ensured by choosing the time step according to the von Neumann stability criteria.

Finite-difference schemes often use a grid of node points that remains undeformed throughout the calculation. Here we allow the grid to deform according to $u(x,t)$, the horizontal advective velocity (i.e. a Lagrangian formulation). The main advantage of a deforming grid is that the horizontal advective terms of Equations 9 and 14 reduce to zero when there is no horizontal motion with respect to the grid itself. A deforming grid also simplifies the subsidence calculation, since an increment of subsidence at a given time step is simply added to the accumulated subsidence because material is being tracked as it moves horizontally. The vertical grid spacing is fixed throughout. Solving for $T(x,z,t)$ is the slowest part of the algorithm. The time spent calculating temperature history is minimized by using the smallest possible number of node points consistent with finite-difference stability criteria.

Lithospheric loading

The temperature structure is used to calculate the density structure of the crust and lithosphere through space and time. Density varies linearly with temperature according to

$$\rho_T = \rho_0(1 - \alpha T), \tag{15}$$

where ρ_T is the density at temperature T and ρ_0 is the density at 0 °C. This changing density structure generates a series of lithospheric loads. There are two important sources of loading, which evolve as functions of space and time. Firstly, the base of the lithosphere rises and lithospheric mantle is replaced by asthenospheric mantle, which slowly cools to become lithospheric mantle. This load is usually negative during extension because lithospheric mantle is being replaced by less dense asthenosphere. When extension stops, the sign of this load changes. The second source of loading is generated at the Moho, where crust is replaced by lithospheric mantle. This load grows during extension and is always positive. The effect of melt generated by decompression of asthenosphere has not been included.

Crustal and lithospheric loads act in concert to deflect the Earth's surface. The total load, $L(x,t)$, is given by

$$L(x,t) = A(1 - 1/\beta(x,t)) - BQ(x,t), \tag{16}$$

where $A = (\rho_m - \rho_c)gt_c$, $B = \alpha\rho_m g$ and

$$Q(x,t) = \int_0^a [T(x,z,t) - T(x,z,\infty)]dz. \tag{17}$$

A and B are constants, which are calculated using the parameters listed in Table 1, but $Q(x,t)$ is a measure of the difference between the perturbed and equilibrium temperature structure and is necessarily a function of $G(x,t)$.

Stratigraphic evolution

Finally, the loading history is used to calculate the subsidence history, $S(x,t)$. The relationship between $L(x,t)$ and $S(x,t)$ depends upon D, the flexural rigidity of the lithosphere. D is often expressed in terms of τ_e, the equivalent or effective elastic thickness (Watts *et al.* 1982). For simplicity, it is assumed that τ_e does not vary through space and time, although this restriction could be relaxed. It is often inferred that τ_e increases as a function of time but there is little direct evidence to support this conjecture within the continents. At a time t_1, the deflection of the Earth's surface, $w(x,t_1)$, is obtained by solving

$$D\frac{d^4w}{dx^4} + (\rho_a - \rho_{fill})gw = L(x,t_1), \tag{18}$$

where ρ_{fill} is the density of the material that is deposited when the Earth's surface is deflected downwards. If $\mathscr{L}(k)$ and $\mathscr{W}(k)$ are the Fourier transforms of $L(x,t_1)$ and $w(x,t_1)$, respectively, then

$$\mathscr{W}(k) = \frac{\mathscr{L}(k)}{(\rho_a - \rho_{fill})g + Dk^4} \tag{19}$$

where k is the wavenumber. If $D \rightarrow 0$, Equation 18 reduces to the Airy isostatic form. The deflection of the Earth's surface is calculated as a function of time and space and is identical to the subsidence history, $S(x,t)$ (White 1994).

Search engines

There are many ways to search for optimal solutions and only a brief discussion is given in this contribution. One important issue is whether to linearize the problem (Parker 1994). Here, the simple and pragmatic approach advocated by White (1994) is summarized. Faulkner (2000) has developed a more formal linearized inversion scheme but the Fréchet derivatives must be calculated numerically, which considerably slows down his algorithm. $G(x, t)$ is parameterized by using M discrete values of G in the space direction, sampled at intervals of δx, and N discrete values of G in the time direction, sampled at intervals of δt. As before, it is necessary to impose smoothing and positivity on G_{ij} in order to stabilize the inversion. Thus a trial function, H, is minimized such that

$$H = \left[\frac{1}{K}\sum_{j=1}^{K}\left(\frac{1}{L}\sum_{i=1}^{L}\left(\frac{S_{ij}^{o} - S_{ij}^{c}}{\sigma_i}\right)^2\right)\right]^{1/2} + \mathscr{P}, \tag{20}$$

where S_{ij}^{o} and S_{ij}^{c} are the observed and calculated water-loaded subsidence, respectively, j is the number of stratigraphic horizons (varying from 1 to K), i is the number of points on each horizon (varying from 1 to L), and σ_i is the palaeobathymetric uncertainty. H is minimized by varying $G(x, t)$, which controls S_{ij}^{c}. $\mathscr{P}$ is a set of weighting factors which ensure that the first and second derivatives of G are smooth and that G is positive. Smoothing criteria are necessarily applied to G rather than to S. Thus

$$\begin{aligned}\mathscr{P} = {} & P_1\left[\frac{1}{A}\sum_{i=1}^{M}\sum_{j=1}^{N}|\log G_{ij}|\right] \\ & + P_2\left[\frac{1}{A}\sum_{i=1}^{M}\sum_{j=1}^{N}\left(\frac{\partial^2 G_{ij}}{\partial t^2}\right)^2\right]^{1/2} \\ & + P_3\left[\frac{1}{A}\sum_{i=1}^{M}\sum_{j=1}^{N}\left(\frac{\partial G_{ij}}{\partial t}\right)^2\right]^{1/2} \\ & + P_4\left[\frac{1}{A}\sum_{i=1}^{M}\sum_{j=1}^{N}\left(\frac{\partial^2 G_{ij}}{\partial x^2}\right)^2\right]^{1/2} \\ & + P_5\left[\frac{1}{A}\sum_{i=1}^{N}\sum_{j=1}^{M}\left(\frac{\partial G_{ij}}{\partial x}\right)^2\right]^{1/2}\end{aligned} \tag{21}$$

where $A = M \times N$ and P_{1-5} are weighting coefficients. The P_1 term ensures that G_{ij} stays positive since this term tends to ∞ as G_{ij} tends to zero. The P_{2-5} terms cause G_{ij} to be smooth with respect to the first and second derivatives through space and time. The results presented here were obtained using $P_1 \sim 10^{-2}$, and $P_{2-5} \sim 10^{-4}$. H is an *ad hoc* function and it is important to check how inversion results vary when different values of the weighting coefficents are used. Experience shows that varying P_{1-5} by several orders of magnitude has negligible effects; the main purpose of the weighting coefficients is to ensure that the observed subsidence is not over-fitted. A more sophisticated approach would optimize the weighting coefficients during inversion.

Powell's algorithm is used to minimize H (Press *et al.* 1992). It is a direction set method which performs successive line minimizations to try to locate the global minima of a misfit function. Consider point **P** in N-dimensional space. A vector direction $\mathbf{u_1}$ is chosen and the function of N variables, $f(\mathbf{P})$, can be minimized in that given direction using a 1-D search engine. Then we choose a different direction and minimize along it, repeating the process until the global minimum is found. Obviously, by changing directions, it is possible to 'spoil' any previous minimization by searching along in a subsequent direction (i.e. when $f(\mathbf{P})$ is minimized along the second vector, $\mathbf{u_2}$, the function may no longer be at a minimum with respect to the first vector, $\mathbf{u_1}$). Powell's method overcomes this problem by choosing a set of conjugate directions, **n**, which are superior to the coordinate vectors $e_1, e_2, \ldots$. This non-interfering direction set defines 1-D search vectors, which yield large decreases in functional value.

Application of two-dimensional scheme

The 2-D strain-rate inversion algorithm has been applied to ~40 sedimentary basins and continental margins located worldwide. Here, this algorithm is applied to two examples. The purpose is to illustrate strengths and weaknesses of the existing 2-D approach. Further details about each basin and its location are given by Bellingham & White (2002).

San Jorgé Basin

This small basin is located just off the east coast of Argentina, several hundred kilometres inboard of the continental margin, and it is a straightforward application of the algorithm (Fig. 3). The basin formed by multiple extensional episodes prior to, and coeval with, the break-up of Gondwanaland (Fitzgerald *et al.* 1990). It is filled with predominantly shallow marine sedimentary rocks, which makes it ideal for modelling since uncertainties in palaeobathymetry are negligible (0–50 m; Faulkner 2000).

Inverse modelling shows that the observed stratigraphic record can be matched for most of the basin's history. Significant misfit occurs at the northern end of the basin where later uplift and denudation have modified the basin's feather edge. The resultant strain-rate history, which was generated for $\tau_e = 0$ km, is complex, with two phases of extension. Bellingham & White (2002) have shown that τ_e is less than about 3 km and that, for values of 0–3 km, the strain rate patterns are quite similar. An earlier intense phase is confined to the centre of the basin and lasts from 140 to 120 Ma. A second phase of extension, lasting from 110 to 90 Ma, has lower strain rates and is more spatially diffuse. Between 150 and 200 km, this second phase is continuous with the earlier event. A range of tests with different compaction models has been carried out, which suggest that this second phase of weak extension is real. After ~80 Ma, strain rate reduces automatically to negligible values, indicating that the gradient of subsidence is consistent with thermal subsidence driven by the preceding extensional event. As before, the offset between peak strain rate and peak cumulative stretching factor is a logical consequence of the horizontal advection of lithosphere away from the fixed left-hand boundary.

Independent evidence for the duration and number of rift periods can be obtained from the history of normal faulting and from the timing of volcanic activity (Faulkner 2000). The accepted view is that rifting commenced during the Tithonian (151–144 Ma), with activity reaching a peak during the Early Cretaceous (142–99 Ma). There are two main phases of normal faulting. The first phase is concentrated towards the centre of the basin where there is excellent evidence for substantial stratigraphic growth across major faults. The second phase of faulting is concentrated between 80 and 140 km (Fig. 3a). The existence and distribution of these phases corroborate the strain rate patterns shown in Figure 3. There does not appear to be any need for strain rate to vary significantly with depth (i.e. evidence for depth-dependent stretching or lower crustal flow).

North Sea Basin

The evolution of the northern North Sea is well understood and this basin has been an excellent testing ground for rifting models over the last 30 years. Here, the inversion algorithm is applied to a profile which crosses the basin at 61 °N (Fig. 4). The western half of this profile between 0 and 80 km shows the classic tilted block geometry of the East Shetland Basin. Maximum subsidence occurs further east in the Viking Graben between 100 and 130 km. The structural development and subsidence history of the North Sea Basin have been thoroughly investigated using a variety of 1-D and 2-D forward-modelling techniques (e.g. Barton & Wood 1984; Marsden *et al.* 1990). The latest phase of rifting occurred in the Late Jurassic (155–145 Ma) and its spatial and temporal distribution is well understood. Interpretational difficulties and a lack of well penetration into the pre-Jurassic section mean that the preceding Triassic extensional episode is less well constrained. For simplicity, only the dominant Late Jurassic extensional event is modelled.

The cross-section shown in Figure 4a is based on a seismic reflection profile which was calibrated with well log information. Bellingham & White (2002) describe how this profile was converted from two-way travel time to depth. The water-loaded results presented here allow easy comparison with published models. There are three important sources of error in calculating water-loaded subsidence. First, the compactional history of a basin is poorly known. Fortunately, synthetic testing shows that a large range of initial porosities and compaction decay lengths can be tolerated without seriously affecting calculated strain-rate distributions (Faulkner 2000). Secondly, decompacted sediment is converted into a uniform water load, which requires assumptions about isostasic response. As in other studies, sediment loads have been replaced by assuming that the isostatic response of the lithosphere can be approximated either by Airy isostasy or by flexure which assumes some elastic thickness. When a flexural response is used, it is essential that the same elastic thickness is used to unload and load the observed or predicted subsidence. Note that τ_e, the elastic thickness, is not explicitly inverted for, although the residual misfit can be plotted as a function of τ_e to identify the optimal value (Bellingham & White 2002). Thirdly, the largest source of uncertainty arises from poor knowledge of palaeobathymetry. In the Cretaceous, water depths are particularly poorly known and could range from 200 to 800 m. Palaeobathymetric errors are formally included within either the 1-D or 2-D inversion schemes (e.g. White 1994). For clarity, these error bars have not been plotted but the necessary details are given in Bellingham (1999). In general, conservative estimates of palaeowater depth have been assigned.

The calculated strain-rate distribution in Figure 4b has picked out the principal rifting episode in the Late Jurassic (155–145 Ma). Strain rate is almost uniform over the East Shetland Basin where domino-style normal faulting occurs. Peak strain rates occur in the Viking Graben proper, where extension contin-

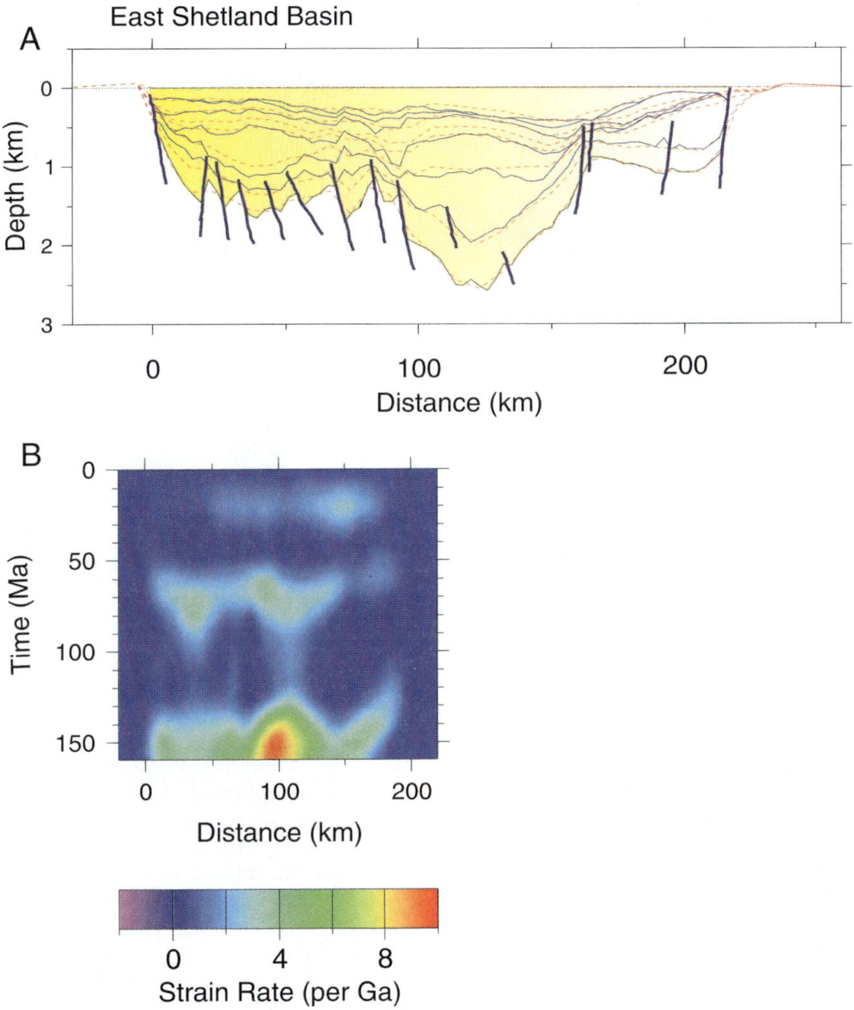

Fig. 4. Regional profile which crosses East Shetland Basin and Viking Graben of northern North Sea (see Bellingham & White (2002) for further details and for location). (**a**) Depth-converted and water-loaded profile. Yellow zone, modelled basin; thin lines, principal stratigraphic horizons (Upper Callovian (159 Ma); Upper Jurassic (142 Ma), Upper Albian (99 Ma), Upper Campanian (71 Ma), Upper Cretaceous (65 Ma), Upper Paleocene (55 Ma), Upper Eocene (34 Ma), seabed at present-day). Dashed lines, best-fitting synthetic horizons generated by inverse modelling. As before, palaeobathymetric errors have been omitted for clarity. Thick lines, normal faults. (**b**) Spatial and temporal variation of strain rate for $\tau_e = 0$ km, which yields the synthetic horizons shown in (**a**). Note Jurassic phase of extension which is more protracted within the Viking Graben (~ 100 km).

ued into the Early Cretaceous. The total amount of horizontal extension across the basin is ~ 25 km, which compares well with the estimate of Marsden *et al.* (1990) of a 22 km horizontal extension (see also Roberts *et al.* 1993). During the Cretaceous and Cenozoic, strain rates are generally negligible, as might be expected. However, two minor events have been highlighted by inversion: during the Late Cretaceous/Paleocene (80–60 Ma) and during the Neogene (20–10 Ma). The inversion algorithm is evidently picking up small increases in the subsidence gradient that cannot be accounted for by thermal subsidence following Late Jurassic rifting. Neither episode can easily be attributed to rifting, although there is some evidence for a mild extension during the Late Cretaceous: localized extension occurred around major basin-bounding faults on the western of the East Shetland Basin and in the Viking

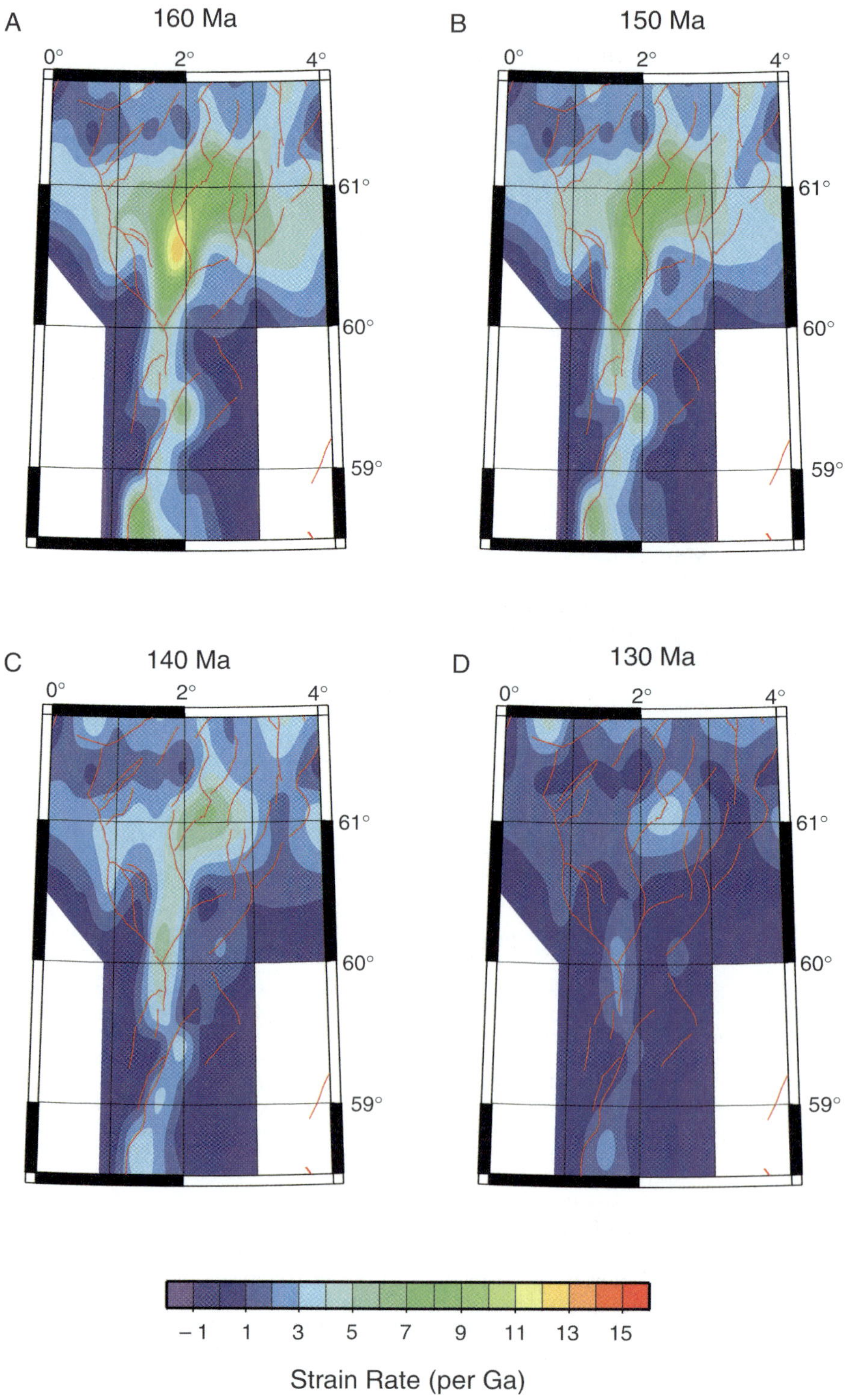

Fig. 5. Four strain-rate maps for northern North Sea calculated by inverting a set of approximately east–west profiles.

Graben. In both cases, a small number of faults cut through the Cretaceous section. This faulting has not previously been attributed to rifting. There is excellent evidence for Mid–Late Cretaceous rifting on the Atlantic margin several hundred kilometres further north, and associated thermal effects appear to have affected the North Sea Basin to the south (Roberts *et al.* 1993). The subsidence anomaly continues into the Paleocene and there is excellent evidence throughout the northern North Sea for anomalous Paleocene subsidence which is usually linked to the evolution of the Iceland plume. The Neogene strain episode occurs further east and is much more localized. Once again, an extensional origin cannot be justified. It may also be associated with the Iceland plume. Thus inverse modelling can be used to extract the detailed pattern of Late Jurassic rifting and to identify the temporal and spatial distribution of anomalous events.

Differences in basin deflection for Airy isostasy and for elastic thicknesses of 1 or 2 km are small. However, if τ_e is greater than 5 km, the residual misfit is significant and calculated strain-rate distributions are geologically less plausible. For a given value of τ_e, strain rate is varied during inversion to achieve the smallest misfit. There is clearly some trade-off between τ_e and strain rate but it is relatively small. The elastic thickness during syn-rift and post-rift phases is inferred to be less than 2–4 km, although variation of residual misfit with elastic thickness is so small that the difference between 0 and 2 km cannot be easily distinguished. The relationship between free-air gravity anomalies and load topography in the frequency domain confirms that $\tau_e < 5$ km (Barton & Wood 1984).

A set of east–west regional profiles which cross the northern North Sea have been inverted and the results used to construct maps of the spatial distribution of strain rate (Fig. 5). These strain-rate maps are only valid if extension occurred strictly parallel to the profiles (i.e. east–west). During the early stages of rifting, deformation is distributed across all of the basin (Fig. 5a). Over a 10 Ma period, strain rate decreases rapidly, especially over the East Shetland Basin. The Viking Graben itself remains active for the longest period. Strain rate maps can be easily converted into maps of heat flux by converting vertical strain rates into temperature gradients.

Conclusions

The main features of a general 3-D inverse algorithm for modelling extensional sedimentary basins and passive continental margins have been outlined. This model is a generalization of 1-D and 2-D models developed by White (1994) and Bellingham & White (2000). At present, only 2-D planform and section implementations exist. In the second half of this contribution, it is shown how the 2-D strain-rate inversion of Bellingham & White (2000) can be used to model stratigraphic cross-sections. This approach can be used to test, and therefore risk, different interpretations. The resultant strain-rate patterns can also be used to calculate the structural and thermal evolution with time (Jones pers. comm. 2003). Model resolution has not been discussed, although this issue forms a central part of the application of inverse theory.

A variety of different forward models could have been chosen as the basis for an inversion algorithm. The general approach concentrates on extracting information from an imperfect and noisy subsidence record which is, nonetheless, probably the best-constrained observation in many extensional sedimentary basins. The essential features of the original stretching model have been incorporated but the detailed, short-wavelength deformation of the brittle upper crust has been ignored. Rougher models, which are more structurally complex and thus include the effects of normal faulting, can be developed by permitting strain rate to vary as a function of depth. A complete dynamical description of lithospheric extension relies on assumptions about driving forces and about lithospheric rheology, which together determine the spatial and temporal patterns of strain and strain rate. The dynamical constraints can be refined by measuring these patterns in a large number of basins.

I am very grateful to P. Clift and J. Turner for helpful reviews. Figures were prepared using Generic Mapping Tools (Wessel & Smith 1995). Department of Earth Sciences Contribution Number 7889.

References

BARTON, P. & WOOD, R. 1984. Tectonic evolution of the North Sea Basin: crustal stretching and subsidence. *Geophysical Journal of the Royal Astronomical Society*, **79**, 987–1022.

BEAVAN, J. & HAINES, J. 2001. Contemporary horizontal velocity and strain rate fields of the Pacific-Australian plate boundary zone through New Zealand. *Journal of Geophysical Research*, **106**, 741–770.

BELLINGHAM, P. 1999. *Extension and subsidence in one and two dimensions, north of 60°N*. PhD thesis, University of Cambridge.

BELLINGHAM, P. & WHITE, N. 2000. A general inverse method for modelling extensional sedimentary basins. *Basin Research*, **12**, 219–226.

BELLINGHAM, P. & WHITE, N. 2002. A two-dimensional inverse model for extensional sedimentary basins. 2. Application. *Journal of Geophysical Research*, **107**, ETG 19. [DOI: 10.1029/2001JB000174]

FAULKNER, P. 2000. Basin formation in the South Atlantic Ocean. PhD thesis, University of Cambridge.

FITZGERALD, M. G., MITCHUM, R. M., ULIANA, M. A. & BIDDLE, K. T. 1990. Evolution of the San Jorge Basin, Argentina. *Bulletin of the American Association of Petroleum Geologists*, **74**, 879–920.

FLESCH, L. M., HOLT, W. E., HAINES, A. J. & SHEN-TU, B. 2000. Dynamics of the Pacific-North American plate boundary in the western United States. *Science*, **287**, 834–836.

GEMMER, L. & NIELSEN, S. B. 2000. SVD analysis of a 3D inverse thermal model. In: HANSEN, P. C., JACOBSEN, B. H. & MOSEGAARD, K. (eds) *Methods and Application of Inversion*. Springer-Verlag, Berlin, Lecture Notes in Earth Sciences, **92**, 142–154.

GEMMER, L. & NIELSEN, S. B. 2001. Three-dimensional inverse modelling of the thermal structure and implications for lithospheric strength in Denmark and adjacent areas of northwest Europe. *Geophysical Journal International*, **147**, 141–154.

HAINES, A. J. 1982. Calculating velocity fields across plate boundaries from observed shear rates. *Geophysical Journal of the Royal Astronomical Society*, **68**, 203–209.

HAINES, A. J. & HOLT, W. E. 1993. A procedure for obtaining the complete horizontal motions within zones of distributed deformation from the inversion of strain rate data. *Journal of Geophysical Research*, **98**, 12 057–12 082.

HOLT, W. E., CHAMOT-ROOKE, N., Le PICHON, X., HAINES, A. J., SHEN-TU, B. & REN, J. 2000. The velocity field in Asia inferred from Quaternary fault slip rates and GPS observations. *Journal of Geophysical Research*, **105**, 19 185–19 210.

JACKSON, J. A., HAINES, A. J. & HOLT, W. E. 1992. The horizontal velocity field in the deforming Aegean Sea region determined from the moment tensors of earthquakes. *Journal of Geophysical Research*, **97**, 17 657–17 684.

KREEMER, C., HAINES, J., HOLT, W. E., BLEWITT, G. & LAVALLEE, D. 2000. On the determination of a global strain rate model. *Earth, Planets, Space*, **52**, 765–770.

MCKENZIE, D. P. 1978. Some remarks on the development of sedimentary basins. *Earth and Planetary Science Letters*, **40**, 25–32.

MCKENZIE, D. 1981. The variation of temperature with time and hydrocarbon maturation in sedimentary basins formed by extension. *Earth and Planetary Science Letters*, **55**, 87–98.

MALVERN, L. E. 1969. *Introduction to the Mechanics of a Continuous Medium*. Prentice-Hall, Old Tappan, New Jersey.

MARSDEN, G., YIELDING, G., ROBERTS, A. M. & KUSZNIR, N. J. 1990. Application of a flexural cantilever simple-shear/pure-shear model of continental lithosphere extension to the formation of the northern North Sea basin. *In*: BLUNDELL, D. J. & GIBBS, A. D. (eds) *Tectonic Evolution of the North Sea Rifts*. Clarendon Press, Oxford, 240–261.

NEWMAN, R. & WHITE, N. J. 1999. The dynamics of extensional sedimentary basins: constraints from subsidence inversion. *Philosophical Transactions of the Royal Society, London*, **357**, 805–830

PARKER, R. L. 1994. *Geophysical Inverse Theory*. Princeton University Press, Princeton.

PRESS, W. H., TEUKOLSKY, S. A., VETTERLING, W. T. & FLANNERY, B. P. 1992. *Numerical Recipes in FORTRAN 77: The Art of Scientific Computing*. 2nd edition. Cambridge University Press, Cambridge.

ROBERTS, A. M., YIELDING, G., KUSZNIR, N. J., WALKER, I. & DORN-LOPEZ, D. 1993. Mesozoic extension in the North Sea: constraints from flexural backstripping, forward modelling and fault populations. *In*: PARKER, J. R. (ed.) *Petroleum Geology of Northwest Europe*. Proceedings of the 4th conference, 1993, London, 1123–1136.

WATTS, A. B., KARNER, G. D. & STECKLER, M. S. 1982. Lithospheric flexure and the evolution of sedimentary basins. *Philosophical Transactions of the Royal Society, London*, **305**, 249–281.

WESSEL, P. & SMITH, W. H. F. 1995. New version of the Generic Mapping Tools released. *EOS, Transactions of the American Geophysical Union*, **76**, 329.

WHITE, N. J. 1994. An inverse method for determining lithospheric strain rate variation on geological timescales. *Earth and Planetary Science Letters*, **122**, 351–371.

WHITE, N., THOMPSON, M. & BARWISE, T. 2003. Understanding the thermal evolution of deep-water continental margins. *Nature*, **426**, 334–343.

WHITE, N. & BELLINGHAM, P. 2002. A two-dimensional inverse model for extensional sedimentary basins. 1. Theory. *Journal of Geophysical Research*, **107**, ETG 18. [DOI: 10.1029/2001JB000173]

Index